Occupa...
Health, and Wellness

Third Edition

Fred Stowell
Project Manager/Senior Editor

Barbara Adams
Senior Editor

Elkie Burnside
Graduate Assistant

Validated by the International Fire Service Training Association

Published by
Fire Protection Publications • Oklahoma State University

Title page photo courtesy of Chief Chris Mickal, New Orleans, (LA) Fire Department.

RECYCLABLE

The International Fire Service Training Association

The International Fire Service Training Association (IFSTA) was established in 1934 as a *nonprofit educational association of fire fighting personnel who are dedicated to upgrading fire fighting techniques and safety through training.* To carry out the mission of IFSTA, Fire Protection Publications was established as an entity of Oklahoma State University. Fire Protection Publications' primary function is to publish and disseminate training texts as proposed and validated by IFSTA. As a secondary function, Fire Protection Publications researches, acquires, produces, and markets high-quality learning and teaching aids as consistent with IFSTA's mission.

The IFSTA Validation Conference is held the second full week in July. Committees of technical experts meet and work at the conference addressing the current standards of the National Fire Protection Association® and other standard-making groups as applicable. The Validation Conference brings together individuals from several related and allied fields, such as:

- Key fire department executives and training officers
- Educators from colleges and universities
- Representatives from governmental agencies
- Delegates of firefighter associations and industrial organizations

Committee members are not paid nor are they reimbursed for their expenses by IFSTA or Fire Protection Publications. They participate because of commitment to the fire service and its future through training. Being on a committee is prestigious in the fire service community, and committee members are acknowledged leaders in their fields. This unique feature provides a close relationship between the International Fire Service Training Association and fire protection agencies, which helps to correlate the efforts of all concerned.

IFSTA manuals are now the official teaching texts of most of the states and provinces of North America. Additionally, numerous U.S. and Canadian government agencies as well as other English-speaking countries have officially accepted the IFSTA manuals.

ISBN 978-0-87939-388-5   *Library of Congress Control Number: 2010931185*

Third Edition, First Printing, August 2010   *Printed in the United States of America*

10   9   8   7   6   5   4   3   2   1

If you need additional information concerning the International Fire Service Training Association (IFSTA) or Fire Protection Publications, contact:

Customer Service, Fire Protection Publications, Oklahoma State University
930 North Willis, Stillwater, OK 74078-8045
800-654-4055   Fax: 405-744-8204

For assistance with training materials, to recommend material for inclusion in an IFSTA manual, or to ask questions or comment on manual content, contact:

Editorial Department, Fire Protection Publications, Oklahoma State University
930 North Willis, Stillwater, OK 74078-8045
405-744-4111   Fax: 405-744-4112   E-mail: editors@osufpp.org

Chapter Summary

Appendices

Table of Contents

List of Tables

Preface

The third edition of the IFSTA *Occupational Safety, Health, and Wellness* manual is intended to provide an overview of programs, policies, procedures, and actions that will establish and maintain a culture of safety within the fire and emergency services. The IFSTA Validation Committee for this manual wanted to provide information for three groups: fire and emergency service administrators, supervisors, and emergency responders. By providing this information, it is hoped that a new culture based on safety, health, and wellness concerns can be instilled in the current and future generations of emergency responders. The basis for this manual is NFPA® 1500, *Standard on Fire Department Occupational Safety and Health Program*. Additional standards are listed in the Introduction and in **Appendix A**. This manual is also a companion to the IFSTA *Fire Department Safety Officer* manual.

Acknowledgement and special thanks are extended to the members of the IFSTA validating committee. The following members contributed their time, wisdom, and knowledge to the development of this manual:

IFSTA Occupational Safety, Health, and Wellness, Third Edition, Validation Committee

Chair

Mark S. Pare
Providence Fire Department
Cranston, Rhode Island

Vice Chair

Scott D. Kerwood
Orange County Emergency Services District #1
Vidor, Texas

Secretary

Robert Capelli
Lutz Volunteer Fire Department
Tampa, Florida

Committee Members

Mary Cameli
Mesa Fire Department
Mesa, Arizona

Steve Ingram
Hammer Training Center
Richland, Washington

Jeff Griffin
Oregon Fire Chiefs Association
Salem, Oregon

Byron Johnson
Alexandria Fire Department
Alexandria, Louisiana

Gregory D. Chesser
Assistant Fire Chief
FE Warren AFB Wyoming

Michael J. Mallory
Tulsa Fire Department
Tulsa, Oklahoma

David Horton
Claremore Fire Department
Claremore, Oklahoma

Dean Martin
Columbia Missouri Fire Department
Columbia, Missouri

Continued

Committee Members (*Concluded*)

Laurent R. McDonald
National Fire Protection Association®
Quincy, Massachusetts

William (Bill) Neville
Neville Associates
Penn Valley, California

Steven J. Schneiderman
Coos Bay, Oregon

Jay Willis
Stillwater Fire Department
Stillwater, Oklahoma

William Zoz
Delhi Twp. Fire Dept.
Cincinnati, Ohio

The following individuals contributed their assistance and comments as reviewers for this manual:

Larry W. Anderson
Fire Department Safety Officers Association
Ashland, Massachusetts

Niles R. Ford
Fulton County Fire Department
Atlanta, Georgia

Darren Olquin
San Ramon Valley Fire Protection District
San Ramon, California

Stephen N. Foley
National Fire Academy
Emmetsburg, Maryland

The following individuals and organizations contributed information, photographs, and other assistance that made completion of this manual possible:

Allen (TX) Fire Department
American Emergency Vehicles (AEV)
Cherry Hill (NJ) Fire Department
Bob Esposito
Federal Emergency Management Agency (FEMA)
Glenwood Springs (CO) "Post Independent"
Globe Manufacturing Company
International Association of Fire Chiefs (IAFC)
International Association of Fire Fighters (IAFF)
Ron Jeffers
Bill Koplitz, FEMA
Rich Mahaney
Mike Mallory
Pat McAuliff
McKinney (TX) Fire Department
Chris Mickal
Rick Montemorro
Ron Moore

Marvin Nauman, FEMA
Mesa (AZ) Fire Department
Mike Nixon
Darren Olquin
Michael Porowski
National Fire Fighter NEAR-MISS Reporting System
Pennsylvania Fire Academy
Portland (ME) Fire Department
Stillwater (OK) Fire Department
City of Stillwater, OK
Tulsa (OK) Fire Department
United States Department of Defense
United States Department of Homeland Security
United States Department of the Interior
United States Fire Academy (USFA)

Additionally, gratitude is extended to the following members of the Fire Protection Publications **Occupational Safety, Health, and Wellness Project Team** whose contributions made the final publication of this manual possible:

Occupational Safety, Health, and Wellness Project Team

Project Manager/Staff Liaison/Writer
Fred Stowell, Senior Editor

Editors
Barbara Adams, Senior Editor
Elkie Burnside, Graduate Assistant

Technical Reviewer
Jeff Fortney, Senior Technical Editor

Proofreader
Lynne Murnane, Senior Editor

Photography
Jeff Fortney, Senior Editor

Production Coordinator
Ann Moffat

Illustrator and Layout Designer
Ruth Mudroch, Senior Graphic Designer

IFSTA Projects Coordinator
Ed Kirtley

Library Researcher
Susan F. Walker, Librarian

Editorial Assistant
Tara Gladden

IFSTA Executive Board

Chair

Jeffrey Morrissette
State Fire Administrator
Commission on Fire Prevention and Control
Windsor Locks, CT

Vice Chair

Paul Valentine
Fire Marshal
Mt. Prospect Fire Department
Mt. Prospect, IL

Past Chairs

Chief Dennis Compton
Mesa & Phoenix, AZ

Chief George Dunkel
St. Helens, OR

Executive Board

Stephen Ashbrock
Fire Chief
Madeira & Indian Hill Fire Department
Cincinnati, OH

Roxanne Bercik
Assistant Chief
Los Angeles Fire Department
Long Beach, CA

Paul Boecker III
Illinois Public Risk Fund
Oswego, IL

Bradd Clark
Fire Chief
Owasso Fire Department
Owasso, OK

Frank L. Cotton
Battalion Chief
Memphis Fire Department
Memphis, TN

David Daniels
Fire Chief/Emergency Manager
City of Renton
Renton, WA

John Hoglund
Director Emeritus
Maryland Fire & Rescue Institute
College Park, MD

John Judd
Chairman, Board of Directors
Institution of Fire Engineers
Moreton in Marsh, UK

Wes Kitchel
Captain
Santa Rosa Fire Department
Santa Rosa, CA

Lori Moore-Merrell
Assistant to The General President
International Association of Fire Fighters
Washington, DC

Randal Novak
Bureau Chief
Iowa Fire Service Training Bureau
Ames, IA

Dena Schumacher
Fire & Life Safety Educator
Champaign Fire Department
Champaign, IL

Executive Director
Chris Neal
Director
Fire Protection Publications
Oklahoma State University
Stillwater, OK

Introduction

Introduction Contents

Introduction

The first two editions of the International Fire Service Training Association (IFSTA) **Fire Department Occupational Safety** manual were published in 1979 and 1991, respectively. Both have been out of print since being replaced by the IFSTA **Fire Department Safety Officer** manual in 2001. In the past few years, IFSTA has become aware of the need to provide a basic safety and health manual for the fire and emergency services. This 3rd edition, under the new title of **Occupational Safety, Health, and Wellness** is intended to fill that need.

Providing a safe work environment for fire and emergency services personnel must be one of the goals of all emergency services organizations. Because fire fighting, rescue, emergency medical activities, and other equally important components of the emergency services profession involve high-hazard situations, it is impossible to remove all risks. That does not remove the responsibility of all personnel to promote the highest standard of safety.

Some members of the fire and emergency services profession and the safety profession now believe that it is possible to protect both the lives and property of the community while providing a safe workplace for emergency responders. This manual provides guidance for creating a safe work environment, establishing an organizational and personal culture that embraces an emergency responder's safety, and providing fire and emergency responders with recommendations for a safe and healthy lifestyle.

To accomplish these purposes, this manual will incorporate a different writing style than previous IFSTA manuals. Rather than writing in the third person without personal pronouns, this text will include the reader. *You* will be addressed directly throughout the manual.

The reason for this shift, if not already obvious, will become so as you complete this manual. Safety is all about *you*. Creating a safe work environment depends on *you*. It is *your* responsibility to follow safety policies and procedures. *You* includes every member of the fire and emergency services profession from the new recruit to the chief of the department, and these responsibilities are described as follows:

- *Individual member* — You are responsible for adhering to the organization's safety and health policies, maintaining a healthy lifestyle, and behaving in a safe manner at all times. You are also responsible for the actions of your fellow workers and team members. This responsibility means that you should correct any unsafe behavior immediately, report unsafe equipment or situations, and recommend changes to policies and procedures that will increase personal safety. You are responsible for your personal safety, health, wellness, and survival.

- *Supervisor* — As a company officer or chief officer, you are responsible for the occupational safety, health, and survival of your team. You must set a positive example by adhering to the organization's policies and procedures, enforcing safety and health policies, and monitoring the safety program. You must stop or correct unsafe actions by your personnel immediately, correct unsafe conditions, report unsafe equipment or situations you personally observe, and forward any notifications of safety issues and recommendations for their corrections to a designated authority as soon as possible.

- *Manager* — As the manager of the organization, you must be aware of accident, injury, and line-of-duty death (LODD) trends, both locally and nationally, that indicate a possible breakdown in the safety and health program. You must provide leadership to the extent your organization is provided with funding, equipment, training, and personnel.

- *Organization* — In this case, you refers to the total membership of the organization. You must create and maintain a culture that makes safety and health issues of prime importance and result in a safe work environment for everyone. That culture must overcome and replace the traditional culture that has existed in the fire service for centuries where firefighters placed themselves at unacceptable and/or unreasonable risk, which resulted in both injuries and fatalities. Management, the membership, and the labor union leadership must work together to achieve the goals of a safe and healthy workplace.

While emergency incident safety has received a great deal of attention over the past few decades, it is only part of the answer to the problem. In fact, the highest number of fire and emergency responder LODDs result from stress-induced cardiac arrests and strokes. To reduce these fatalities and long-term disabilities, it is necessary to have a healthy lifestyle. You must consider that the following factors are things that you can control:

- **What and how much you eat**

- **Your physical fitness regimen**

- **Whether or not you use drugs, tobacco products, or alcohol**

- **How you deal with emotional stress**

If you are going to draw a check from that retirement fund you have been paying into, see your grandchildren graduate from college, or spend quality time with your spouse, you must evaluate your own personal behaviors and make the necessary changes. Achieving a safe and healthy work environment may depend on you making behavioral and cultural changes.

You and the other members of your organization may have to change both personal and professional behaviors. The service we are so committed to must change the culture it has promoted and begin supporting a new environment. The design, maintenance procedures, operation, and replacement of apparatus, equipment, and facilities will have to change to provide greater protection for personnel. The final result, though, should be a reduction in accidents, injuries, and fatalities in the fire and emergency services profession.

NFPA® Requirements

The National Fire Protection Association (NFPA®) professional qualifications standards include safety requirements that apply to each job title or function. While this manual is based on NFPA® 1500, *Standard on Fire Department Occupational Safety and Health Program,* the appropriate job performance requirements (JPRs) of each of the NFPA® professional qualifications will be referenced in the text. **Appendix A** includes a correlation between the NFPA®JPRs and the chapters and pages that address them.

The standards that contain safety requirements include the following:

- NFPA® 471, *Recommended Practice for Responding to Hazardous Materials Incidents*

- NFPA® 472, *Standard for Competence of Responders to Hazardous Materials/Weapons of Mass Destruction Incidents*

- NFPA® 473, *Standard for Competencies for EMS Personnel Responding to Hazardous Materials/Weapons of Mass Destruction Incidents*

- NFPA® 1001, *Standard for Fire Fighter Professional Qualifications*

- NFPA® 1002, *Standard for Fire Apparatus Driver/Operator Professional Qualifications*

- NFPA® 1003, *Standard for Airport Fire Fighter Professional Qualifications*

- NFPA® 1006, *Standard for Rescue Technician Professional Qualifications*

- NFPA® 1021, *Standard for Fire Officer Professional Qualifications*

- NFPA® 1031, *Standard for Professional Qualifications for Fire Inspector and Plan Examiner*

- NFPA® 1033, *Standard for Professional Qualifications for Fire Investigator*

- NFPA® 1041, *Standard for Fire Service Instructor Professional Qualifications*

- NFPA® 1051, *Standard for Wildland Fire Fighter Professional Qualifications*

- NFPA® 1071, *Standard for Emergency Vehicle Technician Professional Qualifications*

- NFPA® 1081, *Standard for Industrial Fire Brigade Member Professional Qualifications*

- NFPA® 1670, *Standard on Operations and Training for Technical Search and Rescue Incidents*

Nonprofessional qualifications standards that have references to safety and health requirements include the following:

- NFPA® 921, *Guide for Fire and Explosion Investigations*

- NFPA® 1403, *Standard on Live Fire Training Evolutions*

- NFPA® 1521, *Standard for Fire Department Safety Officer*

- NFPA® 1561, *Standard on Emergency Services Incident Management System*

- NFPA® 1581, *Standard on Fire Department Infection Control Program*

- NFPA® 1582, *Standard on Comprehensive Occupational Medical Program for Fire Departments*

- NFPA® 1583, *Standard on Health-Related Fitness Programs for Fire Department Members*

- NFPA® 1584, *Standard on the Rehabilitation Process for Members During Emergency Operations and Training Exercises*

One of the basic purposes of IFSTA manuals is to assist fire and emergency services personnel and their departments or organizations in meeting the requirements set forth by NFPA® codes and standards. These NFPA® documents are referred to throughout this manual. References to information from NFPA® codes are used with permission from NFPA®, Quincy, MA 02169. This referenced material is not the complete and official position of NFPA® on the referenced subject, which is represented only by the standard in its entirety.

Purpose and Scope

This 3rd edition of **Occupational Safety, Health, and Wellness** is written for all emergency services personnel. The **Purpose** of this manual is to provide all emergency responders with the motivation, knowledge, skills, and abilities that they need to safely engage in emergency and nonemergency activities on a daily basis. It will also provide assistance to the health and safety officer in charge of implementing the organization's safety and health policy.

The **Scope** of this manual is to provide guidance for meeting the requirements found in NFPA® 1500. This standard contains the minimum requirements and procedures for a safety and health program. It calls upon the fire and emergency services to recognize the importance of member safety and health, establish official safety and health objectives for the department or organization, and provide as safe and healthy a work environment as possible. Included in this manual are discussions of the following topics:

- Risks, hazards, and risk management

- Suggestions for creating a safety culture

- Physical fitness, wellness, and health considerations

- Training safety

- Apparatus safety

- Tool and equipment safety, maintenance, and operation

- Personal protective equipment selection, care, and use

- Emergency scene safety

- Facility safety

- Safety program establishment and implementation

Fire and emergency responder safety is a primary concern of all organizations and individuals engaged in providing emergency services. This manual intends to meet the requirements for specific ranks or positions in the fire and emergency services organization by providing the general knowledge that may not be found in other manuals and textbooks.

At the same time, this manual is a basic textbook for college- or university-level safety and health courses. Some of the Fire and Emergency Services Higher Education (FESHE) learning objectives recommended for the *Principles of Firefighter Health and Safety* course outline are included in this manual.

Resources

The glossary included at the end of the text contains definitions of safety- and health-related terms that will help you understand the material easier.

Additional educational resources to supplement this manual are available from IFSTA and Fire Protection Publications (FPP). These resources include a study guide that will assist readers in mastering the contents of this manual and is available in both hardcopy and electronic formats.

Terminology

IFSTA has traditionally provided training materials that are used throughout the U.S. and Canada. In recent years, the sales of IFSTA materials have expanded into a truly international market and resulted in the translation of materials into German, French, Spanish, Japanese, Hebrew, Turkish, and Italian. Writing the manuals, therefore, requires the use of *Global English:* words and terms that can be easily translated into multiple languages and cultures.

This manual is written with the global market as well as the North American market in mind. Traditional fire service terminology, referred to as *jargon*, must give way to more precise descriptions and definitions. Where jargon is appropriate, it will be used along with its definition. The glossary at the end of the manual will also assist the reader in understanding words that may not have their roots in the fire and emergency services. The sources for the definitions of fire-and-emergency-services-related terms will be the *NFPA® Dictionary of Terms* and the IFSTA **Fire Service Orientation and Terminology** manual.

Key Information

Various types of information in this manual are given in shaded boxes marked by symbols or icons. See the following definitions:

Occupational Safety, Health, and Wellness Sidebar

Sidebars give additional relevant information that is more detailed, descriptive, or explanatory than that given in the text.

Occupational Safety, Health, and Wellness Information

Information boxes give facts that are complete in themselves but belong with the text discussion. It is information that may need more emphasis or separation. They can be summaries of points, examples, calculations, scenarios, or lists of advantages/disadvantages.

Occupational Safety, Health, and Wellness Case History

A case history analyzes an event. It can describe its development, action taken, investigation results, and lessons learned. Illustrations can be included.

Occupational Safety, Health, and Wellness Key Information

Key information is a short piece of advice that accents the information in the accompanying text.

Three key signal words are found in the text: **WARNING, CAUTION,** and **NOTE.** Definitions and examples of each are as follows:

- **WARNING** indicates information that could result in death or serious injury to fire and emergency services personnel. See the following example:

WARNING!
Live-fire training must adhere to the requirements set forth in NFPA® 1403, Standard on Live Fire Training Evolutions (Current edition).

- **CAUTION** indicates important information or data that fire and emergency service responders need to be aware of in order to perform their duties safely. See the following example:

CAUTION
Fire and emergency responders must be familiar with the physiological, emotional, and technological limitations caused by the use of respiratory protection equipment to prevent injury or death.

- **NOTE** indicates important operational information that helps explain why a particular recommendation is given or describes optional methods for certain procedures. See the following example:

NOTE: This information is based on research performed by the National Institute of Occupational Safety and Health (NIOSH).

Risks of a High-Hazard Occupation

Chapter Contents

Divider page photo courtesy of Bob Esposito.

chapter 1

Key Terms

Job Performance Requirements

This chapter provides information that addresses the following job performance requirements of NFPA® 1021, *Standard for Fire Officer Professional Qualifications* (2009):

NFPA® 1021 References

4.7.1

5.7.1

7.7.1

FESHE Learning Objectives

Occupational Safety and Health for Emergency Services
List and describe the components of risk identification, risk evaluation, and incident management.

Risks of a High-Hazard Occupation

Learning Objectives

After reading this chapter, students will be able to:

1. Differentiate between hazards and risks.
2. Describe the all-hazards concept.
3. List the types of hazards.
4. Explain each element of the Three E's of Safety Model.
5. Explain the risk management model.
6. Explain frequency and severity and how they are used to evaluate risk.
7. Compare the risks of fire fighting to other occupations based on the number of fatalities and injuries per 100,000 employees.
8. Explain the human factors that cause or contribute to occupational risk.
9. Explain the environmental factors that cause or contribute to occupational risk.
10. Explain the mechanical factors that cause or contribute to occupational risk.
11. Explain the management and political factors that cause or contribute to occupational risk.
12. Explain political factors that cause or contribute to occupational risk.
13. Explain uncontrollable factors that cause or contribute to occupational risk.
14. Define the term accident.
15. Explain each of the causes of accidents.
16. Describe how multiple causes can contribute to an accident.
17. Describe the consequences of accidents on the individual, organization, and community.

Chapter 1
Risks of a High-Hazard Occupation

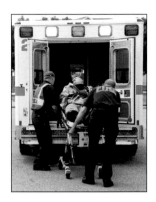

Case History

On February 2, 2010, firefighters responded to a reported unconscious/unresponsive victim at a local motel. The incident turned out to be a suicide by hydrogen sulfide. The hydrogen sulfide was created by the combination of common household cleaning products. The firefighters later learned that this method of suicide has been increasing nationwide in the past several months. The victim had sent letters to family members in advance of the suicide and had posted a hazmat warning sign on the motel room door. While no emergency responders were injured at this incident, it's important to understand that several could have been very easily killed or seriously injured had they not seen the warning signs.

Remember, there is no such thing as a normal response. Always be prepared for the unknown, wear your personal protective equipment, follow established policies and procedures, and be prepared to change your tactics rapidly.

Source: *National Fire Fighter Near-Miss Reporting System*

Fire and emergency responders are members of a high-hazard occupation. A *high-hazard occupation* is one in which people engage in activities or are exposed to conditions that may result in injuries, fatalities, or property losses. As an emergency responder, the individual tasks and activities that you participate in may or may not be hazardous. However, the situation in which you perform those tasks and activities certainly will be hazardous.

For the purpose of this manual, hazards that you face may be classified as external or internal. *External hazards* are those created by situations or conditions that you may not have any control over such as a tornado, hazardous materials release, emergency medical incident, or wildland fire. External hazards may also be the result of a mechanical malfunction of a tool or vehicle that you are operating. Finally, an external hazard may be created by a procedure that must be performed but contains a high level of risk such as high-angle or confined-space rescue or exposure to products of combustion.

You may also be exposed to less obvious internal hazards. *Internal hazards* are the result of lifestyle choices, personal behavior, or physiological conditions that you may or may not be aware of or have any control over. An internal hazard created by a lifestyle choice, for instance, is the decision to use tobacco products, abuse drugs or alcohol, or eat an excessive amount of food that has little or no nutritional value. Examples of hazards caused by personal behavior may be knowingly performing an unsafe act or ignoring safety policies by mentally justifying the action. Finally, ignoring physical symptoms of an illness or not participating in medical evaluations or a physical fitness program can allow existing conditions to go untreated or continue to deteriorate.

To meet the challenges created by external and internal hazards, you must understand not only what the hazards are but also how to prevent, control, or minimize them. This manual provides the foundation for that understanding. This chapter establishes the fundamental concepts about hazard and risk management, including the differences between hazards and risks, a description of the all-hazards concept, types of hazards, hazard-control methods, a basic risk management model, and how fire and emergency services compare to the risks of other occupations. The various factors that contribute to occupational risks are also discussed. After completing this chapter, you will be able to explain the scope, causes, and consequences of accidents in the fire and emergency services and how to reduce their frequency.

Hazards and Risks

In everyday conversation you probably use the terms *hazard* and *risk* interchangeably. For the purposes of this manual, however, they describe two different conditions. The term *hazard* usually refers to a condition, substance, or device that can directly cause an injury, fatality, or property loss. A *risk*, on the other hand, is the likelihood of suffering harm from a hazard. Risk can also be thought of as the potential for failure or loss. In other words, *risk* is the *exposure* to a hazard; a *hazard* is the *source* of a risk. For example, the hazard may be the release of toxic gases into the atmosphere. The risk would be asphyxiation or chronic respiratory disease among other potential risks. The closer a person is to the spill, the greater the exposure and, consequently, the greater the risk.

Many organizations have written definitions for each term. See the information box for definitions of each term.

Definitions: Hazard and Risk

Various organizations, including the American Society of Safety Engineers (ASSE), the National Fire Protection Association (NFPA®), and the International Fire Service Training Association (IFSTA), define the terms *hazard* and *risk* differently. The following definitions are those used by IFSTA, which defines the terms in its **Fire Service Orientation and Terminology** manual as:

- *Hazard* — Condition, substance, or device that can cause injury or loss; the source of a risk.

- *Risk* — Likelihood of suffering harm from a hazard; exposure to a hazard; the potential for failure or loss.

Risks may be divided into the following three types:

- *Natural* — Results of those hazards that are caused by natural acts such as naturally caused fires, lightning, hurricanes, tornados, blizzards, ice storms, floods, earthquakes, avalanches, and droughts **(Figure 1.1a)**.

- *Technological* — Results of practical or mechanical sciences to industry or commerce that are capable of harming persons, property, or the environment. For instance, a release of hazardous gases into the atmosphere by the failure of a safety control system that is outside of human control would be a technological risk **(Figure 1.1b)**.

- *Human-caused* — Results of those hazards that are caused by the acts of humans such as deliberately or accidentally caused fires, hazardous materials releases, explosions, vehicle accidents, and household accidents. Human-caused risks may be intentional or unintentional, but in either case they pose a threat to people and can result in injuries, fatalities, and personal property losses **(Figure 1.1c, p. 14)**.

Figure 1.1a Hurricane Katrina that struck New Orleans and the Texas Gulf coast is an example of a hazard that is caused by a natural act. *Courtesy of Chief Chris Mickal, New Orleans Fire Department.*

Figure 1.1b A hazardous materials fire, spill, or leak resulting from a safety control systems failure is an example of a technological risk. *Courtesy of Rich Mahaney.*

Figure 1.1c Motor vehicle accidents are an example of human-caused risks. *Courtesy of Bob Esposito.*

Hazards can exist without creating a significant level of risk to humans or the environment. For instance, gasoline is considered a hazardous material. As long as it is contained and separated from ignition sources, the possibility of exposure and the level of risk are deemed acceptable.

An example of how exposure affects a hazard is the situation that occurs when you pump fuel into your vehicle's gas tank. As long as the nozzle is properly fitted into the vehicle's filler tube, both you and the environment are exposed to a low, and acceptable, level of risk. If, however, the fuel vapors or the liquid itself are allowed to escape into the atmosphere, you're exposed to a greater hazard and your level of risk increases. If there is an ignition source nearby, the potential for ignition increases the level of risk to an unacceptable point. At the same time, unignited vapors can harm you and the environment.

Risk is the result of an exposure to a hazard; in other words:

$$Risk = Hazard + Exposure$$

By eliminating or controlling the hazard and reducing the exposure, risk decreases. Not all hazards can be eliminated or controlled. We cannot prevent or decrease the ferocity of a hurricane. However we can reduce our exposure to the hurricane by controlling where people live, preplanning evacuation routes, building safe compartments into structures, and providing rescue equipment in areas that are prone to having hurricanes. Each of these actions can reduce the risk to the community.

Historically, the fire service has focused primarily on the hazards and associated risks caused by fires. During the 1950s, the focus widened to include the hazards created by the manufacture, storage, shipment, and use of hazardous materials (or *dangerous chemicals* as they were referred to in that period). As the fire service mission expanded to include confined space and collapsed structure rescue and emergency medical services, including basic and advanced life-support and transportation, the exposure to additional hazards increased the level of risk for firefighters. In the 21st century, the threat of terrorism and weapons of mass destruction (WMDs) create the potential for fire and emergency responders to be exposed to chemical, biological, radiological, nuclear, and explosive devices (CBRN[E]).

Today, the fire and emergency service has embraced an all-hazards concept for defining and responding to hazards and their risks. To understand and apply this concept, you will need to know the following information:

- How the all-hazards concept applies to the community, organization, and individual
- Types of hazards
- Hazard control methods
- How to use an appropriate risk management model
- How to monitor the risk management program
- How to compare the hazards that fire and emergency responders face to those that workers from other occupations are exposed to

All-Hazard Concept

The all-hazards concept provides a coordinated approach to a wide variety of potential incidents. All responders use a similar, coordinated approach with a common set of authorities, protections, and resources. Thus, when emergency responders plan and train for any particular emergency, they have a higher state of readiness to handle a variety of incident types.

The all-hazard concept is used to categorize and prioritize types of hazards, the potential exposure they pose, and therefore the level of risk they generate. The concept may be applied to the community, the fire and emergency services organization, or individual members of the organization. While the community and the organization must deal with external hazards, the individual members are concerned with both external and internal ones.

Community Hazards

From the standpoint of a community, the all-hazards concept refers to the categorizing of all potential hazards that might affect the community. Those hazards are determined by a community risk assessment. This approach determines community risk based on hazards that are identified and placed into two broad categories: natural and human-caused. Community risk is the total sum of all natural and human-caused hazards that a community may suffer.

As an example of the natural hazards and risks a community may face, a river that passes through a city may cause a flooding hazard. Uninhabited areas adjacent to the river could still flood frequently but have no risk associated with them because no one lives in the area. Developed areas within a designated flood plain may have a higher risk because of the possible property loss and potential for human injury, but the frequency of such floods may be limited to 50- or 100-year cycles. Structures that exist near creeks or storm drains that run into the river may have a higher level of risk because the drains may back up with water or be unable to contain large amounts of rain water.

Organizational Hazards

The risks that an organization may face are those created by the hazards faced by the community. Natural and human-caused hazards place fire and emergency services organizations and their members at risk as they provide fire suppression, rescue, structural or topographical stabilization, hazardous materials control, and emergency-scene medical services.

While organizations should not take on tasks that they are not trained, equipped, or authorized to perform, once the community determines that a response is necessary and provides the funding to support it, the organization must prepare for it. Those procedures, such as high-angle rope rescue, may also be hazardous themselves. Equipment and training must be purchased and developed to provide these services in the safest possible manner.

In the example of the community that has a potential flooding hazard, the fire and emergency services organization may prepare to provide services such as water rescue, evacuation, and fire suppression in flooded areas (**Figure 1.2**). To prepare for a flooding incident, the organization would need to perform the following actions:

- Purchase rescue boats, flotation devices, rescue ropes, and other water-rescue equipment.

- Train personnel in water-rescue and boat operations.

- Develop plans for warning residents, evacuating low-lying areas, and transporting victims to evacuation centers.

- Develop procedures for fighting fires or protecting exposures in areas that may be covered with flood waters.

- Consider the possibility that structures might have to be abandoned and allowed to burn.

Figure 1.2 Communities that are located on bodies of water usually purchase water-rescue craft and train personnel for water rescue.

Individual Member Hazards

The external hazards that you the individual will face include the identified community hazards such as an earthquake, the resulting damage such as structural collapse, and the procedures that an organization uses to respond to the hazard such as confined-space operations. The part that you play in the emergency response, training, station activities, and apparatus operation and the tasks that you perform may place you at additional risk.

For instance, entering a confined space to search for and rescue victims will expose you to areas where low levels of oxygen or high concentrations of toxic atmospheres may exist. At the same time, you may be working in an area that is unstable with a potential for additional collapse. Jagged metal, concrete, and other debris will present a potential for injuries.

You must also be aware of the internal or physiological hazards that you will face **(Figure 1.3)**. You should consider the following possibilities:

- Inhalation or ingestion of toxic materials is possible.

- The stress of the moment may cause you to behave in an unsafe manner in an attempt to save a life or mitigate the incident.

- You may be tempted to perform a task that you are untrained or unqualified to do, which could result in injury or death.

- You may not be physically fit to perform the task you are assigned to do.

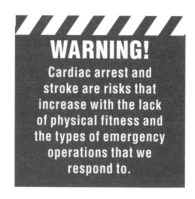

WARNING!
Cardiac arrest and stroke are risks that increase with the lack of physical fitness and the types of emergency operations that we respond to.

Figure 1.3 Firefighters who are engaged in fire suppression activities will be exposed to many types of stress. Rehabilitation sections provide an area for rest, rehydration, and medical monitoring. *Courtesy of Ron Jeffers.*

Hazard Types

Government agencies, nongovernmental organizations (NGOs), the insurance industry, and safety professionals may use a variety of methods for describing hazards by type. One of the many methods used for categorizing hazards is the one used by NFPA® for developing its annual fatality and injury report.

The NFPA® divides hazard types into five categories that include:

1. *Fireground* – fatalities and injuries that occur during fire suppression activities

2. *NonFire* – fatalities and injuries that occur during other types of emergency incidents, including hazardous materials spills and medical, rescue, or extrication incidents

3. *Responding/Returning* – fatalities and injuries that result from motor vehicle or apparatus accidents

4. ***Other On-Duty*** – fatalities and injuries that occur during nonemergency activities such as station maintenance, inspections, or physical fitness programs

5. ***Training*** – fatalities and injuries that occur during training evolutions or courses

These categories are further divided by the type of injury such as:

- Burns
- Smoke/gas inhalation
- Wounds, cuts, or bleeding bruise
- Fractures or dislocations
- Heart attack or stroke
- Strains, sprains, or muscular pain
- Thermal stress

A further subdivision of the injuries is by cause, which includes:

- Overexertion
- Contact with an object
- Fall or jump
- Struck by an object
- Extreme weather
- Exposure to fire products
- Exposure to chemicals/radiation

For the purpose of tracking workplace injuries and illnesses, the U.S. Department of Labor (DOL), through its Bureau of Labor Statistics (BLS) and the American National Standards Institute (ANSI), has also developed a means for categorizing hazards and their causes. These categories include the following:

- Type of task being performed
- Type of injury or illness
- Body part affected
- Time, date, and location of injury/illness

By establishing standardized categories of hazards, the government and employers can effectively gather accurate hazard, injury, fatality, illness, and accident information. Accurate record keeping permits the government and industry to recognize and respond to hazardous trends.

Hazard Control

The traditional approach used by safety professionals for controlling hazards has been referred to as the *Three E's of Safety Model:* Engineering, education, and enforcement **(Figure 1.4)**. These topics are applied to members of the fire service to reduce hazards they face.

- ***Engineering*** — Eliminates or reduces the hazard through the design or modification of an item or process. The concept behind engineering is to create a barrier between the individual and the hazard. The National Fire

Protection Association (NFPA®) standards were created to provide testing and design criteria to reduce the hazard of fire. For example, engineering applies to the following:

— Building construction

— Fire apparatus design

— Personal protective equipment design, construction, and testing

- *Education* — Fire and emergency responders are taught how to alter personal behaviors by applying proper safety procedures when working in a hazardous environment. Fire and emergency responders also teach citizens fire prevention and emergency care that result in reduced emergency responses by the organization.

- *Enforcement* — Implements safety policies and procedures and is the responsibility of management, supervisors, and employees. Management must develop and maintain a uniform safety program for the organization. Program development must include all members of the organization. Management must work to involve employees and gain their acceptance of the program. All officers must strictly enforce safety. The final result must be a program that will form a foundation for a safety culture within the organization with the goal of holding personnel accountable for unsafe behaviors.

Many hazards can be controlled when supervisors implement a safety program by training their subordinates in the application of safety policies and the correct use of procedures. Supervisors must set a positive example for their subordinates by adhering to the safety program. Supervisors must also monitor the program and report its effectiveness or need for improvement.

Employees are responsible for adhering to the policies and applying the procedures. They must also monitor themselves and each other and correct unsafe behavior when they are aware of it.

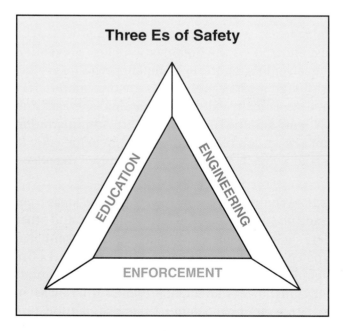

Three Es of Safety

EDUCATION

ENGINEERING

ENFORCEMENT

Figure 1.4 Hazards can be controlled through engineering, education, and enforcement.

Classic Risk Management Model

The classic risk management model process incorporates several components that can be applied to producing a community risk analysis. This model is also used to develop the fire and emergency service organization's internal risk management plan that is required by NFPA® 1500, *Standard on Fire Department Occupational Safety and Health Program*. The classic risk management model as adopted by NFPA® has been successfully used by general industry for decades. The basic risk management model includes the following risk components:

- Identification
- Evaluation
- Prioritization
- Control Techniques Implementation
- Management monitoring

In this section the model is applied to the safety and health of the employees of the organization. Because many injuries and some fatalities occur during training evolutions or physical fitness sessions, it is important to apply the risk analysis to all nonemergency activities as well as emergency activities in order to prevent or reduce the number of incidents.

Development of an agency's risk management model is the responsibility of the chief of the department or organization. The actual implementation of the model is often delegated to the health and safety officer (HSO) or a health and safety committee. You must remember, however, that risk management is every member's responsibility, especially officers. In any case, the final responsibility lies with the officer in charge of the organization.

Identification

To identify the risks, the HSO or health and safety committee compiles a list of all emergency and nonemergency operations and duties in which the organization participates. Ideally, the HSO or health and safety committee should take into consideration a wide range of possible conditions or potential events, including major disasters and concurrent multiple events.

Many sources are available to assist with this identification process. The first (and possibly the most effective) is the organization's loss prevention data, which consists of annual fire-loss reports by occupancy type, loss value, incident frequency, etc. Although most organizations are too small to rely on their own databases for a statistically valid trend, national averages and trends are available from NFPA® and the United States Fire Administration (USFA). National data is not always complete, accurate, or up to date due to collection inconsistencies and a time lag of 1 to 2 years required to collect, analyze, and publish it.

The HSO or health and safety committee should seek input and ideas from organization personnel, trade journals, professional associations, and other emergency service providers such as hospital emergency room personnel to identify potential risks. When using information provided by other area fire and emergency services organizations, the HSO or health and safety committee should consider local circumstances that might present a different set of emergency and nonemergency operations and duties. Other risk identification sources include the following;

- Risk management plans developed by local industry and companies that produce, process, transport, or store hazardous materials
- Vulnerability analyses
- U.S. Environmental Protection Agency (EPA) and local, state, and regional emergency management plans
- Local Emergency Planning Committees (LEPC)

Evaluation

Once the HSO or health and safety committee identifies the risks, these can be evaluated based on frequency and severity **(Table 1.1)**. *Frequency,* referred to by the U.S. Occupational Safety and Health Administration (OSHA) as *incidence rate,* addresses the likelihood of occurrence.

Typically, if a particular type of incident such as injuries related to lifting has occurred repeatedly, it will continue to occur until a job hazard or task analysis has been performed to identify the root causes and effective control measures have been implemented. In this example, the HSO or health and safety committee must develop and implement guidelines that outline proper lifting techniques and physical fitness requirements or provide mechanical aids for lifting.

Severity addresses the degree of seriousness of the incident and can be measured in a variety of ways such as lost time away from work, cost of damage, cost of and time for repair or replacement of equipment, disruption of service, or insurance and legal costs.

Prioritization

Taken in combination, the results of the frequency and severity assessments help to establish priorities for determining action. Any risk that has both a high probability of occurrence and serious consequences deserves immediate action and is considered a high-priority item. Nonserious incidents with a low likelihood of occurrence are lower priorities and can be placed near the bottom of the action-required list.

Control Techniques Implementation

Once the HSO or health and safety committee prioritizes the risks, it is time to apply risk control measures. Several approaches can be taken in risk control, including the following:

- ***Risk elimination*** — One approach is removing the hazard completely. Although it is not always possible in the fire and emergency service, some hazards can be removed through local laws

**Table 1.1
Risk Analysis Based on Frequency and Severity**

	Severity
Frequency	**Low** → **High**
High	Incidents of high frequency/low severity Incidents that fall into this category are of moderate concern **Example:** high frequency of carpel tunnel injuries in office staff **Action:** evaluate workplace ergonomics and budget for ergonomic keyboards, desks, and chairs	Incidents of high frequency/high severity Incidents that fall into this category require immediate action **Example:** high frequency of fireground fatalities due to cardiac arrest **Action:** immediately initiate annual medical evaluations, monitored physical fitness program, smoking cessation program, and nutrition program
Low	Incidents of low frequency/low severity Incidents that fall into this category are of the least concern **Example:** singular incident of elbow strain during annual softball game **Action:** none required	Incidents of low frequency/high severity Incidents that fall into this category are of moderate concern **Example:** singular incident of broken arm while changing tire chains on apparatus **Action:** analyze incident to determine if procedure, equipment, or education would prevent future incidents. Budget accordingly

or ordinances. For instance, mandating that condemned or structurally unsound buildings be demolished within a given time period can remove the risk that a fire may occur requiring a response.

- *Risk spread* — When the hazard cannot be eliminated, then its effect can be minimized by spreading the risk around. For instance, if one person has a greater risk of back injury from lifting a patient, requiring that a minimum of three people lift any victim will reduce the potential for injury to anyone of them.

- *Risk avoidance* — Avoid the activity that creates the risk (which is the best risk-control choice). In a fire and emergency services organization, this approach frequently is impractical. *Examples:*

  — Lifting a stretcher presents a serious back injury risk, but personnel cannot avoid this risk and still provide effective service. Training in the use of safe lifting techniques and/or the use of mechanically assisted lifting equipment would be more acceptable solutions.

  — Including a policy prohibiting smoking by new candidates when they are hired would reduce the potential for lung cancer and other smoking-related illnesses among members.

- *Risk transfer* — Transfer the risk to someone else or purchase insurance. Transfer of risk may be difficult if not impossible for a fire or emergency services organization. *Examples:*

  — Contracting the cleanup and disposal of hazardous waste would transfer risks to a private contractor who accepts the liability of the risks associated with those activities.

  — Purchasing insurance transfers financial risk only and does nothing to affect the likelihood of incident occurrence. Buying fire insurance for the station — while highly recommended to protect the assets of the organization — does nothing to prevent the station from burning. Insurance is no substitute for effective control measures such as installing an automatic sprinkler system.

- *Control measures* — Reduce the likelihood of incident occurrence or mitigate its severity. Risk reduction is the most common method used for the management of risk. Effective control measures include the following:

  — Safety, health, and wellness programs

  — Training and education programs

  — Safety features incorporated into the design of facilities, equipment, and apparatus

  — Well-defined standard operating procedures or guidelines (SOPs/SOGs)

Once control measures have been implemented, they need to be evaluated to measure their effectiveness. For example, changes in station apparatus bay design and apparatus-backing procedures have been very practical **(Figure 1.5)**. The risks associated with backing apparatus into station bays are well documented. An expensive solution would be the replacement of old single-door stations with new drive-through stations. Simple solutions are as follows:

- Improved driver/operator training
- Stripes painted on the apparatus bay floors to guide the rearward movement of vehicles
- Policy that requires a second person standing at the rear of the apparatus to guide the backing operation

Typical control measures instituted to control incident scene injuries include the following:

- Use of accountability systems
- Use of full-protective clothing
- Mandatory respiratory-protection plans
- Training and education sessions
- Health and wellness SOPs or SOGs
- Implementation of the National Incident Management System - Incident Command System (NIMS-ICS)
- Implementation of a preincident planning program for high-hazard occupancies

Figure 1.5 When it is not possible to build a station with a drive-through bay, policies and training must be provided to prevent accidents involving apparatus backing into the station. *Courtesy of McKinney (TX) Fire Department.*

Management Monitoring

The last component in the model process is risk management monitoring. This component ensures that the system is dynamic and facilitates periodic reviews of the entire program or plan. This works to identify any problems that occur in the process and revise or modify the program. The intent of the risk management model is to develop a strategy for reducing the inherent risks associated with fire and emergency services activities.

Table 1.2
Fatalities per Profession in 2000

Profession	Fatalities
Truck drivers (includes all types of delivery services	852
Construction workers	288
Pilots	230
Timber cutters	105
Firefighters	103
Extractive occupations (mining)	69
Roofers	65
Fishermen	52
Structural steel workers	47

Source: Bureau of Labor Statistics (BLS) and U.S. Federal Emergency Management Agency (FEMA) based on 1,096,900 firefighters in the U.S.

Risk Comparison with Other Occupations

Firefighters have long held the belief that fire fighting was the most dangerous of all professions. Based on the BLS for 2000, the most hazardous jobs are truck driving, construction, flying (pilots), and timber cutting. See **Table 1.2** for a compilation of the most hazardous jobs in the U.S. in 2000.

Regardless of how the professions are compared and what the ranking is, fire fighting must still be considered a high-hazard occupation. Ironically, much of the risk caused by hazards can be controlled. Understanding the causes and contributing factors to the risk is a start toward reducing the number of fatalities, injuries, and workplace illnesses.

Factors Contributing to Occupational Risk

Multiple factors cause or contribute to occupational risks created by hazards. While some are uncontrollable, most can be controlled, either eliminating the risk or reducing its effect. The risk management process mentioned previously can be used to control some of these factors. The *Three E's of Safety Model* may also be applied to them. Contributing factors include the following:

- Human
- Environmental
- Mechanical
- Management
- Political
- Uncontrollable

Human

Although humans are exposed to the risks created by hazards such as fires, tornados, or mechanical failures, humans also cause or contribute to occupational risks. Human risks may be the result of personal or organizational factors. Personal factors can be controlled by the individual. Some of these factors are the same as those that cause accidents as discussed in the Accidents section that follows. Organizational factors are the responsibility of management. Some human factors may also be found in the Management section.

Personal Factors

Unsafe behavior, improper attitude, lack of physical and medical fitness, and poor communications are personal factors that can be altered and controlled by the individual. Through his cartoon character and alter ego Pogo, the cartoonist Walt Kelly once said, *We have met the enemy and he is us.* Considering that 50 to 95 percent of all accidents are the result of human error, this statement could very well apply to safety.

Behavior. Knowingly or intentionally behaving in an unsafe manner can frequently result in injuries, fatalities, property damage, or at the very least a near miss. Taking unnecessary risks is unjustified and only places the emergency responder, and sometimes the victim, at greater risk. To clarify what an acceptable level of risk at an emergency scene is, the Phoenix (AZ) Fire Department established the following set of criteria:

1. We will risk ourselves a lot, within a structured plan, to save a savable life.

2. We will risk ourselves a little, within a structured plan, to save savable property.

3. We will not risk ourselves at all to save lives or property that is already lost.

Adopting this philosophy and applying it as individuals and as an organization can reduce risk and eliminate behavioral actions that could increase the risk of a hazard. Equally important is the adoption of a personal policy of avoiding unsafe acts to reduce risks.

Attitude. As fire and emergency responders, we may have developed a *bulletproof* attitude; that is, we think that nothing can harm us. Not only do we think we are personally immune to injuries and risks, we think our personal protective equipment (PPE) will protect us from anything. This attitude has contributed to many injuries and some fatalities as firefighters have advanced farther into structural fires simply because they believed their PPE would protect them. We even have the attitude that PPE designed to protect us from heat will also protect us from biological or chemical contamination when, in fact, PPE is not tested or certified for those hazards. We must change our attitude in the following ways:

- Accept that we cannot *leap tall buildings in a single bound* and that we can be injured.

- Have a safety-first attitude, knowing and recognizing our own limitations and the limitations of our PPE.

Physical and medical fitness. NFPA® statistics make it very clear that poor physical condition is a firefighter's greatest risk. A lack of physical and medical fitness is the result of complacency: We are satisfied with ourselves and believe that we are immune to diseases and injuries. We deny that we may have a problem or lack the physical ability to perform our duties. In some cases, we do not listen to our own bodies. We ignore symptoms until a disease becomes incapacitating. We also believe that our bodies and how we treat them is our business and not the responsibility of the organization or government. To overcome this self-image, we must change our attitude about ourselves and make changes in our daily lives. Some of these changes may include the following:

- Participate actively in a regulated exercise program.

- Eliminate the overuse or abuse of tobacco, alcohol, or drugs.

- Follow accepted nutritional standards by eating the proper quantities and types of food.

- Have regularly scheduled medical examinations.

- Report symptoms or changes in normal physical conditions to a doctor immediately.

- Learn to identify and cope with stress-causing activities

Communications. As a human factor, *communications* means the ability to speak with clarity and accuracy. You must be able to state what the situation is during a crisis. You must also be able to listen and understand others as they state the situation and give orders. Inaccurate status reports can cause the incident commander to make the wrong decision and commit resources to a hazardous situation. Some personal behaviors that can improve emergency scene communications are as follows:

- Following a communication model **(Figure 1.6)** that includes speaking clearly, concisely, and slowly and using clear-text messages can ensure that information is properly transmitted.

- Repeating the message to the sender ensures that you as the receiver are understood.

- Practicing this communication model during training evolutions will help to identify and correct any problems.

Organizational Factors

Other human factors that are the responsibility of the organization, such as training and crew readiness and to some extent fitness shortcomings, will require a change in the organization's culture. The organization must provide appropriate programs to ensure that all personnel are prepared to perform the duties and tasks assigned to them.

Training shortcomings. Providing the appropriate level of training for members of the organization is the responsibility of management. Applying the training and maintaining the level of training is the responsibility of the individual. Attempting to perform tasks that the individual is not trained for can contribute to the risks and hazards. For instance, personnel trained in wildland fire suppression but not in structural fire suppression are not qualified to perform an interior structural attack. Nor are they qualified for

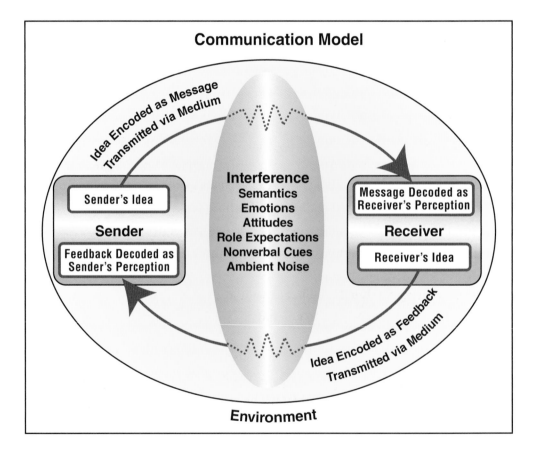

Figure 1.6 Communication Model

a confined-space rescue operation if they lack the training for such a task. In short, never attempt any activity that you are not trained, qualified, or certified for. Officers must never order a responder or unit to attempt a hazardous task that they are not trained to perform.

Fitness shortcomings. A lack of physical or medical fitness also contributes, in a major way, to injuries and fatalities. Attempting to lift heavy objects or victims can result in back injuries and even cardiac arrest. Maintaining proper physical fitness is your responsibility. Knowing your physical and medical limitations is also up to you. You must be familiar with and adhere to the organization's physical fitness program, maintain the correct weight-to-height proportions, and follow the medical evaluation schedule appropriate to your age. Organizational fitness programs can not only provide communities with more capable firefighters, but can reduce fire protection costs by keeping men and women on the job.

Personal and crew readiness. Readiness includes many of the human factors listed earlier. Individuals must be ready for any potential hazard through the proper attitude, training, and physical fitness. Crew readiness is the responsibility of the unit leader or company officer. That officer must ensure that all members of the unit are trained and qualified to perform the tasks that may be assigned to them. A lack of readiness on either the personal or crew level can result in inefficient and ineffective activities that can increase the hazard and risk that individuals may face. Readiness is also the responsibility of the organization that must provide adequate staffing, equipment, and training for its personnel.

Environmental

The environment can be a primary hazard (such as a hurricane) or it can be a contributing factor to a hazard such as the effect that terrain can have on a wildland fire. The environment consists of weather, traffic, road, and incident site conditions along with terrain features. While the presence of environmental factors may contribute to a hazard, a lack of infrastructure can also increase a hazard. For instance, a lack of adequate river crossings that can support the weight of fire apparatus can increase the response times for emergency vehicles thereby allowing the incident to grow worse.

Weather, Traffic, and Road Conditions

Humans have no real control over weather. However, preparation through early warning and planning can reduce the effect of weather as a primary hazard and a contributing factor. Traffic and road conditions can be controlled through planning on the part of the community, organization, and individual.

Weather. An adverse weather condition may be the primary hazard that requires an emergency response. These conditions may result in broken gas lines, downed power lines, cave-ins, fires, structural collapse, or trapped victims. They may also prevent or slow emergency responses to fires, medical situations, or rescues. These conditions may also destroy portions of infrastructures that responders need to perform their duties such as roadways, water supply systems, and radio communications systems. Some conditions can restrict visibility during responses or at emergency incident scenes.

Temperature inversions can keep smoke or toxic gases close to the ground and prevent fires from properly ventilating. Electrical storms can disrupt radio communications. Examples of weather-related incidents include the following:

- Floods
- Ice, dust, or electrical storms
- Tornados
- Earthquakes
- Tsunamis
- Hurricanes
- Land or mudslides
- Heavy rain or fog
- Temperature inversions
- Lightning
- Wildfires
- Temperature extremes

Traffic. Traffic conditions may be the result of poor infrastructure or urban planning that has not progressed with population spread. Traffic can slow or even prevent an emergency response and increase the risk to emergency responders working at a road or highway incident. Planning the best travel routes, knowing the times that roadways may be blocked by trains or raised drawbridges, and observing points and times of traffic congestion can lessen possible response delays.

Roads. Road condition factors can be mitigated through planning by the organization and the community administration. However, the company officer and crew members are responsible for keeping track of conditions within their response area and reporting any deficiencies. Road conditions generally refer to the following factors **(Figure 1.7)**:

- Width of roadways
- Condition of road surfaces
- Lack of traffic-control devices
- Lack of emergency-response lanes
- Lack of water supplies along roadways
- Limited weight-carrying capacity of bridges and road surfaces
- Obstructions such as railroad crossings or physical barriers
- Road construction
- Travel against one-way traffic direction

Terrain Features

The *terrain* features or the surface characteristics of the land in the immediate area of the incident can contribute to the hazard. The terrain can directly affect the direction and spread of a wildland or forest fire or the spread of a liquid or gas release into low-lying areas. The movement of the tectonic plates of the earth creates earthquakes, causing damage to structures of all types. The terrain may also restrict access to aircraft or railroad crash sites or some sides of structures.

Figure 1.7 Narrow roadways with obscured curves must be considered when planning responses. Firefighters must also exercise caution when travelling on these roadways during emergency responses.

Incident Site Conditions

Conditions at an incident site can contribute to the hazard you are called to eliminate or control. Besides the weather, road, and traffic conditions and terrain features in the immediate area, other factors include the following:

- *Immediate physical area* — Barriers and obstructions in the immediate area around the incident can contribute to the hazard by preventing or delaying access. These barriers may be natural such as trees, creeks, or terrain features or human-made features such as setbacks, blast barriers, fences, or walls. Some of the most common barriers that slow access to residences are security grills and bars on windows and doors. Not only do these barriers prevent access, they also prevent emergency egress of both citizens and firefighters.

- *Structural conditions* — These conditions may be the result of the incident such as a structural collapse caused by an explosion or the result of the design or construction of the structure such as the use of lightweight construction, firewall penetrations, or unprotected floor openings. Previously burned structures may lack structural integrity and be susceptible to early collapse in case of subsequent fire.

- *Structural contents* — These contents may be the primary cause of the incident or contribute to the growth of the hazard. For instance, the contents of a warehouse that stores ignitable liquids may catch fire, exposing the structure to extreme temperatures and rapid burning. A structure containing a printing process may have a fire in the bulk paper storage area that spreads to the ignitable liquids storage area, causing the liquids to add to the intensity of the fire. Content storage may restrict firefighter access or may overload upper floors making them susceptible to early collapse **(Figure 1.8, p. 30)**.

- *Adjacent exposures* — These exposures may contribute to the hazard by adding directly to the hazard such as flammable or combustible liquid storage tanks adjacent to a burning building. They may also require additional resources to protect them such as those required to protect houses from a spreading wildland fire.

- *Actions of others* — Probably the most difficult aspect to prepare for are the actions of victims and others at an incident scene. Responding to a domestic violence call that may have resulted in injuries places responders in a potentially dangerous situation. Victims that are under the influence of alcohol or illegal drugs may turn violent while responders are attempting to provide assistance.

Figure 1.8 Warehouse fires involving large quantities of materials can expose the structure to high temperatures. *Courtesy of U. S. Federal Emergency Management Agency, Bill Koplitz, FEMA Photographer.*

Infrastructure Conditions

The lack of infrastructures such as roadways, water supplies, and utilities can contribute to a hazard. Without roads or bridges, it is very difficult to access wildland or forest fires in remote areas with fire apparatus. This can contribute to the growth of a wildfire, increasing the risk to firefighters. A lack of piped water-supply systems means that water has to be transported by water tenders and transferred to pumpers or water tanks to fight fires in remote structures. Water transporting has often led to accidents due to poor road conditions, lack of operator training, or improper apparatus design. Even the lack of radio repeaters or transmissions towers creates barriers to radio communications at emergency incident scenes.

Mechanical

Mechanical factors may be the cause of a hazard or contribute to it. For instance, the failure of a liquid transfer pump may cause a spill of ignitable or hazardous liquids. Likewise, the failure of a foam fire-suppression system may contribute to the spread of a fire involving the spill.

From the standpoint of firefighter safety, a mechanical failure involving apparatus or equipment can create a life-safety hazard for firefighters or delay a response to the incidence. The failure of personal alert safety system (PASS) devices has contributed to the deaths of a number of firefighters over the past decade. Hydraulic failures in aerial platforms have delayed or prohibited the deployment of elevated fire streams, which then contributed to the length of time it took to control a fire.

Some of these mechanical failures can be traced to a lack of preventive maintenance programs. Preventive maintenance reduces the possibility of mechanical failure by ensuring that apparatus and equipment are inspected on a regular basis. While the lack of maintenance can be considered a contributing factor to mechanical failure, it is also a management issue **(Figure 1.9)**.

Mechanical failure can also occur when apparatus and equipment are improperly used or abused. Using the wrong type of tool for a job can result in damage or excessive wear to the tool. Apparatus brakes and tires can receive uneven wear from improper driving, turning, and stopping by driver/operators. Overloading an aerial apparatus has often led to ladder collapse or vehicle rollover.

Improper use may also be considered both a management issue and a human issue. Management is responsible for establishing procedures and providing training for the proper use of apparatus and equipment. Individuals are responsible for adhering to those procedures and using them within the limitations of the apparatus or equipment design.

Management

The organization's management or administration is ultimately responsible for risk management and providing a safe workplace for its employees. Therefore, a failure on the part of management can cause a hazard or contribute to it. Management factors include the following:

Figure 1.9 Proper apparatus preventive maintenance can increase the life and reliability of vehicles.

- *Safety culture* — Management should establish a culture that rewards and promotes safe actions. Without a strong safety culture, an organization can experience a cultural drift where employees do not adhere to safety policies or procedures, taking shortcuts that result in injuries, fatalities, or equipment damage. Praising an unjustified dangerous act can result in a dangerous precedent being established.

- *Organizational discipline* — A lack of strong leadership that is committed to a safe working environment sends a message to employees that safety does not matter or is not important. Employees are not held responsible for unjustified, unsafe acts and management does not correct unsafe conditions.

- *Clear mission, goals, and objectives* — An organization must include safety in its mission statement, goals, and objectives to formally establish safety as part of the organizational culture and structure. Including safety will also prevent an organization from engaging in operations that it is not trained, equipped, or responsible for, activities that can contribute to the hazard and increase the risk to personnel.

- *Planning* — A lack of planning contributes to hazardous operations as personnel and units perform unnecessary tasks, do not coordinate efforts, or operate outside the control of the incident commander. Planning includes the application of the National Incident Management System-Incident Command System (NIMS-ICS), developing operational plans for all potential hazards, and developing preplans for all structures and facilities in each unit's response area with special attention to unusually high hazardous occupancies. Planning includes the development of safety programs, policies, and procedures for all aspects of the department **(Figure 1.10, p. 32)**.

Figure 1.10 The safety committee is composed of members of the department, representatives of the labor union, and the department's health and safety officer.

- *Required programs* — Management is responsible for developing, implementing, evaluating, and monitoring safety-related programs. Some of these are mandated such as the respiratory protection, training, and hazardous materials awareness programs, while others are *good industry practice* such as physical fitness, health and wellness, and apparatus preventive-maintenance programs.

- *Incident management* — Management is responsible for the implementation of an incident management system that includes NIMS-ICS. The use of ICS decreases the hazards and risks of all emergency scene operations. Along with ICS, hazards can be reduced through proper emergency scene decision-making, a skill that all fire officers must be taught and able to apply.

- *Resources* — Management is responsible for providing the resources required to provide the fire and emergency services. Without sufficient funds, apparatus, equipment, materials, personnel, and training will not be available when needed. Based on a community risk assessment, an organization must have or know where to get the resources needed for each hazard. For instance, a storage facility might require 500 gallons (1 892.71 liters) of foam concentrate to control the worst possible fire. If the organization only maintains a supply of 50 gallons (189.2 liters), then it may not be able to control the hazard effectively. In those cases where the risk presented by a particular hazardous situation outweighs the resources available, management must make it clear that operations must be limited to those that can be safety conducted with the available resources.

- *Policies and procedures* — Management can cause or contribute to hazards by not establishing and enforcing policies and procedures. For instance, it is a documented fact that the use of seatbelts when operating or riding in motor vehicles can save lives. If management fails to establish a seatbelt policy or enforce an existing one, all personnel are put at risk.

Political

Political factors that cause or contribute to hazards may be internal or external to the organization. Political considerations based on the perception of members of the organization can create either action or inaction, depending on how the person perceives the situation.

Internal politics may consist of a lack of administrative support or the perception that an individual will get in trouble if a certain decision is made. This type of situation may be the result of any of the management factors mentioned previously. Other internal political factors may include pressures from labor groups or the wording found in the labor/management agreement. An individual may feel that making a decision could conflict with the agreement or create conflict with the labor leadership.

External political factors include influence from elected/selected officials, outside government agencies, and the public in general **(Figure 1.11)**. Elected officials may (and do) have their own agendas and priorities, usually defined by the political base that supports them. Safety issues may not be a major concern for them, prompting them to support issues like filling potholes in streets and building recreation centers at the expense of appropriate fire apparatus or training programs.

Figure 1.11 Political officials make decisions that may not be in the best interests of the fire and emergency services organization.

Appointed officials such as city managers may make decisions based on keeping their jobs. Therefore, pleasing some politicians may prevent them from addressing or supporting issues of safety or risk management. Political influence can result in inadequate or unenforceable building, zoning, or fire codes that permit hazards to exist or go uncorrected.

Agencies of the state/provincial and federal government can contribute to hazards faced by the local emergency responders. An unfunded federal mandate is one obvious problem. By placing requirements on the local organization, the government forces the organization to provide a service or program that local government is unable to fund. The local organization must then decide which services must be cut to fund the new mandate.

Finally, public interference can cause a hazard or contribute to it. Direct interference may exist in the form of civil unrest, which includes the intentional burning of property and the obstruction of firefighters in the line of duty. It may also include political activities that oppose proposals that are intended for the public good. Tax protests and changes in local ordinances that reduce or eliminate fire and emergency services programs are examples of this type of activity.

Uncontrollable

Uncontrollable factors are those activities that you or your organization may not be able to predict or prevent. You could argue that there are no uncontrollable factors if effective intelligence gathering and planning are applied to all types of situations. You may also ague that uncontrollable factors are part of each of the previously stated factors such as human, management, political, environmental, and mechanical. You could build a strong case for either argument.

For the purpose of this manual, we propose that uncontrollable situations can exist but that they may be mitigated or reduced through proactive decisions and actions. Uncontrollable situations may include the following:

- *Criminal negligence* — Failure to use reasonable care to avoid consequences that threaten or harm the safety of the public and are the foreseeable outcome of acting in a particular manner. For instance, driving while intoxicated could be considered criminal negligence. If you commit an act that qualifies as criminal negligence, it is controllable. You can decide not to knowingly place yourself and others in danger. If someone else commits such an act, you probably will not have any control over it.

- *Intentional or malicious activities* — Consciously and knowingly behaving in a manner that is intended to inflict pain and suffering. These acts may be illegal such as operating a methamphetamine (meth) lab **(Figure 1.12)**. As in criminal negligence, it is controllable if you are the party who behaves in such a way. If you are not, then it may not be controllable. Responding to a structure fire in a hidden, illegal meth lab would place you and your crew in an uncontrollable hazard.

- *Terrorism and acts of war* — Definitions:
  - *Terrorism:* Premeditated, politically motivated violence perpetrated against noncombatant targets by subnational groups or clandestine agents, usually intended to influence an audience (according to the Central Intelligence Agency [CIA]). Only the best of intelligence agencies can thwart acts of terrorism.

  - *Act of war:* Military attack by one nation-state against another. As we discovered on September 11, 2001, terrorists are difficult to identify, locate, and control. Acts of war are also difficult to control, especially those that are perpetrated without a formal declaration of war such as the attack on the U.S. Pacific Fleet at Pearl Harbor on December 7, 1941.

- *Accidental hazardous materials releases* — Unintentional discharge (including a spill on the soil, in a body of water, or into the air) of any substance or material that possesses an unreasonable risk to the health or safety of persons and/or the environment. Once a hazardous material such as a toxic gas is released into the atmosphere, it may be completely uncontrollable. Allowing it to dissipate and disperse into the air may be the only possible response. Hazardous materials, however, can be controlled. Proper procedures and engineering devices (control valves, leak-detection systems, and warning systems) can prevent releases.

Figure 1.12 Criminal acts such as arson or the operation of illegal drug labs increase the risk to firefighters.

- ***Natural occurrences or extreme weather conditions*** — Natural occurrences or weather conditions such as hurricanes or earthquakes that we cannot stop from occurring. We can, however, control their effects on us and our communities. Planning our response to such extreme weather conditions can prepare us for the possible occurrence and the consequences. Installing warning devices that can provide an adequate warning and permit evacuation is one approach. Stocking evacuation centers and our own homes with provisions are simple solutions to the aftermath of these emergency situations.

Accidents

Over the past few years, there has been a controversy among members of the safety profession over the use of the term *accident*. Some have argued that the term properly describes unplanned events that result in injuries, fatalities, or property damage. Others have argued that accidents do not exist and that the proper term should be *incident* or *event*.

Currently, however, the term *accident* is still in popular use and the preferred term used by NFPA®. The words accident and incident appear in the ASSE *Dictionary of Terms Used in the Safety Profession,* 4th edition. Therefore, it is only appropriate that the term *accident* be used in this manual.

NFPA® defines an accident as follows:

> *. . . an unplanned event that interrupts an activity and sometimes causes injury or damage; a chance occurrence arising from unknown causes; an unexpected happening due to carelessness, ignorance, and the like.*

ASSE defines the term as follows:

> *. . . an unplanned and sometimes injurious or damaging event which interrupts the normal progress of an activity and is invariably preceded by an unsafe act or unsafe condition or some combination therefore. An accident may be seen as resulting from a failure to identify a hazard or from some inadequacy in an existing system of hazard controls.*

As you can see, *an accident must be unplanned, unexpected, and unwanted and sometimes result in injury, loss of life, damage, or unforeseen consequences.* However, an accident may also result from human failure or system failure; that is, you may cause an accident by failing to use the proper procedure for operating a water tender on a rain-slick roadway. The accident might result in a vehicle rollover that can injure you, your passengers, or other citizens and destroy the apparatus and delay the water needed at an incident. Similarly, an automatic shutoff on an electrical system might fail, resulting in an electrical fire in a structure.

Accidents may be further categorized by the terms *major, minor,* or *near miss.* A major accident may result in the hospitalization of an individual, destruction of a vehicle, or even a fatality. A minor accident may result in a bruise, paper cut, or scraped paint on a vehicle. A near miss results in no injury, fatality, or damage but could have had these consequences.

While major accidents are always reported and investigated and minor accidents are sometimes reported, near misses are rarely reported. In the fire and emergency services, near-miss reporting is gaining acceptance. This aspect is discussed in Chapter 2, Culture Change in the Fire Service.

The following sections describe the causes and consequences of accidents and how to reduce the occurrence of accidents. Some of the information may parallel the information provided previously under the Hazards and Risks section.

Workplace Casualties

According to the National Safety Council (NSC), there are 13 unintentional-injury deaths and about 2,560 disabling injuries every hour during the year.

Source: National Safety Council (NSC), *Injury Facts® '05*, page 31.

Causes

According to the National Center for Health Statistics in its *National Vital Statistics Report,* accidents were the fifth leading cause of death in the U.S. in 2002. As indicated in the ASSE definition, accidents may be caused by an unsafe act, an unsafe condition, unsafe personal factors, or unsafe physical factors. As mentioned previously, approximately 88 percent of all accidents are the result of unsafe acts and unsafe personal factors, while 10 percent are the result of unsafe conditions **(Figure 1.13)**. Unsafe physical factors (Acts of God) constitute the remaining 2 percent.

Unsafe Acts

The ASSE *Dictionary of Terms Used in the Safety Profession,* 4th edition, defines an *unsafe act* as:

> . . . *a behavioral departure from an accepted, normal, or correct procedure or practice which, in the past, has produced injury or property damage, or which has the potential for doing so in the future; an unnecessary exposure to a hazard; conduct that reduces the degree of safety normally present in an activity.*

For example, an unsafe act is standing on the tailboard of an apparatus while it is in motion. Another example is to enter an atmosphere that is known or suspected to be immediately dangerous to life or health (IDLH) without wearing the proper respiratory-protection equipment.

Unsafe acts are considered to be human-caused accidents because they are the result of human behavior. The behavior may be unintentional due to inadequate training or knowledge or intentional due to knowingly acting in an unsafe manner and justifying the action. Approximately 15 percent of all accidents in the workplace are the result of human error or behavior.

Human-caused accidents may also include the lack of management policies, programs, procedures, or training. Management-related problems are responsible for a high percent of workplace accidents.

Figure 1.13 Correcting unsafe conditions, such as this standing water, can prevent 10 percent of all accidents.

Unsafe Conditions

The ASSE definition for an *unsafe condition* is as follows:

> *... any physical state which deviates from that which is acceptable, normal, or correct in terms of its past production or potential future production of personal injury and/or damage to property; any physical state which results in a reduction in the degree of safety normally present.*

For example, an unsafe condition might be the lack of adequate wheel chocks to hold engine apparatus on inclines while pumping. The failure to properly maintain a mechanical system can result in an accident for example, the loss of engine power during pumping operations, the break in a hoseline, or the loss of hydraulic pressure on an aerial platform can all result in unforeseen consequences.

Unsafe Personal Factors

An *unsafe personal factor* is defined by ASSE as follows:

> *... a mental or bodily characteristic that permitted or occasioned an unsafe act which contributed to an accident or which has the potential of doing so.*

For example, being physically unable to control a pressurized hoseline or raise a ladder may result in an accident. Being mentally distracted and ignoring proper safety procedures may also be considered an unsafe personal factor.

Unsafe Physical Factors

Finally, an *unsafe physical factor* (may also be called an *Act of God*) is defined by ASSE as follows:

> *... an environmental or physical state or condition, of a physical hazard, that caused an accident or could do so.*

Accidents that are caused by the environment include weather- and topographic-related incidents, flash floods, earthquakes, lightning storms, mudslides, or a multitude of other natural occurrences. Even with modern weather forecasting and warning systems, environmental accidents are still unpredictable and can have a variety of consequences.

Combinations

While a single cause may be identified for an accident, there are usually multiple causes from those listed. For example, a firefighter driving a baffleless water tender while responding to a reported house fire during a rainstorm. The driver exceeds the posted speed limit, enters a curve, loses control of the vehicle, and the truck rolls over. Multiple causes could include:

- *Unsafe acts* — Operating the vehicle at a speed that was greater than existing conditions allowed contributed to the accident.

- *Unsafe conditions* — Without baffles in the water tender's tank, the water sloshed, causing the apparatus to lose its center of balance in the curve resulting in the rollover.

- *Unsafe personal factors* — The driver's mental focus on arriving quickly at the reported house fire contributed to his lack of attention to the existing conditions and caused him to exceed the speed limit.

- *Unsafe physical factors* — The rainstorm, which reduced visibility and created the slick surface on the roadway, contributed to the accident.

Being physically unable to control a pressurized hoseline or raise a ladder in less than desirable circumstances may result in an accident.

Consequences

When an accident occurs, there will be multiple effects on the individual involved, the organization, and the community. The accident may have direct and indirect consequences in each of these areas.

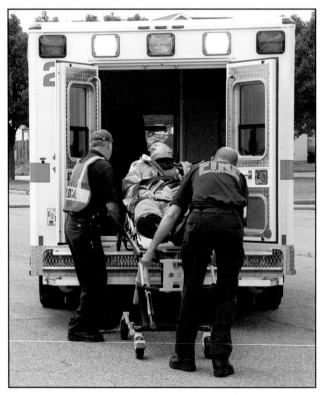

Figure 1.14 Emergency scene injuries can result in the need for immediate transportation to a hospital emergency room.

Individual

The individual may suffer an injury that is minor or life-threatening. Many injuries are so minor that they go unreported. Others require immediate medical care or hospitalization. Some result in permanent disability and loss of employment **(Figure 1.14)**.

Fatalities may also result from the accident. In the fire and emergency services, approximately 100 firefighters have died in the line of duty annually over the past decade. Many of these deaths could have been prevented.

Accidents may also result in exposures to chemical, biological, or radiological materials that can have long-term affects on health. Exposure to asbestos, carbon monoxide, radioactive materials, and someone with hepatitis can all contribute to diminished health and result in a shortened lifespan.

Finally, accidents can result in psychological problems such as increased stress, emotional impairment, lowered morale, chronic depression, and loss of confidence. These problems can also have an effect on close family members and friends.

Organization

Accidents can have a variety of direct and indirect effects on an organization. They can be financial, human, and psychological.

Financial affects include the personnel costs if the accident results in physical injury to an employee. NSC estimates that a non-lost-time injury can cost an organization approximately $7,000. An accident that results in a disabling injury can cost approximately $38,000. A workplace fatality can cost as much as $1,110,000. According to the National Institute of Standards and Technology (NIST) report titled *The Economic Consequences of Firefighter Injuries and Their Prevention*, the average cost for nonfatal injuries in the fire service was $34,100 per incident.

Included in personnel costs are overtime pay, worker's compensation, medical and injury insurance, and staffing costs. Training replacement personnel may also result from the accident. The organization suffers from the loss of a trained employee, either temporarily or permanently.

If the organization is cited by government agencies such as OSHA, then fines may occur. In addition, changes in mandated safety programs, policies, and procedures will also add costs to the organization.

If the accident involves citizens and the organization is considered liable, then other costs are incurred. Increases in insurance premiums, medical costs, and legal fees can also occur.

Apparatus and equipment repairs or replacement may also result from an accident. If an apparatus is destroyed, an organization may not be able to replace it immediately or at all. Some fire protection agencies are self-insured and will not receive an insurance payment to purchase a new vehicle. Funding must be provided from another source and may result in some services being reduced or eliminated.

The organization will also suffer psychological damage from an accident. If fatalities result, either members or citizens, the organization may suffer from a loss of internal morale. At the same time, the image of the organization may suffer if the accident is shown to be the result of organizational failure. A Critical Incident Stress Management (CISM) program should be maintained by management for the benefit of personnel to reduce the effects of psychological damage.

Increased health insurance costs or a decrease in insurance benefits may also have a direct effect on the individual. Loss of income, increased medical costs, and other associated expenses can add to the stress caused by an accident.

Community

The effects of an accident also include financial and psychological aspects. The community may have to adjust priorities in order to pay for the replacement of equipment or increased liability or worker's compensation insurance premiums. Although financial assistance is sometimes available from other levels of government, this aid may not cover the total loss and may be delayed. For example, the winter ice storms that struck Oklahoma in January 2007, were estimated to have caused $39 million in damages in 23 counties. Federal aid, however, rarely reimburses 100 percent of the damages or costs incurred by

Costs of Accidents

In 2000, the National Safety Council (NSC) issued a summary of the cost of workplace accidents in the U.S. in 1999. The costs were divided into the following four categories:

Wages and lost productivity	$ 64 Billion
Medical costs	20 Billion
Administrative	34 Billion
Fire and vehicle damage	4 Billion
Total Costs	$122 Billion

Add the cost of lost-time days, $124 Million, and the overall total was approximately $123 Billion.

local government. The difference between the actual loss and the amount provided by the federal government must be absorbed by the state or local government or the public.

In 2004, the NSC estimated that the cost to society for unintentional injuries due to accidents was $574.8 billion. This figure includes the total of all costs, including insurance, medical, lost wages, productivity, work days, training, etc.

The community may suffer a loss of industry and jobs from an environmental disaster when facilities or infrastructures are destroyed or workers are displaced. For instance, a grain silo that is destroyed may reduce the number of people who travel to the community when delivering grain, thereby reducing the local sales of clothing, food, or gasoline. With the loss of businesses, the community loses tax revenue that it needs to rebuild.

The psychological affect on the community can be devastating. Confidence in the government and the agencies that people depend on to protect and help them can be lost. The prime example of the loss of confidence is Hurricane Katrina in 2005 **(Figure 1.15)**. Not only was New Orleans devastated but much of the Gulf Coast also suffered. All levels of government failed to provide what the residents expected. However, some local exceptions were able to provide the expected support.

A community can also suffer a major psychological effect by the loss of a single structure. The loss of structures such as government buildings, historic landmarks, or places of worship that are central to the life of the community can have a major effect.

Small accidents can have an effect on the community as well. The family, coworkers, and friends of emergency responders who are injured or killed will suffer major distress from an incident. Litigation that may result from an accident can also have an effect on the community, in some cases causing people to take sides in the issue.

Reduction

It should be obvious that a proactive approach to safety can reduce the occurrence of accidents and, therefore, the consequences. The responsibility for a proactive response rests with the community, the organization, and with you, the individual.

Figure 1.15 Citizens living along the Gulf Coast lost confidence in the ability of the government to help them following Hurricane Katrina in 2005. *Courtesy of U. S. Federal Emergency Management Agency, Marvin Nauman, FEMA Photographer.*

The community must be willing to expend time, resources, and funds for a community risk assessment. Once the assessment is complete, the community must establish the necessary programs and implement the recommendations. Certainly environmental accidents cannot be prevented. However, preparing for them through planning, training, and communications can reduce their effect.

The organization is responsible for having a safety program. Establishing a safety program such as the one outlined in NFPA® 1500 is less expensive than paying the costs associated with an accident. This topic is addressed in Chapter 11, Safety and Health Program Administration.

Finally, you are responsible for your own safe behavior. As mentioned before, 88 percent of all accidents are caused by unsafe acts, which include the following actions:

- Ignoring safety policies or procedures
- Modifying safety equipment
- Modifying safety procedures
- Attempting tasks that you are not trained or equipped to perform
- Taking risks
- Taking shortcuts

- Allowing a sense of urgency to replace common sense
- Ignoring warning signs or indicators of a hazard
- Being overconfident
- Failing to recognize and analyze your own near misses
- Participating in unsafe horseplay
- Working while under the influence of drugs, narcotics, or alcohol
- Ignoring signs or symptoms of illness

You are also responsible for the actions or inaction of those you work with. If you recognize unsafe behavior, you should correct it. Just because the incident safety officer has the authority to stop unsafe behavior at an emergency scene does not relieve you of the same responsibility.

Summary

You should now have an understanding of the hazards that you will encounter in the fire and emergency services. The specific hazards you face will be determined by where you live, the types of activities that occur in your community, and how you act. The risks these hazards create will determine how you respond to the hazards.

By understanding the risk management model and applying it, you can reduce the level of risk to yourself and your organization. You should also be able to recognize why accidents occur, what causes or contributes to them, and how to prevent them from occurring. The information in this chapter should motivate you to act in a safe manner, act as a positive role model for others, and correct unsafe actions or conditions when you see them.

Review Questions

1. Discuss the differences between the five categories of hazards used by the NFPA®.
2. What three aspects of potential incidents does the all-hazard concept explore?
3. Discuss the difference between hazard and risk.
4. What types of factors can cause or contribute to occupational risk?
5. Explain how multiple causes can contribute to an accident.
6. What are possible organizational consequences of an accident?
7. Discuss the elements that make up the Three E's of Safety Model.
8. How are frequency and severity used to evaluate risk?

Cultural Change in the Fire Service

Chapter Contents

Divider page photo courtesy of Bob Esposito.

chapter 2

■ FESHE Learning Objectives

Occupational Safety and Health for Emergency Services

Describe the history of occupational health and safety.

Identify occupational health and safety programs for industry and emergency services today.

Cultural Change in the Fire Service

Learning Objectives

After reading this chapter, students will be able to:

1. Describe the current fire and emergency services culture.

2. Explain how the current culture influences an organization's safety efforts.

3. Explain how to recognize the need for cultural change in an organization.

4. Identify barriers to change.

5. Describe methods used for changing an organization's culture.

6. Select an effective change model for improving the safety culture.

7. Explain the results that a change in the safety culture will have in an organization or individual.

8. Describe the safety initiatives that can help in causing a change in the safety culture.

Chapter 2
Cultural Change in the Fire Service

Case History

While responding to an alarm called in by an alarm-monitoring company, the dispatch office notified the responding units that it was a confirmed false alarm. Because the fire department had no cancellation policy, units proceeded to the incident scene in nonemergency manner to confirm the alarm. Upon arrival, it was determined that there was no one present with a key to open the residence. When the owner was notified, he stated that if his wife's car was in the garage, she was home. After seeing a car in the garage, the company officer did a complete 360 survey. It was determined with information from the owner that this very well could be a medical emergency. At this point, a firefighter noticed that a woman was laying face down in the garage and not moving. The crew used forcible entry through a garage window on the side of the structure. One firefighter gained entry as another followed through the window. At this time the interior firefighters, with no respiratory protection on, relayed to the exterior firefighter that the vehicle inside the structure was still running and breathing apparatus was needed. The firefighters inside tried to open the garage door, but realized the power was cut to the garage. The interior firefighters finally opened the garage door manually. At this time the scene was deemed a suicide by carbon monoxide (CO) poisoning. All three firefighters at this point were inside the structure assessing the patient. The firefighters were inside a CO-rich environment with no ventilation and no respiratory protection. Although they only spent a few minutes inside the CO-rich environment, they could have been permanently injured or killed. All the firefighters exhibited the side effects of CO poisoning including headaches that lasted up to two days after the incident.

The desire to rescue a victim should not prevent you from protecting yourself. Always take the time to assess the situation, wear the correct type of personal protective equipment, and be aware of your surroundings and the clues that indicate potential hazards. In a safety culture, you must think before you act.

Source: *National Fire Fighters Near-Miss Reporting System*

If the fire and emergency services are to reduce the number of line-of-duty fatalities, injuries, illnesses, and property loss, we must change our culture. The traditional culture that placed an emphasis on heroism at all cost must give way to one that rewards safety. The change must be complete and permanent to ensure that *everyone goes home* and that we do not fall into a *cultural drift* that takes us back to a time when casualties were an accepted part of our profession.

Individual change is not easy. Examples of personal behavioral changes that can take time and resources are losing weight or stopping smoking. Cultural change within an organization or profession is even more difficult because it can mean going against tradition. In the fire and emergency services, tradition has molded the professional image that we hold so dear.

We take pride in the fact that we expose ourselves to danger in order to protect the citizens of our community. The citizens also hold us in high regard as firefighters and emergency responders. These internal and external images can create barriers to the change of our culture.

To create an organizational and professional safety culture, we must take aggressive actions such as the following to cause the change:

- Understand what our current culture is and on what it is founded.
- Recognize the need for change and the barriers that may inhibit it.
- Understand the various models for change that can help us make the change.
- Recognize the results of the change and know the benefits it will bring.
- Be aware of the various programs and initiatives that we can use to support the change to a new safety culture.

This chapter provides a brief look at the history on which the current culture is based, the cultural changes that are currently active in the fire service, and sample change models that can be used to facilitate change. Finally, a variety of safety-related initiatives that are intended to reinforce the need for changing the culture are described.

Current Fire and Emergency Services Culture

The current culture of most fire and emergency services organizations is based on the history of the fire service in general. From the paramilitary organizational structure and rank structure to the can-do attitude, the fire and emergency services has a strong historical image. Changes have been difficult and have taken years, if not decades, to become established in the fire service.

To understand the current culture and how it affects issues of safety, you must understand the past and its influence on the fire service. This section contains the following elements:

- Discusses the general history of the fire and emergency services and how a cultural image emerged
- Introduces the changes in service delivery and its effect on the culture
- Shows how demographic changes, including a culturally diverse and aging population, have created a need for change
- Shows how the current culture results in unsafe actions and creates unnecessary costs to the community, organization, and individual

History of the Fire and Emergency Services

Organized fire protection dates from the Roman Empire around 24 B.C.E. (Before the Common Era). As people crowded into Rome, constructing shanties made of combustible materials and cooking on open fires, destructive fires became prevalent. This led Emperor Caesar Augustus to form the *familia publica*, a group of 600 servants stationed by the gates of the city for the express purpose of fighting fires.

In 6 C.E. (Common Era), following another disastrous fire, Emperor Augustus instituted the *Corps of Vigiles*. This group would protect Rome from fire for the next 500 years. The Corps was a night patrol of slaves who checked for fires and alerted the town if a fire was discovered. The Vigiles, correctly dressed and equipped with buckets and axes, also fought fires and enforced fire prevention measures. Rome was divided into districts and protected by about 7,000 Vigiles. The Corps was organized by a ranking system of officers and ordinary firefighters. This organization was similar to that used by today's fire and emergency services organizations.

As civilization spread through Northern Europe and people congregated in cities, fires continued to devastate society. Small cooking fires could result in the destruction of villages, towns, and cities. Following the fires, inhabitants would rebuild and start over.

Local communities would pass laws requiring the use of certain types of building materials such as slate roof tiles instead of thatched straw, but no real attempt to form fire-fighting organizations existed. Some communities enacted laws that required *able-bodied men* to respond to an alarm and gave them authority to pull down burning buildings or use gunpowder to destroy buildings to form a fire break.

The first attempt at forming official fire-fighting units occurred following the Great London fire of 1666 **(Figure 2.1)**. City authorities recommended many new proposals to improve fire-fighting methods, but their suggestions were not taken. The theory of the *hand of God* in fire disasters had such a hold on citizens' mentality that their thoughts never turned to skilled fire brigades or when they did, the cost was considered too high.

Figure 2.1 Following the Great London fire of 1666, recommendations were made for improving fire-fighting. These recommendations were rejected and not implemented.

As a result of the London fire, the first fire insurance company was formed. The creation of the fire insurance company was to have a major effect on the fire service of Britain. In turn, those traditions and ideas were in turn spread to North America.

To protect their interests, British insurance companies formed fire brigades that would respond to fires in structures that the company insured. The properties of the insured were marked with a metal *fire mark* that was associated with the owner's insurance company. No attempt would be made by a fire brigade to extinguish a fire in a building not insured by their employers. However, they would protect their insured buildings if they were exposed to the primary fire.

In 1717, Boston established America's first fire department. Because of the terrible fires that the city had experienced, having organized fire protection was vital to its existence. The following year, the first mutual aid fire society was formed. This event was the beginning of the colorful age of the volunteer firefighter who was to play an important role in American history. The mutual aid fire societies operated independently and volunteered assistance to the regular Boston municipal firefighters.

In 1736, Benjamin Franklin established the first fire organization in Philadelphia, the Union Volunteer Fire Company, patterned after the Boston mutual aid fire societies. The Union Volunteer Fire Company responded to any fire calls in its vicinity, unlike the Boston mutual aid fire societies that responded only to members' properties. During this same period, American insurance companies were established that formed fire brigades to protect their member's homes and businesses **(Figure 2.2)**.

Figure 2.2 The fire mark was symbolic of early Colonial America and used to indicated that a structure was protected by a certain volunteer fire company.

Because towns had no permanent firefighters and volunteers received no pay, insurance companies or town treasuries would authorize pay to the first-arriving firefighters who put water on a fire. This incentive set in place the fierce rivalries that developed among volunteers in later years. Thus the desire to be *first due* or be the *first to get water on the fire* became a tradition that firefighters still attempt to achieve.

Cultural Image Emerges

The creation, development, and competitive spirit of the American volunteer firefighter system did several things for the fire service. The zeal to beat rival companies led manufacturers and firefighters to work continuously at developing new and better apparatus, equipment, and methods of fire fighting. Firefighters practiced making fast responses to fires, drilled to save time in putting engines to work, and experimented with the *bunking-out* system (sleeping at the station).

It was also during these early years that firefighters emerged as heroes. Their daring rescues and heroic stands before fires were captured in song and verse as well as in the drawings of Currier and Ives **(Figure 2.3)**. Young boys across America wanted to be firefighters when they grew up. The firefighter's exciting and colorful job was of major interest to the townspeople who would gather to watch firefighters in action. Competitive pumping contests would bring cheers and encouragement for local favorites.

Early volunteers were selected carefully and underwent a thorough character investigation and trial period before the company accepted them. Most of the members were men of means — merchants, manufacturers, and profession-

Figure 2.3 The cultural image of the heroic firefighter dates to the earliest traditions of the fire service as shown in this historical print.

als — who kept their fire companies as socially exclusive as a private country club. To be a firefighter was to be *somebody*, and everybody who was *anybody* usually joined.

Volunteers were expected to endure extremes of weather; the hazards of fire, smoke, and falling debris; and physical exhaustion. Every fire was a challenge, and the volunteers reveled in fighting it. The status of belonging to a fire company, the teamwork, and the pride in the beautiful machinery made membership in the company very desirable.

The heroic image of firefighters carried over into times of national crisis. At the beginning of the American Civil War, members of many volunteer fire companies in New York City volunteered to fight for the Union as the 11th New York Volunteer Infantry, Fire Zouaves, and the 2nd New York Fire Zouaves. Firefighters from Lancaster and Philadelphia, Pennsylvania, also formed regiments to support the Union cause **(Figure 2.4)**. In the Confederacy, firefighters from Atlanta, Georgia, and Norfolk, Virginia, exchanged fire-fighting equipment for muskets and bayonets. The pride of belonging to a fire company was a strong bond even in times of war.

The influence of the military image does not end with the American Civil War. The relationship between the fire service and the military may be found in the lexicon of terms that describe what firefighters do: We *fight fires*, wear *helmets*, *attack* the seat of the fire, and take *casualties*. We place a great deal of importance on taking care of our own and being ready to rescue a fallen firefighter. Our uniforms have a similar appearance to military dress uniforms, as does our rank system and the terms we use to describe our unit structures.

Based on the history of the fire service, certain characteristics can be found throughout it. The following characteristics have created the culture that has and will continue to result in firefighter fatalities and injuries:

Figure 2.4 During the American Civil War, volunteer firefighters laid down their axes and nozzles and picked up muskets and bayonets to fight for the Union.

- Pride
- Heroism
- Duty
- Loyalty
- Sacrifice
- Teamwork
- Competition
- Machismo
- Volunteerism
- Apathy
- Bulletproof
- Affiliation

Certainly, not all of these characteristics are negative and some, like teamwork, can be the basis for changing the culture. Some have changed as the demographics of the fire service have changed. Others have changed because the hazards have changed. However, the current organizational culture has costs that must be reduced or eliminated.

Diversity of Service Delivery

As mentioned in Chapter 1, Risks of a High-Hazard Occupation, hazards and their associated risks have changed over the past half century. The all-hazards concept now means that firefighters must be prepared to face multiple types of hazards. And those hazards require a far different approach than the ones used to *fight fires*.

While fire fighting can never be considered simple or easy, the centuries of fire fighting have established approaches that are fairly consistent. Water, or some type of extinguishing agent, is applied to the fire until it is extinguished.

Protective equipment is designed to repel heat and smoldering debris as well as water to prevent steam burns. Tools such as pike poles have long been used for the task of pulling down walls and ceilings to expose burning structural members.

With the advent of the all-hazards concept, firefighters must take time to perform the following actions:

- Determine the type of hazard and the level of risk it poses to them. For instance, has the fire damaged the structural integrity of the building, making it too hazardous to enter the structure?

- Select the best approach to dealing with the hazard. For instance, should we elect to go inside or adopt an defensive mode to protect exposures?

- Select the appropriate type of equipment required to control or mitigate the hazard at an acceptable level of personal risk. For instance, on a brush fire, should we use aircraft, should we drop water or chemicals, should we use dozers to construct a fireline or use torches to set backfires?

The days of rushing into the electrical vault to extinguish a fire, no matter the risk of electrocution, or diving into a swollen river to make a rescue without proper equipment and training are long over. Taking time to evaluate the hazard decreases the level of risk and increases our ability to deal with it effectively.

In the years following World War II, the manufacture, storage, transportation, and use of hazardous materials has increased exponentially. Fire and emergency responders regularly respond to chemical spills, hazardous vapor leaks, and fires involving plastics, flammable metals, and other materials that are difficult to extinguish and produce hazardous vapors when burned.

The increasing risks caused by these materials has resulted in changes in response strategy, improved personal protective clothing and respiratory protection, and federally mandated training in hazardous materials response. Greater degrees of knowledge and certification have led to the creation of special units that are trained and equipped to respond to these situations.

The next change in types of service came in the form of emergency medical responses **(Figure 2.5)**. A variety of factors led to this change, including the following:

- Increased costs associated with providing prehospital care and transportation

- An aging population that has experienced life-threatening medical conditions that require immediate attention

- Development of new contagious and fatal diseases

- The increasing reliance on emergency-room care rather than a primary care doctor

As firefighters have expanded their roles as first responders to medical emergencies and taken on greater emergency medical responsibilities, the need for personal protective equipment (PPE) that is certified to protect against bloodborne and airborne pathogens has increased. Not all of our traditional structural fire-fighting PPE is certified for that level of protection. Additional technical training and certification has also become necessary for emergency medical responders.

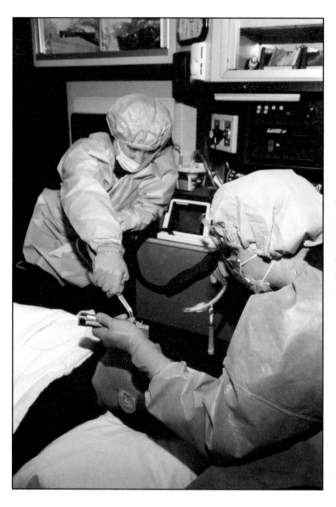

Figure 2.5 The trend toward fire departments providing emergency medical services has been increasing over the past 20 years. *Courtesy of Captain Mike Nixon, Portland (ME) Fire Department.*

The fire service's assumption of a primary role in rescue from buildings severely damaged from earthquakes, tornados, floods, etc. gave rise to the Urban Search and Rescue (USAR) programs creating a whole new genre of equipment, training, procedures and operational safety issues.

Water rescue has evolved from an incidental of fire service functions to a full-blown specialty function with its own equipment, training, and safety issues.

While firefighters have always responded to natural disasters, the increase in the population and the expansion of urban and suburban communities into areas that are prone to natural occurrences has increased the number of responses. Wildland/urban interfaces, the areas where forests and residences coexist, are becoming more prevalent as are the fires in these areas. These fires increase the need to understand the unique safety issues involving this area of fire fighting and learn new skills (e.g., wildland fire behavior, managing aircraft safety, etc.) to protect ourselves as well as our citizens.

Finally, the 21ˢᵗ Century has witnessed an increase in global terrorism. weapons of mass destruction (WMDs), once the concern only of the Defense and State Departments, are now considered hazards for all first responders, including hospital emergency room personnel. Firefighters must be trained in the recognition and response to such devices, equipped to handle them, and able to mitigate the results of their use. The risks include both the direct effect of WMDs and the psychological affect on the individual who has responded to such an incident.

Cultural Differences

As hazards have changed over the past half century, so have members of the fire service. In 1950, the majority of fire departments were composed of white males who were born in the 1920s and 1930s. Therefore, their *worldviews* were limited and influenced by their ethnicity, gender, and generational values. Today, the fire and emergency services are a reflection of the demographics of North America: multiethnic, multigender, and multigenerational, which results in a multicultural atmosphere.

Figure 2.6 The composition of fire departments often reflects the local population and includes firefighters of all racial, ethnic, and cultural backgrounds.

Ethnicity

Because the fire service was originally fraternal, it traditionally attracted men who had similar ethnic backgrounds. The New York City Fire Department was populated by men of Irish decent, while departments in Pennsylvania attracted Germans.

Shifts in populations that resulted from the Great Depression (1929–1940), the Dust Bowl (1933–1940), and World War II (1941–1945) changed the demographics of the U.S. and Canada. Populations became ethnically diverse as people moved to find employment and improve their lives. In addition, a new wave of immigration following World War II resulted in an influx of people from East Asia and Central Europe.

Finally, the integration of African Americans into mainstream culture beginning in the mid-1950s has changed the composition of the fire service. Beginning with the integration of the armed forces in 1948 and public schools and accommodations in 1955 along with the Civil Rights movement of the 1960s, African-Americans have continued to become a larger part of most fire and emergency services organizations **(Figure 2.6)**. The fire and emergency services reflect this shift in ethnic composition. With this shift comes a change in cultural values of an organization. Views on authority, work ethics, and loyalty to an organization change as new members join the organization and gain rank and influence.

Gender

One of the major changes in the fire and emergency services is the addition of females into career organizations. Women have long played an active role in volunteer fire departments, but only in the past 35 years have they been hired by career organizations. Currently, over 6,200 females hold positions in career organizations **(Figure 2.7)**.

As departments hired female firefighters, the image of the all-male fire service changed. The presence of females has altered the daily living conditions in the station, changed the attitude of male firefighters toward women, and provided a more nurturing atmosphere when dealing with victims. Studies of female workers have also shown that they are more safety conscious than their male counterparts. Also, women tend to think of different ways to accomplish a task. Where men will often use brute strength, women will try to discover how to get the job done by other methods that are sometimes safer than just using raw strength.

Figure 2.7 While some volunteer departments have included females since World War II, this is a more recent change in career departments.

The presence of women in the fire service has helped to match a diverse composition of the community, which is composed of 53 percent women. At emergency incidents, female firefighters are able to relate to female victims to help allay fears. Many times a woman victim feels another woman understands how she is feeling. Nurturing and compassion help on many emergency incidents as well, especially when children are involved.

Generational

The composition of fire departments is always changing. As older members retire, younger recruits are hired to take their places. However, with the advent of pension plans that permit members to retire after 20 years, the average age of them has decreased. No longer is it necessary to work until you are 65 years of age in order to draw a retirement check.

The differences in generations, or age groups, are determined by personal life experiences. Things that are important to a person born in 1930 are vastly different from what is important to someone born in 1990. See the information box for the cultural characteristics of various generations of Americans.

Generational Diversity

North America is composed of people who have been born between the late 1920s and the present. The population has been divided into the following four distinct groups that have varying characteristics:

- Traditionalists
- Baby boomers
- Gen Xers
- Dot.coms, millenniums, or nexters

Continued on page 56

Generational Diversity (*Concluded*)

You should remember that these categories are artificial and very general. Members of each group are *not* identical. The influence of family, history, tradition, and media create individuals within one group that have strong ties to the characteristics of the previous generation.

Traditionalist

Traditionalists were born between 1920 and 1944. They have the life experiences of the economic boom and bust of the 1920s, the Great Depression and the Dust Bowl (for some) of the 1930s, and the tragedy and horror of World War II in the 1940s. Their values include loyalty, patriotism, hard work, and education. They tend to be fiscally conservative and place a high value on institutions such as universities, corporations, and religion.

Baby Boomer

Currently, this group composes the largest portion of the native-born population in the U.S. They were born between 1946 and 1964. They are idealistic and place a high value on fairness, equality, hard work, and competition. However, they also have a history of questioning authority and wanting to know why something is important. The leadership and upper ranks of fire and emergency services organizations are composed of members of this group. They also have the life experiences of war, social turmoil, and economic recession and affluence. Baby boomers tend to place a high value on education, family, and personal leisure time.

Gen Xer

These are the children of baby boomers, and they were born between 1961 and 1980. They require personal flexibility and thrive on feedback from supervisors. However, they do *not* like constant supervision and prefer to work independently. Gen Xers are also considered entrepreneurs. They try to balance work and leisure time in their lives. Because of the mobility of their parents, the high divorce rate in the U.S., and the rapid expansion of urban America, they are used to change. This group composes the majority of fire and emergency services organizations.

Dot.com, Millennium, or Nexter

This age group could be considered the future of society. They were born after 1980 and have the following characteristics:

- Raised on technology
- Enjoy visual stimulation (computer games and television)
- Are optimistic
- Appreciate diversity
- Ready to work and learn but usually want instant gratification in the form of tangible results for their efforts
- Have broad worldview
- Accept nontraditional families and lifestyles
- Have high expectations

Since the beginning of the new century, they have been exposed to terrorism and war and the uncertainty that both bring with them. They are experiencing what the traditionalists did in World War II and the baby boomers did during the Vietnam War. In addition, those individuals who were activated as part of military reserve and National Guard units have been faced with drastic changes in their personal and family lives. This generation will continue to fill the ranks of the fire and emergency services and bring with them these experiences and values.

As the members of each generation advance through the ranks, their world-views influence the decisions they make for themselves as well as the departments. For instance, a fire chief in 1940 would not have been concerned with firefighters smoking cigarettes in the station. Today, the use of cigarettes or other tobacco products is banned in most fire stations and facilities. Members of each generation may have different opinions about the effects of smoking and the authority of an organization to control what some may argue is a personal choice.

As young members join the fire service, they accept the current level of safety as normal or expected. Therefore, new personnel will adhere to the safety policies and use safety equipment and procedures more readily than older veterans. Also, the younger members see the effects that lower, nonexistent, safety standards or the failure to enforce them have had on older members and do not want to be placed in the same positions.

Costs Resulting from the Current Culture

Any attempt to quantify the exact cost of fire and emergency services' fatalities, injuries, and job-related illnesses is very challenging. The fire service, as well as each level of government and the insurance industry, has done an inconsistent job of collecting data related to the types and cost of injuries and illnesses.

These inconsistencies became apparent in 2004 when the National Institute of Standards and Technology (NIST) commissioned an analysis of the economic consequences of firefighter injuries. The results of the report, which are included in NIST GCR 05-874 and available from the NIST web site, are based on the best possible estimates for the years 1992–2002.

Firefighter injuries and job-related illnesses result in both *direct* and *indirect costs.* Direct costs usually include all costs related to treating and compensating for the injury or illness. Components of direct costs according to the NIST Report are as follows:

- Lost wages to the firefighter that exceed disability payments
- Overtime wages, above the cost of the injured firefighter's wages, to substitute for the firefighter
- Medical expenses
- Psychological counseling for pain and anguish suffered by the firefighter, the firefighter's family, and (occasionally) the firefighter's coworkers
- Time spent by the firefighter, supervisors, and others in investigating the incident and writing an injury report
- External investigations
- Disability/retirement income (when firefighters cannot return to work for years or retire on disability short of normal career lengths)
- Litigation expenses

Indirect costs are those associated with preventing or mitigating the injury and include the following:

- Additional staff sent to emergency incident scenes for preventing and dealing with firefighter injuries (staff to meet the two-in/two-out rule, for rapid intervention teams, or to serve as safety officers)

- Firefighter protective equipment, including respiratory protection
- Maintenance of protective equipment (cleaning and inspecting protective equipment after each use)
- Training for firefighter safety (for example, how to escape in extreme situations or buddy breathing)
- Administrative costs of insurance
- New safety technology such as sensors that are being developed to locate lost firefighters
- Health and safety officers
- Union time and fire-management time spent on negotiations related to safety
- Safety aspects of firefighter personnel record keeping

The NIST report determined that there is no single all-encompassing method for determining the annual injury costs for firefighters. The results of the study, based on five analysis methods, indicate that the cost of firefighter injuries ranges from $2.8 billion to $7.8 billion annually.

By comparison, the cost of safety training ranges from $65 million to $146 million annually. The cost of wellness and fitness programs was estimated to be $75 million. Obviously, investments in training and fitness are much less than the costs generated by actual injuries. Other cost estimates determined in the NIST report may be found in subsequent chapters of this manual.

Recognition and Acceptance of Cultural Change

We still see evidence of a lack of focus on preventing firefighter injuries. Too often photographs in our industry magazines show firefighters operating in an unsafe manner, e.g., without appropriate PPE, on roofs without appropriate safety precautions, with no second means of egress, no safety man with the roofcuter; firefighters attempting to make ice rescues without appropriate safety equipment; and hose streams being directed into ventilation holes with firefighters still inside the structure.

NFPA® and U.S. Fire Administration (USFA) statistics make it clear that lack of physical conditioning resulting in fatalities caused by cardiac arrest is still our service's most deadly risk. Weekly we see stories and pictures of fire apparatus accidents, often due to poor driving or faulty equipment, that result in the second highest number of firefighter fatalities **(Figure 2.8)**.

It should be evident from the previous sections that the fire and emergency services must change in order to create a safety culture that is effective. The top three issues that result in unnecessary injuries and fatalities year after year are: 1) lack of physical conditioning, 2) failure to consider risk versus value of building/life hazard, and 3) failure to insist on safe and sane driving.

The fire service must not continue to accept the current level of fatalities, injuries, illnesses, and their associated financial burdens. To create change, we must progress in the following ways:

- Recognize that we are participating in an unsafe culture.
- Accept the need for change.

Figure 2.8 Not all apparatus accidents result in fatalities, although property damage may occur. *Courtesy of Chief Mike Mallory, Tulsa (OK) Fire Department.*

- Recognize the barriers to change.

- Remember that we exist to remove the problem, not make it worse by injuring ourselves or others through unsafe acts.

Once these change tasks have been accomplished, we can select a method for making the change and implement it. These tasks can be used for organizational or individual change.

Recognition of Counterproductive Cultures

The current culture, which is based on tradition, heroism, peer acceptance, and public image, must be recognized for what it really is: a *counterproductive culture*. By adhering to the current culture, job-related fatalities, injuries, and illnesses will continue at the current level rather than decreasing.

In order to change this culture, we must first recognize that it exists, both in our own organizations and in the fire and emergency services as a whole. **Table 2.1, p. 60,** shows the characteristics and built-in activities of an unsafe and counterproductive culture.

Once these characteristics or symptoms are recognized, it is necessary to establish the need for change, which can be accomplished by linking each symptom to the negative results they can cause. For instance, the cost of not enforcing the minimum physical fitness and weight requirements can be shown in such costs as the following:

- Sick leave

- Special-sized protective clothing

- Worker's compensation claims for cardiac-related medical conditions

- Damage to station furniture

- Increase in medical insurance premiums

- Need to shift tactical focus from incident control to providing care to a firefighter who needs assistance at an incident, thereby delaying control of the incident and increasing damage

Table 2.1
Counterproductive Culture Symptoms and Built-in Activities

Unsafe Characteristics or Symptoms	Statistically Based Symptoms	Built-in Counterproductive Activities
• The attitude that *we have always done it this way*. • Safety is for wimps. • Self-sacrifice is a job requirement. • You get paid to take risks.	• Increase in job-related injuries • Increase in workers compensation claims • Increase in near-miss incidents • Increase in the use of sick days • Increase in equipment and apparatus repairs • Occurrence of a job-related fatality	• Fail to enforce minimum physical fitness requirements and testing. • Hold the employee solely responsible for an accident. • Use fault-finding techniques rather than fact-finding ones when trying to determine the cause of an accident. • Use threats or discipline to correct unsafe behaviors. • Reward unsafe behavior by underreporting accidents, injuries, and near-misses. • Fail to reward supervisors for enforcing safety policies, procedures, and behaviors. • Fail to hold personnel accountable for knowingly making unsafe choices. • Lack of management leadership in creating and enforcing a safety culture. • Fail to make safety an important part of every training experience

Additional documentation can include the number of personnel who are considered to be obese based on national standards. Obesity, which can be largely controlled by proper nutrition and exercise, can contribute to other diseases such as diabetes and high blood pressure.

Once the cost of a counterproductive culture has been quantified, the cost of prevention should be calculated. As mentioned previously in this chapter, NIST has attempted to provide general cost estimates for developing a safety culture. Its analysis can be used to create a local estimate that can be compared to the local cost of an unsafe culture.

Acceptance of the Need for Change

The next task we must accomplish is to accept the need for change. Because the fire service organization is composed of individuals, the realization that change is necessary must start with you. Chapter 1, Risks of a High-Hazard Occupation, provided the statistics that underscore the need for change: approximately 100 firefighter fatalities and over 100,000 injuries per year. Add to these figures the number of job-related illnesses that go unreported, and the need for change becomes evident.

But, these are just statistics and average firefighters tend to ignore them, believing that they will *not* become part of these numbers. This optimistic attitude, sometimes referred to as *being bulletproof,* justifies our actions and can cause us to behave in unsafe ways.

The truth is that we can all become one of those statistics. The cause, like those hazards and risks listed in Chapter 1, may be external or internal. To keep from becoming a statistic, we must be proactive and work to change the very culture that is threatening us. Reading this manual and mentally deciding that change is needed in your organization, your profession, and yourself are the first steps.

Barriers to Change

Unfortunately, there will always be barriers to change. Some of these barriers are psychological (internal) while others are generated by the organizational culture (external). An understanding of the various types of barriers will help you recognize and overcome them before they derail the change process. Change can also fail if the organization uses inappropriate methods to initiate and promote change.

Psychological (Internal) Barriers

Internal barriers are the result of psychological conflicts and beliefs within the individual. These barriers can prevent the individual from making personal changes in lifestyle or behavior or in accepting the need for organizational change. These barriers **(Figure 2.9, p. 62)** are described as follows:

- *Fear* — Remains one of the main personal barriers to change; may be in the form of any number of anxieties including the following:
  - Loss of power
  - Loss of skills
  - Loss of income
  - The unknown

  Supervisors may resist change because of fear of losing control and an overload of current tasks and responsibilities that may be caused by a change.

- *Peer pressure* — Causes an individual to behave in a manner that is acceptable to others within the group. The influence of a social group on an individual can create a major barrier to change. At the same time, peer pressure can have a positive effect on an individual who is showing inappropriate behavior. Although peer pressure is an external influence, the reaction by the individual is internal.

- *Lack of awareness* — Causes an individual to misunderstand the need for change; may also misinterpret information that is provided to justify a change.

- *Distrust* — Causes lack of trust in one's own ability to implement a change and in others to provide accurate information and support to implement the change; result of low self-esteem that can cause paralysis and keep any change from occurring.

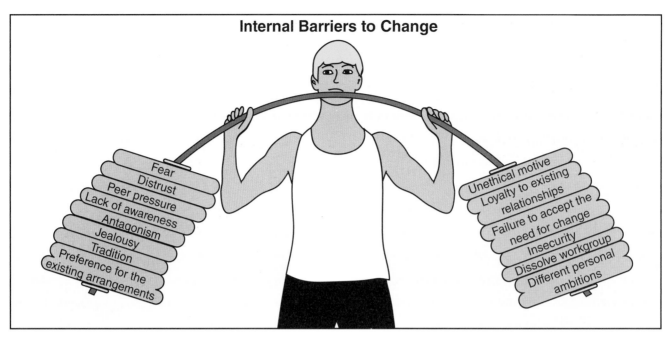

Figure 2.9 Internal Barriers to Change.

- *Antagonism* — Means that active opposition to change by others causes the individual to compromise even though the compromise might not be the best solution to the problem; may be the result of ego, personality differences, or jealousy.

- *Jealousy* — Causes an individual to act irrationally and block the suggestions of others. The decision-making ability of a committee can be crippled if one or more members are jealous and respond in a negative manner.

- *Unethical motive* — Causes an individual to seek personal gain, enhanced self-image, and personal protection at the expense of others. Bad decisions can be generated when the reason for making them is unethical. Eliminating this barrier requires an insistence that decisions be motivated by and based on an ethical foundation. Personal agendas need to be subordinated to the good of the organization and solution of the problem.

- *Tradition* — Bases lack of change on the history of the organization and individual loyalty to that history. Change is considered in direct opposition to *the way we have always done things*. Change may devalue the old ways and call them into question.

- *Loyalty to existing relationships* — Forms a strong bond that creates a barrier to change; may take two forms: (1) loyalty to the views of a respected superior who does not agree with the change or (2) loyalty to a supervisor who may see change as a threat to the unit composition.

- *Failure to accept the need for change* — Fails to see a need for change and rejects any argument for change; basically, *things are fine the way they are, don't confuse me with the facts.*

- *Insecurity* — Causes an internal feeling that an individual is incapable or afraid of handling a situation; generally founded on a lack of self-esteem.

- ***Preference for the existing arrangements*** — Prefers the comfort of the status quo or how things are arranged at present; common trait among most people.

- ***Dissolve workgroup*** — Fears that change will require the unit to be dissolved; similar to the desire to keep the status quo and to stay with the current supervisor.

- ***Different personal ambition*** — Conflicts with the individual's personal agenda; for instance, a proposal that officer candidates hold a college degree may be a conflict to someone who wants to promote but does not have the educational requirements.

Organizational (External) Barriers

External barriers that exist outside the individual may be created by the organizational culture or structure and are described as follows **(Figure 2.10)**:

- ***Structural inertia*** — Exists in any organization and may prevent not only change but any progressive movement. Tradition, status quo, and the organizational structure can all create inertia.

- ***Existing power structure*** — Interprets the proposed change as a threat to the group's influence and power; may be the labor union, a specific division, or officers of a certain rank. The members of these groups may band together to oppose or even stall changes.

- ***Resistance from workgroup*** — Opposes changes because of the perceived threat to workgroup status quo. For instance, fire-suppression personnel may oppose performing fire and life-safety inspections because of the time required to perform them.

- ***Failure of previous change initiative*** — Attempts previously to institute change may have failed for a number of reasons. Each of these reasons must be analyzed and corrected to ensure that the current initiative succeeds.

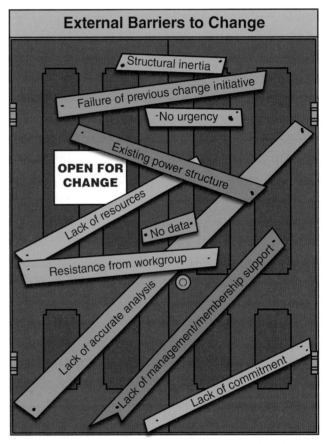

Figure 2.10 External Barriers to Change.

- ***Lack of urgency*** — Neglects to create a sense of urgency for changes; without a sense of urgency, members of the organization will not see the value of the proposed changes and will lose interest in adopting the proposal.

- ***Lack of data*** — Includes the lack of accurate, sufficient, or timely information and can result in making no decision, making the wrong decision, or accepting a compromise that is not adequate to completely resolve the problem through change.

- ***Lack of accurate analysis*** — Leads to an unacceptable decision even when data is accurate, recent, or sufficient. Every effort must be made to look at the total picture of the problem based on all available information to determine the correct relationship of the various parts.

- ***Lack of resources*** — Delays change if necessary resources are not available. The lack of finances, personnel, time, skills, and equipment or the inability to enforce a change can prevent change from occurring. A lack of resources may require that the change be implemented in phases or postponed until the required resources are available.

- *Lack of management/membership support* — Dooms change to failure from the start. Lack of support (such as in the area of allocation of resources) will simply result in a good intention that does not become an action. Support from upper management or the membership is built through the use of team building, open communication, and the empowerment of the membership. Depending on the issue or initiative, building support can start from the bottom up or the top down.

- *Lack of commitment* — Leads to ineffective change decisions. The personal commitment by the *change agent* and those affected by the change are required in order to have effective decisions. The use of oral and written communication skills and public statements of commitment create the image of commitment necessary to ensure acceptance of the change decision.

Inappropriate Change Management

Change is often resisted because of failures in the way it is presented by the organization or change agent. Inappropriate change management includes the following (**Figure 2.11**):

- *Failure to explain the need for change* — Failure to communicate adequate justification for the change to the leadership, membership, or stakeholders prevents them from understanding the need. Change must be based on facts and supported by accurate statistics.

- *Failure to provide information* — Lack of accurate information destroys the credibility of the change agent and prevents the leadership or membership from making an informed decision.

- *Failure to consult, negotiate, and offer support and training* — Failure to consult with the membership and their labor representatives (stakeholders) prevents participants from knowing what changes are needed and why. When necessary, the organization must negotiate to overcome resistance and provide adequate resources to support the change. Training, both in the change process and the elements of the change, must also be provided.

- *Lack of involvement in the process* — Failure to involve the membership and stakeholders in the change process will alienate the people who are affected by the change and those who can influence acceptance of it.

- *Failure to build trust and a sense of security* — Lack of trust by the membership and stakeholders as well as a feeling of insecurity can prevent change from occurring. If members believe that the organization is taking advantage of them or that their jobs are in jeopardy, they will resist any attempts at change.

- *Poor employee relations* — Paternal attitudes toward employees and treating them or their concerns disrespectfully or ignoring requests and suggestions can doom any attempt at change.

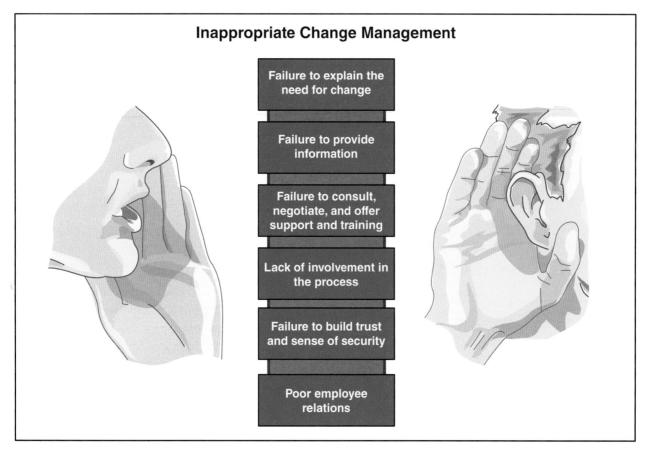

Inappropriate Change Management

Failure to explain the need for change

Failure to provide information

Failure to consult, negotiate, and offer support and training

Lack of involvement in the process

Failure to build trust and sense of security

Poor employee relations

Figure 2.11 Inappropriate Change Management.

Overcoming Barriers to Change

Some simple steps that you can take to overcome the change barriers, both as an individual and as an organization, include the following:

- *Communicate* — Explain the need, process, results, and benefits of the proposed change. A sense of urgency must be stressed to the membership.

- *Involve* — Keep members and stakeholders in the process.

- *Inform* — Keep the members and stakeholders informed of the process, progress, and results of the change.

- *Support* — Provide analysis, resources, leadership, and moral support to the change process.

- *Commit* — Provide strong commitment to the change process and the actual change.

- *Relate* — Explain how the change relates to the organization, membership, and stakeholders as well as to the tradition and history of the organization.

- *Create* — Establish an atmosphere that is receptive to change in the organization; must be based on mutual trust, ethical conduct, and job security.

Methods for Changing the Culture

Changing the organization's culture, and specifically its safety culture, requires planning based on a proven model. Many change models are available that have been used successfully in business, industry, and the public sector. The models can also be adapted to making personal changes. The following models are described in this section:

- Five-step change
- Social-change
- Social-marketing
- Lewin change
- Kotter's eight-step change

Five-Step Change Model

One common change model that has been used by the fire service for many years is based on the five steps of the planning process model used in many organizations. The purpose of this five-step planning model (or any other similar program-planning model) is to provide planners with a systematic approach to decision-making. While not new, this model is familiar to most firefighters.

The model can be applied to either emergency or nonemergency situations including, as in this case, causing organizational change to occur. The change model can be applied to the individual as well as the organization. The five-step model includes the following steps:

Step 1: *Identify* — Select a problem that requires a response. The problem that we are considering in this manual is the high fatality and injury rate in the fire service in general. Locally, the problem may be high lost-time injury rates or near-miss incidents.

Step 2: *Select* — Choose the appropriate response to the problem, which requires establishing the goals, outcomes, and objectives required to meet them. Nationally, the fire service has set a goal of reducing fatalities by 25 percent in five years and 50 percent in 10 years. On the local level, the goal may be to reduce lost-time injuries by a similar amount.

Step 3: *Design* — Determine the steps required to meet the goals, outcomes, and objectives previously selected. On the national scene, the *Everyone Goes Home Initiative* states 16 objectives to meet the stated goals. Locally, the steps may include developing courses in proper lifting techniques or the implementation of a physical fitness program.

Step 4: *Implement* — Perform the selected activity that will mitigate the problem. In this step the program is placed into effect. Nationally, the implementation of the *Everyone Goes Home Initiative* began with an awareness campaign to convince local fire departments to adopt the 16 objectives. In the local example, personnel would be required to attend a training course in proper lifting techniques and participate in the fitness program.

Step 5:   *Evaluate* — Determine the effectiveness of the activities in meeting the goals or outcomes. This step includes monitoring the various elements of the program and evaluating them to determine how effective they are. Some alterations in the elements may be necessary to ensure that the goals are met. For instance, in the local example given above, the fitness program may result in an increase in injuries due to improper implementation or the lack of stretching exercises before exercising.

The five-step change model can be represented as a circular path or as a flowchart. The results of the evaluation step may indicate the need for some changes in the activities. When changes are identified, a new change cycle begins again.

Social-Change Model

Developed by members of the higher education community in the 1990s, the social-change model is regarded by some as the leadership model for the future. It is a value-based model of leadership that places service at the core for social change. This model has been used in university student affairs organizations and nonprofit organizations that focus on social problems and the need for change.

The model's purpose is to make change for the betterment of others through leadership. The goals of the social-change model are as follows:

- Promote in the individual self-knowledge and an understanding of one's own interests, talents, and values.

- Increase leadership competence in order to cause positive cultural change in an institution, community, or society.

Corresponding to and interacting with the goals are seven critical values that are subdivisions of the model's three areas:

1. Individual — Critical values:

— *Consciousness of self and others:* Awareness of values, emotions, attitudes, and beliefs that motivate people to action

— *Congruence:* Similarity of thoughts, feelings, and actions toward others

— *Commitment:* Personal investment of time and energy for the duration of the project

2. Group — Critical values:

— *Common purpose:* Shared goals and values defined by the active participation of members of the group

— *Collaboration:* Cooperation based on the application of mutual trust as a means of empowering others and self

— *Controversy with civility:* Acknowledgment that group members will inevitably hold different views and differences must be addressed in a civil values-based, respectful manner

3. Community/society — Critical value:

— *Citizenship:* Acknowledgment that the members of the group have both individual rights and responsibilities to the community

Social-Marketing Model

Another model similar to social change is *social marketing.* It is used widely in third-world countries for health initiatives. It has been very successful in changing deep-rooted cultural beliefs and behaviors in those areas.

Social marketing was created in the 1970s when it was realized that the same marketing principles that were being used to sell products to consumers could be used to sell ideas, attitudes, and behaviors. The social-marketing concept seeks to influence social behaviors to benefit the target audience and the general society rather than the seller of the ideas. This technique has been used in Africa in the fight against sexually transmitted diseases (STDs), acquired immune deficiency syndrome (AIDS), water purification, and in the U.S. to alter beliefs about drug abuse, heart disease, and organ donation.

The model can be successfully applied to an individual, a small group, an organization, or a society. Resources are applied to each step until it has been completed. The social-marketing model contains the following seven elements to behavioral change:

1. ***Knowledge*** — Provide information about the problem to members of an audience so they become aware of a need for change. The audience must gain the following knowledge:

   — Know a problem exists

   — Know there is a practical solution or alternative

   — Personalize the problem, which causes them to own the problem

   — Understand costs of inaction, both personal and organizational

   — Recognize the benefits of action

2. ***Desire*** — Describe the benefits of the change so members of an audience are able to visualize how the change will improve their lives, their organizations, and their society. This vision of a new, improved future will cause them to desire the change. Because desire is an emotion, advertising agencies have founded their sales campaigns on it, creating consumer demands for products that the consumer perceives will make their lives better, happier, and more fulfilling.

3. ***Skills*** — Demonstrate the steps that are required to accomplish the change or new behavior to the audience to create a visual roadmap from the current condition to the desired condition.

4. ***Optimism*** — Demonstrate support of the leadership to create optimism. The chief of the department, officers, labor representatives, and political officials must actively support and adhere to the change. Otherwise, members of the audience will give up and believe that the change will not be effective.

5. ***Facilitation*** — Provide resources (funds, training, equipment, time, personnel, and knowledge) so that members of the audience will be able to enact or implement the change. The fire service has done a very good job of providing personal protective equipment (PPE) designed to protect firefighters as they approach and extinguish fires of all types. However, the fire service has been less successful at changing firefighter attitudes and personal behaviors that lead to cardiac arrest due to poor nutrition and lack of physical fitness.

6. *Stimulation* — Provide humans stimulation (or the boot in the backside) to motivate the individual or organization to accept change. Firefighters are often complacent with their current conditions and are creatures of habit who may not respond to all the knowledge, desire, support, and resources created in the previous steps. Stimulation or motivation may take one of the following two forms:

— *Direct, personal threat:* May take the form of a proposed layoff or reduction in rank or compensation.

— *Inspirational call to action:* Can be the improvement in personal health and/or in the quality of life

7. *Feedback and reinforcement* — Give continuous feedback on the results of the change. Audience members must know that the change is working and their efforts have been worthwhile. At the same time, they must be permitted to provide feedback to the leadership on how they perceive the change succeeding or failing. The importance of the new conditions and benefits of the change must be reinforced to the participants as well. This feedback and reinforcement demonstrate that the leadership is still in support of the new behavior or condition.

Lewin Change Model

One model that has been available in the private sector since the 1950s is the Lewin Change Model **(Figure 2.12)**, which consists of the three following phases:

1. *Unfreezing* — Managers demonstrate to workgroups the difference between the current level of performance and the desired or new level.

2. *Moving* — Workgroups learn new behaviors, values, processes, or procedures.

3. *Refreezing* — Changes become permanent.

Because change is a continuing process, it must be monitored to determine the effectiveness of the new process or procedure. It is important to remember that initial implementation of a change and institutionalizing that change are usually two different things.

Since the development of the Lewin Change Model, the business environment has itself changed and become more technical and fast-paced. Likewise, the environment in which the fire and emergency services operate has changed in a similar fashion. Therefore, a new model for change has been created to facilitate most forms of change facing an organization; it is based on the following five steps:

Step 1: **_Recognize the need for change_** — Clearly declare the need for change and establish objectives. Consider the effects the change will have on other parts of the organization.

Step 2: **_Identify resistance and overcome it_** — Identify potential resistance to the change and determine and implement the best method for overcoming it.

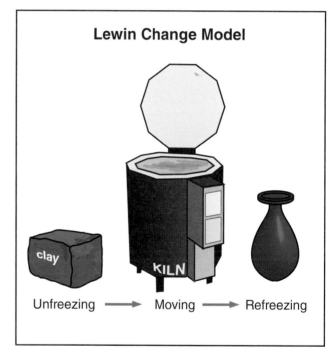

Figure 2.12 Lewin Change Model.

Step 3: ***Plan the change interventions*** — Recognize that a variety of change agents or interventions exist that can help in implementing the change. Some of those interventions are as follows:

— *Training and development*: Use to develop skills, behaviors, and attitudes that will be used in the workplace; may include technical skills, interpersonal skills, or communication skills to name a few.

— *Teambuilding:* Focus on how to get the job done to minimize exposure to injury while accomplishing the task. The team approach (one of the most widely used intervention forms) allows for user involvement in defining the needed change and how to implement it.

— *Sensitivity training*: Teach people to recognize their own and their fellow crew members' hazardous behaviors, the effects they have on others, and how to alter or improve them.

— *Job design*: Change or alter the types of tasks members of the organization do. It includes simplification, rotation, or expansion of the tasks to improve efficiency or effectiveness.

— *Direct feedback*: Use an outside agent such as a manufacturer's representative to train members in the use of new equipment or techniques. Consider a review by an outside agent (state fire training academy, another fire agency, consultant, etc.) of basic functions, e.g., ladder raising, hose pulling, emergency driving, gurney lifting, etc., focused on injury prevention.

— *Survey feedback*: Use a written questionnaire or brain-storming sessions designed to gather data from members of the organization. Analyze the data and determine recommended changes.

— *Process consultation*: Focus on how people interact to get a job accomplished. Have an outside observer analyze the relationships within the group and make recommendations for change to the organization.

Step 4: ***Implement the change*** — Use the appropriate change agent, and put the change into operation.

Step 5: ***Control the change*** — Enforce, review, monitor, and analyze the change model. Take correction action if change objectives are not met.

Kotter's Eight-Step Change Model

Professor John Kotter is a recognized authority on managing change. In the past decade, he has created an eight-step model for implementing and handling change. In each step, Kotter acknowledges a key principle that relates to people's response and approach to change. Because change is personal, people must see, feel, and then act on it. The model's eight steps are as follows:

Step 1: ***Create a sense of urgency*** — Motivate personnel and create real, relevant, and attainable objectives.

Step 2: ***Build a guiding team*** — Select team members who have the appropriate knowledge, skills, and abilities and who are committed to the change process.

Step 3: ***Establish the vision*** — Allow the team to establish a simple vision and strategy. Establish attainable and measurable objectives that support the vision's ultimate goal.

Step 4: ***Communicate the vision to everyone who will be affected*** — Involve as many people as possible, including internal and external customers and stakeholders. Communicate the essential information in a simple and concise manner. Appeal and respond to needs of the participants.

Step 5: ***Empower the team*** — Remove obstacles, permit constructive feedback, and provide support from leaders. Reward and recognize progress and achievements of the team.

Step 6: ***Create short-term objectives*** — Set objectives that are easy to achieve, and establish manageable numbers of initiatives. Complete each objective before attempting the next.

Step 7: ***Don't declare victory too soon*** — Encourage determination and persistence in the team. View change as a constant process, not one that ends at the completion of the project.

Step 8: ***Institutionalize the change*** — Reinforce the value of successful change. Make the change process part of the culture.

Results of a Change in Culture

Implementing the desired change and institutionalizing it may be difficult, but the results will be worth the effort. The results will benefit all levels of our profession at the personal, organizational, community, and professional levels.

Personal Benefits

If you consider the relationship between the personnel, organization, community, and profession to form a pyramid, you will realize that the personnel who form the base will gain substantial personal benefits from a safety, health, and wellness culture. By adopting this culture on a continual basis, you and your associates will have a safer work and home environment **(Figure 2.13, p. 72)**.

You will be assured that the organization you work for has your best interest at heart and, policies and procedures will be grounded in safety. You will also be assured that your supervisors are making decisions based on a full awareness of the hazards that you face and the risks they can create.

These benefits will not be strictly the result of changes at the top of the organization or in the community. They will also require that you are fully supportive of the new culture and make internal behavioral changes as well. You must change unsafe behaviors, have a proactive attitude toward safety, and relate ethically to subordinates, peers, supervisors, and those in authority. You must be willing to take responsibility for your actions and base your actions on the safest and healthiest choices possible.

It is a very emotional experience to visit the National Fallen Firefighters Memorial at the National Fire Academy at Emmetsburg, Maryland. The names listed there often commemorate acts of heroism that resulted in the ultimate sacrifice. Too often however, postincident analysis reveals that the death could have been avoided through application of reasonable precautions, e.g., main-

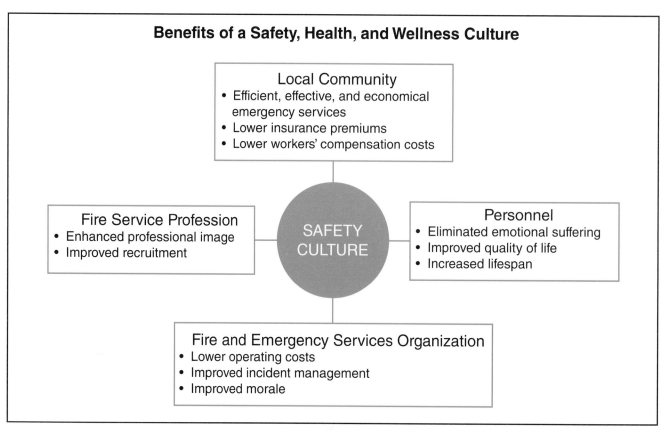

Figure 2.13 Benefits of a Safety, Health, and Wellness Culture.

tain good physical condition, have regular physical examinations, stay off the roof or do not stay under the roof of a building that has experienced long-term extensive fire, drive defensively, etc. The majority of firefighters would rather see their names on a service pension check than listed on a memorial. The primary benefit to active participation in a safety, health, and wellness culture is the ability to live a long and active life free of disabilities. An added benefit is the elimination of emotional suffering for the individual and their family.

Organizational Benefits

The fire and emergency services organization will also benefit from the shift to a safety-based culture. Organizations will experience the following benefits:

- *Financial* — The organization will save money through reduced insurance costs, apparatus and equipment repair and replacement costs, and liability payments to victims. The organization will also save money by not being cited and fined by the Occupational Safety and Health Administration (OSHA) for infractions of federal safety requirements.

- *Operational* — Benefits will include improved emergency incident site management, reduced or eliminated fatalities and injuries, and increased membership participation and morale. *Others:*

  — Recruitment, both for career and volunteer personnel, can be facilitated as the organization projects an image of being safety-conscious and employee-centered.

  — Labor/management relations will be improved based on a mutual partnership for safety.

— Training will be more consistent with a focus on the crew resource management approach to safe operations (See Chapter 3, Behavioral-Based Safety).

- *Morale* — The organization will be able to adopt safety, health, and wellness not strictly as a program but as a core value that will determine strategic goals, establish objectives, and influence decision-making both at emergency incidents and in nonemergency activities. Professional development at all levels of the organization should include safety as a focus of each training session, course or program.

Community Benefits

Our communities will benefit in multiple ways. The service they receive will be more efficient, effective, and economical. Apparatus and equipment maintenance costs will be less when accidents are eliminated or reduced. Fire insurance premiums will be reduced as services improve, and funds once used for worker's compensation benefits and high health insurance premiums can be shifted to improve Insurance Services Office (ISO) rating requirements. Worker's compensation premiums can be reduced as the *experience modifier* is reduced (see information box for an explanation).

Worker's Compensation Experience Modifier

Public and private organizations purchase worker's compensation insurance policies from commercial insurance companies for their employees. Premium rates are based on an *experience modification rate* (EMR). The experience modifier is a formula that compares the organization's accident and injury record over a three-year time period with average data for all organizations in a similar industry such as fire departments. The median is given a rating of 1.0. Companies that have an EMR greater than 1.0 pay a higher-than-average premium and those less than 1.0 pay a lower than average premium. A low EMR is important for private industry as it drives their direct overhead costs and gives those with a lower EMR a competitive advantage. Many public departments are self-insured by their government agency. This means that any worker's compensation loss not budgeted for must be taken from contingency funds or other previously approved and funded budget items.

Professional Benefits

The fire service will always have the image of heroism and sacrifice. The public will continue to look at us with admiration and respect for the job we do. Shifting our culture to one based on personal safety will enhance our professional image with the public. An enhanced professional public image helps to increase the attraction of the career fire service to potential applicants and support for fiscal needs of the agency. The increased emphasis on safety also helps to attract volunteers who associate safety with a high level of professionalism.

Safety Initiatives

In the past decade, the fire and emergency services have begun to actively embrace the concept of a safety culture. Part of this movement has been the result of unacceptable losses such as the 14 deaths at Storm King Mountain, Colorado, in 1994. At the same time, the recognition that our annual line-of-duty deaths (LODDs) have continued to remain at the same level while our PPE has improved and our fire responses have decreased. Finally, the realization that other professions such as the aviation industry have managed to create successful safety cultures has motivated us to seek similar answers.

A number of initiatives have been developed to assist both organizations and individuals to change their cultures. Among the initiatives, concepts, and resources that are available are the following:

- *Safety Stand Down* — Joint venture between the International Association of Fire Chiefs (IAFC); the International Association of Fire Fighters (IAFF); the Volunteer and Combination Officers Section of the IAFC; and the Safety, Health, And Survival Section of the IAFC to focus attention on safety-related issues.

- *Everyone Goes Home Initiative* — Sponsored by the National Fallen Firefighters Foundation, the initiative contains 16 objectives intended to reduce fire and emergency responder casualties. See **Appendix B** for the list of 16 Initiatives.

- *Front-loading the incident* — Emergency incident strategy that advocates concentrating command resources early in the operation in order to gain early control of the situation. The concept was advocated by Phoenix (AZ) Fire Chief Alan Brunacini.

- *Federal government mandates* — Requirements developed by the U.S. Department of Homeland Security (DHS) and OSHA, including requirements for respiratory protection training programs, confined-space requirements, hazardous materials training, national credentialing, and implementation of the National Incident Management System-Incident Command System (NIMS-ICS).

- *Candidate Physical Abilities Test (CPAT)* — Physical fitness test for use in hiring developed jointly by the IAFF and the IAFC. The test is based on job-related physical tasks; can also be used as the basis for annual fitness evaluations.

- *Crew Resource Management (CRM)* — Program based on the proven approach used in commercial aviation to reduce aircraft accidents; attempts to reduce accidents, injuries, and fatalities through *error management*. Chapter 3, Behavioral-Based Safety, is dedicated to describing the CRM approach to safety.

Ethical considerations can also be used to help create change. Simply put, *doing the right thing* can go a long way to convince members of an organization that change is necessary. This approach is primary in most values-based leadership.

Summary

The fire and emergency services have a long history that is based on heroism and sacrifice for the good of those we serve. This tradition has resulted in a counterproductive culture that has resulted in continued fatalities and injuries in the line of duty. Only by changing this culture to one that embraces safety, health, and wellness can we affectively reduce those casualties, improve our public image, and provide a safe workplace for ourselves and our peers.

Review Questions

1. What characteristics describe the current fire and emergency services culture?

2. Discuss how internal and external barriers can prevent change in the safety culture of an organization.

3. What are the differences between the social-change model and the social-marketing model for change?

4. How does the cultural image of fire and emergency services affect organizational safety efforts?

5. What potential personal and organizational benefits can a change in safety culture bring about?

Behavioral-Based Safety

Chapter Contents

Key Terms

Job Performance Requirements

This chapter provides information that addresses the following job performance requirements of NFPA® 1001, *Standard for Fire Fighter Professional Qualifications*

NFPA® References

6.3   6.4.1

This chapter provides information that addresses the following job performance requirements of NFPA® 1021, *Standard for Fire Officer Professional Qualifications*

NFPA® References

| 4.1.1 | 4.2.1 | 4.2.2 | 4.7 | 4.7.1 |
| 6.7 | 6.7.1 | 7.6.2 | 7.7 | 7.7.1 |

This chapter provides information that addresses the following job performance requirements of NFPA® 1071, *Standard for Emergency Vehicle Technician Professional Qualifications*

NFPA® References

6.2.3.3   4.4.2   5.4.3

This chapter provides information that addresses the following job performance requirements of NFPA® 1051, Standard for Wildland Fire Fighter Professional Qualifications

NFPA® References

| 1.3.12 | 1.3.16 | 5.5.3 | 8.2 | 8.2.1 | 8.2.3 |

FESHE Learning Objectives

Occupational Safety and Health for Emergency Services
Describe the responsibilities of individual responders, supervisors, safety officers, and incident commanders, safety program managers, safety committees and fire department managers as they relate to health and safety programs.

Behavioral-Based Safety

Learning Objectives

After reading this chapter, students will be able to:

1. Describe the similarities between flight crews and fire and emergency service crews.
2. Explain each of the five factors that crew resource management (CRM) is intended to change.
3. Describe each of the five components of the communication model.
4. Explain interference as it relates to the communication model.
5. Describe how the sender can improve communications.
6. Describe how the listener can improve communications.
7. Describe CRM communications tasks.
8. Explain eight indicators of situational awareness loss.
9. Explain the DECIDE Model of decision-making.
10. Describe social skills required for teamwork.
11. Explain the four skills that are important for a leader to possess.
12. Describe disruptive attitudes that increase the risk to crew safety.
13. Describe workload management.
14. Describe each of the three layers used in error management.

Courtesy of Cherry Hill (NJ)
Fire Department.

Chapter 3
Behavioral-Based Safety

Case History

South Canyon (Storm King Mountain) Fire, July 1994

On July 2, 1994, during a year of drought and at a time of low humidity and record high temperatures, lightning ignited a fire 7 miles west of Glenwood Springs, Colorado. The fire was reported to the Bureau of Land Management (BLM) on July 3 as being in South Canyon, but later reports placed it near the base of Storm King Mountain. The fire began on a ridge, which was paralleled by two canyons or deep drainages, called in this report the east and the west drainages. In its early stages, the fire burned in pinyon-juniper fuel type and was thought to have little potential for spread.

Figure 3.1 A simple memorial commemorates the deaths of 14 firefighters at Storm King Mountain near Glenwood Springs, Colorado. *Courtesy Glenwood Springs (CO) "Post Independent."*

Over the next 2 days the South Canyon Fire increased in size, the public expressed more concern about it, and some initial attack resources were assigned. On the afternoon of July 4 the District sent two engines. Arriving at 6:30 p.m. at the base of the ridge near Interstate 70, the crew sized up the fire but decided to wait until morning to hike to the fire and begin firefighting efforts.

The next morning, a seven person BLM/Forest Service crew hiked $2\frac{1}{2}$ hours to the fire, cleared a helicopter landing area (Helispot 1) and started building a fireline on its southwest side. During the day an air tanker dropped retardant on the fire. In the evening the crew left the fire to repair their chainsaws. Shortly thereafter, eight smokejumpers parachuted to the fire and received instructions from the Incident Commander to continue constructing the fireline. The fire had crossed the original fireline, so they began a second fireline from Helispot 1 downhill on the east side of the ridge. After midnight they abandoned this work due to the darkness and hazards of rolling rocks.

On the morning of July 6 the BLM/Forest Service Crew returned to the fire and worked with the smokejumpers to clear a second helicopter landing area (Helispot 2). Later that morning, eight more smokejumpers parachuted to the fire and were assigned to build the fireline on the west flank. Later, ten Prineville Interagency Hotshot Crew members arrived, and nine joined the smokejumpers in line construction. Upon arrival, the remaining members of the hotshot crew were sent to help reinforce the fireline on the ridgetop.

Continued on page 80

Case History *(Concluded)*

At 3:20 p.m. a dry cold front moved into the fire area. As winds and the fire activity increased, the fire made several rapid runs with 100-foot flame lengths within the existing burn. At 4:00 p.m. the fire crossed the bottom of the west drainage and spread up the drainage on the west side. It soon spotted back across the drainage to the east side beneath the firefighters and moved onto steep slopes and into dense, highly flammable Gambel oak. Within seconds a wall of flames raced up the hill toward the firefighters on the west flank fireline. Failing to outrun the flames, 12 firefighters perished **(Figure 3.1, p. 79)**. Two helitack crew members on the top of the ridge also died when they tried to outrun the fire to the northwest. The remaining 35 firefighters survived by escaping out the east drainage or seeking a safety area and deploying their fire shelters.

Within 3 hours of the blowup, an interagency team was forming to investigate the entrapment of the South Canyon Fire. In the next few days the team investigated the fire and fatality sites and began a series of 70 interviews with witnesses. On July 22, with the interviews and much of the investigation report completed, the team adjourned. On August 9-11, the team reconvened to review a draft of the completed report in preparation for its publication.

The Investigation Team determined that the direct causes of the entrapment in the South Canyon fire were as follows.

- Fire behavior

- Strategy and tactics

- Escape routes and safety zones were inadequate for the burning conditions that prevailed. The building of the west flank downhill fireline was hazardous. Most of the guidelines for reducing the hazards of downhill line construction in the Fireline Handbook (PMS 410-01) were not followed.

- Strategy and tactics were not adjusted to compensate for observed and potential extreme fire behavior. Tactics were also not adjusted when Type I crews and air support did not arrive on time on July 5 and 6.

- Safety Briefing and Major Concerns:
  - Given the potential fire behavior, the escape route along the west flank fireline was too long and too steep.
  - Eight of the 10 Standard Firefighting Orders were compromised.
  - Twelve of the 18 Watch Out Situations were not recognized or proper action was not taken.
  - The Prineville Interagency Hotshot Crew (an out-of-state crew) was not briefed on local conditions, fuels, or fire weather forecasts before being sent to the South Canyon Fire.
  - The "can do" attitude of supervisors and firefighters led to a compromising of Standard Firefighting Orders and a lack of recognition of the 18 Watch Out Situations.
  - Despite the fact that they recognized that the situation was dangerous, firefighters who had concerns about building the west flank fireline questioned the strategy and tactics but chose to continue with line construction.

Source: Bureau of Land Management article, U.S. Department of the Interior website

Achieving a safe work environment is a multilevel effort. The effort must exist in all areas of a fire and emergency services organization to be successful. As mentioned previously, the responsibility for safety belongs to the administration, the supervisor, and the individual.

The key to creating a safe work environment, however, is not in the type of equipment that is purchased or the policies that are implemented. Statistics show that injuries increase because the emphasis on safer equipment may cause people to become overconfident and place themselves in greater harm. Stronger policies requiring the reporting of injuries has resulted in more injuries being documented. Therefore, creating a safe environment should be based on the behavioral change of personnel. You, your peers, and your supervisors must change how you perform your tasks and how you make decisions. Working safely must be the primary mission as you go about your duties in the station or on scene.

Behavioral-based safety finds its roots in risk management. Over the past century, social scientists and safety professionals have taken various approaches to creating a safe work environment. Currently, the two in use by fire and emergency services are Crew Resource Management (CRM) and High Reliability Organization (HRO). This chapter will provide you with an overview of both concepts and the evolution of risk management that they are based on.

History of Risk Management

The evolution of risk management and safety occurred over a relatively short time. Prior to 1932, and the research by Herbert William Heinrich, the limited focus of the safety profession was on interrupting injuries through engineering. This usually resulted in the design of barriers that were intended to put a barrier between the employee and the risk. Machine guarding is a good example of a safety barrier. A metal guard was installed on a machine to prevent human contact with a moving part that could result in an injury. In the fire and emergency services, the addition of layers of heat resistant fabric to protective clothing was intended to place a barrier between the firefighter and the heat source.

Herbert Heinrich introduced many of the modern day safety theories that are still considered state of the art thinking. After reviewing thousands of accidents, Heinrich came to the conclusion that approximately 88 percent of workplace accidents were caused by *unsafe acts*. His review of accident reports indicated that supervisors typically blamed the employees for being injured and failed to conduct detailed investigations and rarely determined the *root cause* of the accidents. When he published *Industrial Accident Prevention, a Scientific Approach* in 1932, Heinrich laid the foundation for what has come to be known as *behavioral-based safety*.

Accident Causes as Determined by Heinrich
- 88 percent caused by *unsafe acts*
- 10 percent caused by *unsafe conditions*
- 2 percent *unavoidable*

As a means for determining the cause of an accident, Heinrich developed what he called the Domino Theory. This theory is illustrated by standing a series of dominos up on end close together in line. When the first one is pushed, it strikes the next causing it to strike the following one until all the dominos in the line have collapsed. While the theory can consist of an infinite number of dominos depending, on the cause/effect relationship you are trying to demonstrate, Heinrich determined that there were only the following five components of an accident:

1. ***Ancestry and social environment*** — Unsafe behavior is the result of inherited traits or attitudes (ancestry) or on acquired traits learned from peers, family, or instructors (social environment).

2. ***Fault of the person*** — Personal unsafe traits, regardless of the source, are the reason for unsafe acts or conditions.

3. ***Unsafe act or hazardous condition*** — Unsafe acts by people and hazardous conditions caused by people are the direct causes of accidents.

4. ***Accident*** — When an accident occurs, it typically results in an injury, illness (due to exposure or stress), fatality, or damage to property. A near miss, an accident that did not occur but could have, should be included in this factor when analyzing accidents and safety prevention.

5. ***Injury, illness, fatality, or property loss*** — The final domino in the series; when this one falls, there is pain, suffering, and financial loss.

One of the most important safety perspectives Heinrich developed was the importance of the Human Factor in accident prevention. Heinrich further believed that people's errors often result in a near miss and sometimes resulted in an injury. His theory further developed what has become known as the *Accident Pyramid,* which suggested that hundreds of near misses occurred prior to the actual disabling injury **(Figures 3.2 a and b)**. He suggested that, if changes were to be made and accidents reduced, these near misses should be investigated to the same detail as injury-producing accidents. For years, fire and emergency services have focused on fatalities as illustrated by the fallen

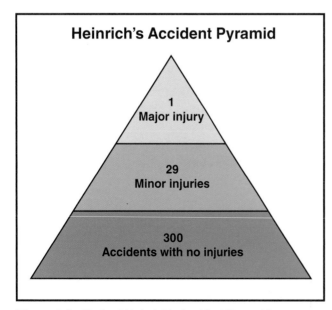

Figure 3.2a Herbert Heinrich's Accident Pyramid developed in the 1920s.

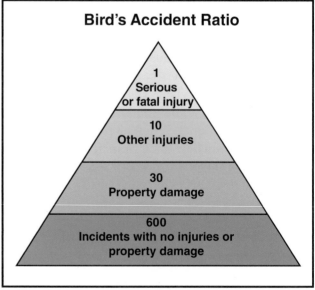

Figure 3.2b Frank Bird's Accident Ratio created in the 1960s.

fire fighter reporting system. In fact many training agencies display the number of fallen firefighters for the current year prominently on the classroom bulletin board. Several leading fire service organizations have recently embraced the near-miss concept and developed a near-miss reporting system that has provided a forum for departments and personnel to tell their near-miss incidents to others. A renewed focus on near misses should result in fewer injuries with a subsequent reduction in fatalities.

It is important to note that these *human errors* are usually the result of good people working from a flawed base of knowledge. Frequently employees think that working fast is more important than working safe. This attitude promotes shortcuts. This belief that the work is so important that personal safety is counterproductive and that it is acceptable to take high risks in a low reward situation, placing themselves and others at risk, is misguided. It is difficult for some to grasp that personal safety does not cost valuable seconds; it maximizes the person's ability to work efficiently and longer in dangerous situations. When supervisors begin to understand that safety is human driven, the obvious need to proactively manage work through a risk management system becomes a reality.

In the early years of the fire and emergency services, safety was primarily the job of the training division or training officer. The training officer balanced teaching individuals to be aggressive in attacking a fire and how to work safely. The very idea of teaching people to work safely in a high-risk environment seemed like an impossible task. Other than limited personal protective equipment (PPE), experience, and gut intuition, training officers had little to work with **(Figure 3.3, p. 84)**.

Even today, the federal government through agencies such as the Occupational Safety and Health Administration (OSHA) recognizes that regulations do not cover all work activities in the fast-moving, ever-changing environment of an emergency scene. Because of this, training safely is a challenging and critical part of the fire service. Besides OSHA, safety organizations have existed for over a century. The development timeline of safety organizations is shown in the following information box.

Development of Safety Organizations
- 1896 — National Fire Protection Association (NFPA®)
- 1912 — National Safety Council (NSC)
- 1914 — American Society of Safety Engineers (ASSE)
- 1938 — American Conference of Governmental Industrial Hygienists (ACGIH)
- 1939 — American Industrial Hygiene Association. (AIHA)
- 1970 — Occupational Safety & Health Administration (OSHA)
- 1970 — National Institute for Occupational Safety and Health (NIOSH)

Today's training officers have many tools to work with, from significantly improved PPE to thermal imaging cameras. The training officer's job has shifted from solely task and tactics to a "safety first" work environment. Even so, with all of the new technology, training alone cannot change our risk-taking

Figure 3.3 Many fire departments assign their Health and Safety Officer the duty of providing safety training to members of the organization.

culture. Cultural change, and therefore behavioral change, must accompany safety-oriented training. This revolution in our traditions will require a single unwavering focus throughout the entire fire and emergency services.

A significant step in accident reduction was fostered by OSHA with the requirement for an increase in employee involvement through health and safety committees. The need for employee involvement was based on the following three factors:

1. Employees are exposed to hazards daily.

2. Group input can create more effective solutions to problems.

3. Employees are more likely to support programs where input has been provided.

As of 2007, there were 26 states or territories managing their own federal OSHA-approved state programs with the understanding that they will be as effective, if not more so, in their establishment and enforcement of occupational safety regulations. Most states adopt updated OSHA federal regulations within six months of any rule change. In line with this concept, many states have adopted specific safety committee requirements that are more stringent than federal OSHA regulations.

One of the early efforts to increase employee involvement was the suggestion box. Over the years, safety suggestion boxes have been used with limited effectiveness. With the development of labor representation in the fire and emergency services, they have been replaced with the concept of labor-management committees, commonly referred to as *safety committees*.

The National Fire Protection Association (NFPA®) recognized the importance of safety and the creation of safety committees when they developed NFPA® 1500, *Standard on Fire Department Occupational Safety and Health Program*. The standard required the establishment of a safety committee as well as the position of Health and Safety Officer (HSO) in fire and emergency service organizations.

While safety committees are one of the most effective tools to promote a health and safety program, many safety committees are frequently inappropriately used or underutilized. A primary function of traditional safety committees has been to review accidents and injuries that have occurred since the previous safety committee meeting. This historical examination concept was intended to help prevent future accidents and injuries. Approaching accident prevention from this reactive perspective helped succeed in making labor and management reflect on some of the risks associated with fire and emergency services. However, accident review is an after-the-fact approach and, in some cases, promoted a *fix-blame* approach to accident prevention rather than a *fix-the-problem* approach.

In cases of well-run and supported health and safety committees, the knowledge obtained can serve to augment the fire and emergency service's safety functions. Many health and safety committees have assisted departments with the adoption of additional effective safety and training practices.

Some local fire and emergency services organizations place the health and safety committee function under the human resources division, while others have chosen to place it under the direction of the operations or finance division. Any of these divisions can successfully support the health and safety committee function effectively.

In some jurisdictions, the safety committee and responsibility are external to the fire and emergency service organization. The jurisdiction may choose to locate it in the authority's Human Resources Department or Administration. In these cases, close coordination between the safety committee and the fire and emergency service's personnel must be maintained.

The most successful health and safety committees use the behavioral-based safety model. The behavioral-based safety concept is based on having personnel committed to the process through personal involvement. All members of the organization must feel that they have had input into the development of safety policies and procedures that they must adhere to. Elements of an effective health and safety committee are provided in the following information box.

An Effective Safety Committee
- Receives active administrative support for the committee
- Receives active labor organization support for the committee
- Meets on a regular schedule
- Shares results with the organization's membership
- Ensures continuity by rotation of 50 percent of members
- Conducts or monitors postincident analyses
- Generates three to five goals for each year
- Routinely evaluates its overall effectiveness
- Reviews all of its near misses as well as accidents
- Monitors the progress of safety inspection results
- Helps develop a climate of positive safety peer coaching
- Helps develop an overall health and wellness programs
- Helps evaluate the effectiveness of safety-in-training programs
- Helps evaluate the effectiveness of health and safety programs

A natural progression for the safety committee for many fire and emergency services organizations was to evolve to *strategic safety plans*. This process has been used extensively since the 1990's in many emergency services organizations throughout the country. Strategic plans help focus the safety committee's efforts on prevention through identification of specific goals that support and enhance the safety and risk management activities of the department.

This proactive approach permits the safety committee to become an integral component of the department's risk management program. This team effort creates an environment that moves away from fixing blame to an environment that solves problems. Safety committee members are permitted to evolve from being safety cops to being part of a collaborative team whose efforts are focused on obtaining positive results. Safety committees that operate within this type of environment rarely do not have difficulty obtaining members willing to serve. In the end, strategic plans produce win-win results.

In the late 1990's, the fire and emergency services became interested in the Crew Resource Management (CRM) concept used by the airline industry. It was considered the state of the art in fire service safety risk management. Although CRM is recognized as a best practices, some industry leaders have begun to embrace a safety concept used in many high-risk industries, the military, and some airlines known as High Reliability Organization (HRO).

Safety has been a separate function in many jurisdictions. Management would define what the task and tactics are, and the safety committee would examine safety issues related to the task. HRO's combine and coordinate tasks, tactics, and safety into one cohesive function. One leader in developing a fire and emergency services HRO model is Tualatin Valley Fire & Rescue in Beaverton, Oregon.

Whether your department uses a strategic safety plan, CRM, or HRO approach to a safety management system, an important ingredient that helps make safety work is timely, enthusiastic management support. Mere acceptance on the part of management is not enough to change our pervasive risk-taking culture. Anything less than enthusiastic support will produce less than optimum results.

Strategic Safety Plans

Strategic plans map the course of an organization over an indefinite future that is divided into definite time components. The plan attempts to take into account the external factors that will affect the organization such as changes in the economy, demographics, service requirements, hazards, and technologies. Strategic plans are based on trends and expectations of the future environment, but external factors are somewhat unpredictable and some unforeseen events will occur. Therefore, strategic planning, while relying on the best judgment of the planning committee based on analysis of the available information, must be flexible and responsive to change.

The concept of strategic planning can be effectively applied to the organization's safety needs. The safety committee uses the same process to analyze the safety trends of the organization and develop responses.

Strategic plans are often created through a process called *SWOT analysis* **(Figure 3.4)**. It may be best to have an experienced facilitator lead the SWOT analysis. Letter descriptions are as follows:

- *Strengths* — Identify the strengths of the organization's safety program. Identifying strengths is an important first step for a strategic plan.

- *Weaknesses* — Identify safety program weaknesses.

- *Opportunities* — Recognize the opportunities to improve the level of safety by reducing the number of near-misses, accidents, injuries, or fatalities.

- *Threats* — Identify unfavorable internal and external threats posed by lack of training, lack of funding, and unsafe behaviors.

Strategic safety plans should contain specific contents. These contents include the following:

- Mission Statement
- Vision
- Values
- Internal Strengths and Weaknesses
- External Threats and Opportunities
- Goals and Objectives

The plan must be supported by the administration and the employees, implemented evenly and fairly, and reviewed periodically. If the review indicates that the plan must be altered to meet new threats, then the development process must begin again.

SWOT Analysis	
Strengths What does the organization do well? What resources does the organization have to draw upon? What do citizens see as your strengths?	**Weaknesses** What needs to be improved? What resources does the organization lack? What do citizens see as weaknesses?
Opportunities What opportunities are open to the organization? What trends could the organization take advantage of? How can the organization use its strengths to take advantage of the opportunities?	**Threats** What threats are confronting the organization? What trends are a threat to the organization? What threats do the organization's weaknesses expose it to?

Figure 3.4 Example of a SWOT Analysis.

Crew Resource Management

In order to achieve the goals of the *Everyone Goes Home Initiative* and reduce line-of-duty deaths (LODDs) by 25 percent over the next 5 years and 50 percent over the next 10 years, the fire and emergency services must change the way it operates. One approach that has been adopted by the International

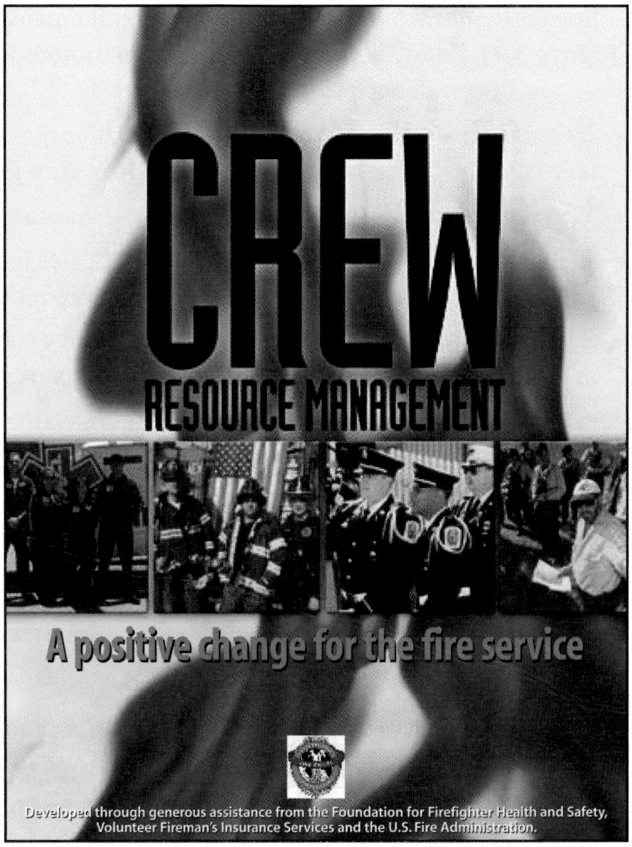

Figure 3.5 The International Association of Fire Chiefs publish their version of crew resource management in this document and make it available on their website. *Courtesy of IAFC.*

Association of Fire Chiefs (IAFC), and recommended by various authors, is the Crew Resource Management (CRM) Model **(Figure 3.5)**. This model was originally developed by the commercial airline industry to reduce airliner accidents and has now been adopted by the U.S. Department of Defense (DoD).

While the primary goal of CRM in the fire and emergency service is to reduce LODDs, the model can also be applied to reducing all types of injuries and accidents in both emergency and nonemergency situations. In fact, it can also be applied by an individual as a means of making behavioral decisions. This section provides a background of CRM, factors that CRM address, and the application of CRM concepts to activities of the individual, unit, and organization.

CRM Background

The roots of CRM in the commercial airline industry can be found in the tragic airliner crashes of the 1960s and 1970s. Crash investigations into numerous airliner disasters during those decades indicated that human error on the part of the flight crew was a contributing factor in an incident **(Figure 3.6)**. In fact, one study indicated that flight crew errors occurred in 50 percent of the incidents studied while air traffic controller errors contributed to 35 percent. A similar analysis of naval aviation accidents indicated that 59 percent were the result of human error. These investigations resulted in the National Aeronautics and Space Administration (NASA) developing an aircrew workshop in 1979 to address the specific elements of human behavior that had been identified as contributing to these accidents: decision-making, lack of interpersonal communications, and leadership.

Figure 3.6 Investigations into civilian and military aircraft crashes have determined that flight crew errors are a contributing factor in the incident. *Courtesy of U. S. Department of Defense.*

The basic CRM Model used for training in the commercial airline industry as well as military aviation has continued to evolve since 1979. According to one source, disasters have dropped from approximately 20 per year to as few as 1 or 2 per year. Other organizations that have adopted CRM have also noticed an improvement in their operational safety records. The U.S. Coast Guard (USCG), for instance, has recorded a 74 percent drop in injury accidents since implementing CRM.

In the International Association of Fire Chiefs (IAFC) *Crew Resource Management* publication, the authors describe the similarities between flight crews and fire and emergency services crews. Similarities include the following:

- *Crew structure* — Both crews have a leader/follower organization or hierarchy.

- *Teamwork* — Both crews work together toward a common goal.

- *Crew composition* — Both crews may be composed of personnel who are permanently assigned to work together. They may also be assembled on short notice from a variety of sources.

- *Workload* — Both crews may experience long hours of mundane activities that are punctuated by periods of intense, high-stress activities.

Modern aircraft cockpits are equipped with multiple safety warning devices, redundant controls, and in some cases computers that can literally fly the aircraft without a pilot's immediate attention. Like the commercial airline industry and aviation in general, the fire and emergency service has attempted to create a safe work environment through the use of technology such as:

- Personal protective clothing and equipment are constructed with fire-resistant materials.

- Chemical fire-extinguishing agents that decrease suppression time, provide a barrier between flame and combustible materials, and reduce the probability of reignition.

- Personal alert signaling devices improve the chances of locating an incapacitated firefighter.

- Radio communications equipment help keep personnel in contact with the command post and telecommunications center as well as each other.
- Improved engineering of fire apparatus to improve braking and stability.

While the commercial airline industry and the military have seen the desired results of improved technology in the form of reductions in accidents, injuries, and fatalities, the fire and emergency services continues to experience unacceptable levels of injuries and fatalities on an annual basis.

CRM Factors

A comparison between airliner accidents that led to the creation of CRM and fire and emergency service incidents involving fatalities and injuries indicates that the same five factors are present in both occupations. The five factors that the commercial airline industry has attempted to change with CRM are as follows:

1. Communications between crew members

2. Situational awareness

3. Decision-making

4. Teamwork

5. Implementation barriers

In addition to these five factors, fire and emergency service CRM training should include techniques for workload and error management. Training emergency personnel to effectively and correctly overcome the negative aspects of each of these factors will result in a change in the work environment. Just as the commercial airline industry has improved its safety record, the fire and emergency service organizations will too.

Because implementation of the CRM Model requires change on the part of the organization's membership, one of the change models introduced in Chapter 2, Cultural Change in the Fire Service, can be used. For CRM to be successful though, the administration must support it, resources must be provided, and members and stakeholders must become involved and accept it.

Communications

Communication model — The process that all people use when interacting with others on a daily basis.

The CRM Model attempts to correct one of the major problems uncovered in the analysis of airliner accidents: lack of communication among flight crew personnel. To correct this problem, crew members must use an effective *communication model*. A communication model is the process that all people use when interacting with others on a daily basis **(Figure 3.7)**. The basic communication model consists of the following five components:

1. *Sender* — Person who initiates the message. The sender develops the idea; translates it into oral, written, or nonverbal symbols; and sends it to the receiver.

2. *Message* — Content or idea as it is contained in the medium.

3. *Medium* — Vehicle that carries the message to the receiver. It may be oral (sounds), written (words on paper or in electronic format), or nonverbal (physical gestures or facial expressions among others).

4. *Receiver* (also known as *Listener*) — Person or persons who is/are the recipient(s) of the message and responsible for interpreting it based on personal knowledge and bias.

5. *Feedback* — Confirmation from the receiver to the sender that the message was received and understood. Effective communication is not complete without feedback. Feedback may use any of the mediums suitable to assure the sender that full understanding on the part of the receiver has occurred. When feedback occurs, the participants exchange the roles of sender and receiver, and the feedback contains a new message.

A sixth component may or may not be present in the communication model: interference. *Interference* is a barrier to effective communications and causes a breakdown in the process. Interference may be external or internal to the participants in a conversation; descriptions are as follows:

- *External interference* — Includes all audible and visual distractions in the environment **(Figure 3.8)**. *Examples:*

  — At an emergency incident, sirens, apparatus or power-tool engine noises, radio static, and noise created by a fire are just some of the audible distractions possible.

Figure 3.7 Accurate and effective communication between emergency responders is essential for safety at emergency incidents.

  — Visual distractions are anything that take the listener's attention away from the sender. The sight of injured victims, massive devastation, flashing lights, and the presence of the media are a few examples.

Figure 3.8 Communicating at an incident is complicated by audible and visual distractions such as flashing lights, engine and equipment noise, and the sounds created by the incident itself. *Courtesy of Chief Chris Mickal, New Orleans (LA) Fire Department.*

- *Internal distractions* — Includes items unique to the receiver. Generally speaking, internal distractions may be more difficult to overcome than external distractions. *Examples:*
  - Personal prejudice toward the sender or message
  - Hearing deficiency
  - Lack of knowledge of subject under discussion
  - Inflated ego on the part of the receiver
  - Lack of self esteem on the part of the sender

Successful communications depend on the sender and receiver overcoming both external and internal interference. The external interference may be the easiest to control by moving the conversation to a quiet, less distracting area, using headphones to block external noise, or turning off audible warning devices.

Internal distractions may be more difficult to overcome because they require some behavioral changes, training, or the use of hearing aids in the case of hearing loss. Two ways of overcoming internal interferences are to apply common communications traits and communication tasks suggested in the CRM Model.

Common Communications Traits

Internal distractions result in the receiver misunderstanding or not receiving a message. The result can be tragic and fatal. To prevent such a result, both the sender and the receiver must improve their speaking and listening skills. Through research, speech and interpersonal communications professionals have identified a series of common communications tasks that can aid the sender and receiver. Those tasks are listed in **Table 3.1**.

CRM Communications Tasks

CRM is based on the successful application of the basic communication model. The CRM Model further suggests that the message sent between the sender and receiver must be respectful, clear, concise, and complete. To further expand and improve on the communication model, CRM gives each member of a crew the following additional responsibilities in the communications process:

1. *Inquiry* — In situations where safety is an issue, it is the ability and opportunity to ask *why*. The purpose is not to question routine tactical issues but to provide the ability to reconcile any difference between what is occurring and what should be occurring. A crew member is *not* questioning authority however. (Care must be taken when using this task as it can easily be overused in emergency situations.)

2. *Advocacy* — Ability to state one's position, based on facts as you perceive them. It may be helpful to communicate those perceived facts during advocacy. Advocacy may help to increase situational awareness, improve understanding, and avoid a disaster.

3. *Listening* — Use of effective listening skills as described previously.

4. *Conflict resolution* — Ability of each member of a crew to stay focused on the situation and how to best resolve conflict. Conflict is normal because

Table 3.1
Tasks for Improving Communications

Sender	Receiver
To ensure that a message is understood and heard, the sender must perform the following tasks: • *Establish a frame of reference for the receiver* — Sender and receiver must both understand the subject of the message. • *Provide a clear and complete message* — Message must contain all the information required to comprehend the situation or accomplish the mission. • *Omit any subjective or biased information* — Sender must adhere to the facts and eliminate any opinions that might alter the understanding of the message. (This "task" may not be invariably applicable to the fire service as "opinions" are at times very useful until more complete "facts" are available.) • *Understand the importance and effect of nonverbal communication* — Nonverbal element of communication carries up to 93 percent of the message. As little as 7 percent of the message is carried by words that are spoken. • *Repeat the message to ensure that it is understood* — Repeating and paraphrasing the message helps to increase the receiver's opportunity to hear and understand the message, provide correct feedback, and act on the information. (This task may not be invariably applicable to the fire and emergency service as the situation does not always allow time for repeating.) • *Be respectful* — Never speak in a disrespectful manner. Insulting, humiliating, or defaming the receiver is the best way to prevent effective communication.	The receiver is equally responsible for ensuring that the communication process is complete and successful. The receiver must perform the following tasks: • *Use effective listening skills* — Essentially, this task means that the receiver must be mentally ready to listen, listen to the complete message before providing feedback, and maintain eye contact with the sender (when possible). • *Put aside preconceived notions* — The receiver must listen to the actual message and not allow personal bias affect the message. • *Properly interpret nonverbal communications* — If the verbal and nonverbal messages do not agree, the receiver asks the sender to clarify the message. • *Ask for clarification of the message* — Through feedback, the receiver tells the sender that the message was not understood. Ask for more information or to have the message simplified or rephrased. • *Be respectful* — Show respect for both the message and the sender.

humans have individual points of view. Conflicts are not about who is right or wrong but what is right. The focus should be on the resolution of safety issues and not tactical decisions.

5. ***Feedback*** — Confirmation that the message is fully understood and the communication process is complete. CRM depends on good feedback by the receiver. If the message was not understood, then the sender restates the message and the process continues until confirmation is received.

With these responsibilities, crew members are empowered to have an active role in the communications process. They are no longer just receivers but active contributing partners. You must be fully aware that the supervisor's authority is not being questioned. Crew members are allowed to question decisions when the decision does not seem to fit the situation and may result in a life-threatening condition.

When considering communication in the context of the Incident Command System (ICS), it is critical to remember that it is a two-way street; i.e., subordinates must provide their superiors with up-to-date information on their situation. In many if not most incidents, the company or squad leaders will have views of the situation very different for those managing larger groups or geography such as Division or Group Supervisors. In effect, the small-group leaders will be the eyes and ears for the Division/Group Supervisors.

Situational Awareness

Being aware of your surroundings and what is going on is called *situational awareness*. Errors occur when we lose situational awareness. If we are too focused on getting to the seat of the fire, we may not notice the loss of structural integrity or the change in fire behavior that indicates that flashover is approaching.

Situational awareness is sometimes referred to as a process that exists on the following three levels:

Level 1: *Perception* — Perceive the situation around us.

Level 2: *Comprehension* — Apply our knowledge and past experiences to our perception and develop an understanding of the meaning of the situation.

Level 3: *Application* — Take our understanding of the situation and apply it to the future, thereby predicting how and when the situation will change and what action is appropriate on our part.

CRM requires that all members of a crew maintain situational awareness and communicate any changes they perceive in their environment or the conditions or behaviors of other crew members to the crew leader and each other. The loss of situational awareness creates an opportunity for errors to occur and improper decisions to be made. Eight factors that may lead to the loss of situational awareness include the following:

- *Ambiguity* — Information received is confusing or unclear. For instance, the building is two stories at the front, but three floors at the rear.

- *Distraction* — Loss of focus of the original mission without appropriate rationale. For instance, attempting exposure protection before rescuing a confirmed victim.

- *Fixation* — Too focused on a single element of the situation to the exclusion of all others. This indicator includes personal concerns such as financial or family problems. For instance, raising an aerial into electrical wires while focused on a rescue situation on upper floors.

- *Overload* — Tasks or information overwhelm us or we attempt to perform all the tasks ourselves. For instance, a single company attempting to carry out fire attack, search and rescue, and ventilation at the scene of a high-rise fire.

- **Complacency** — False sense of comfort based on a misconception of the hazard, risk, or situation sometimes based on past, seemingly similar experience (e.g., "we come here twice a week on false alarms, so this must also be a false alarm").

- **Improper procedure** — Policies or procedures are violated or ignored without justification, e.g., a first-arriving company commencing initial operations without giving a size-up would be a violation.

- **Unresolved discrepancy** — Two or more pieces of information do not agree. For instance, the visual indicators of smoke coming from a structure do not match our knowledge of the building's contents.

- **Lack of comprehensive hazard surveillance** — Crew members become so fixated on one detail that they ignore everything else. An example would be focusing on a vehicle accident and ignoring the hazard of the highway traffic.

In the CRM training provided by the USCG, proper situational awareness depends on performing the following actions:

- Maintain effective communications.
- Recognize and make others aware of any deviation from standard operating procedures (SOPs) or policies.
- Monitor crew member performance.
- Provide information in advance of an operation or mission.
- Identify any potential problems or existing hazards.
- Communicate the desired course of action.
- Communicate the mission's status continuously.
- Evaluate the situation for any changes continually.
- Clarify expectations of crew members continually.

Decision-Making

The ability to make correct, well-informed decisions is important in nonemergency situations and critical in emergencies. While you may have ample time and resources to seek the views of others in a nonemergency situation, that will not be the case during an emergency.

The IAFC provides an example of another decision-making model developed by Dr. Ludwig Benner, Jr. of the National Transportation Safety Board (NTSB). The model uses the acronym *DECIDE* to symbolize each of the steps in the process as follows:

- **D**etermine the problem.
- **E**valuate the scope of the problem.
- **C**onsider available options for mitigating the problem.
- **I**dentify the most appropriate option.
- **D**o the most appropriate option.
- **E**valuate the effectiveness of the actions.

Dr. Benner's model is an expansion of the five-step model described in Chapter 2, Cultural Change in the Fire Service. His model separates some of the elements of the individual steps and adds the evaluation of the effectiveness of the implementation of the decision. Either model is appropriate and can be used effectively.

U.S. Coast Guard Version of the *DECIDE* Model

The USCG also uses the acronym *DECIDE* to describe a variation of the decision-making model. This model has been used successfully in the aviation industry. This version consists of the following actions:

- **Detect change** — Determining that there is a difference between what is happening and what should be happening is necessary.

- **Estimate its affect on the situation** — Evaluating how the change will affect the situation and if the crew really needs to react to the change is necessary. Crew members must determine the answers to the following questions:
    - Who is affected?
    - What situation is affected?
    - Where is the problem?
    - When did the problem occur?

- **Choose an objective** — Selecting an objective may be the most important step in this process because the final outcome will be determined by this selection. An objective may be intermediate or the final outcome itself.

- **Identify alternatives** — Having a variety of alternatives helps to ensure that the best response is made by the crew. It also provides contingencies in the event the first choice fails or the situation changes.

- **Do the best alternative** — Implementing the best possible choice will help ensure success. The best choice should adhere to established policies and procedures and minimize the risk to the safety of the crew.

- **Evaluate the decision** — Providing constant feedback to the crew on the effectiveness of the decision is necessary. It also provides an idea of what may occur as the situation changes or stabilizes.

When applied to the CRM process, decision-making moves from an individual task performed by the crew leader to a group effort involving all crew members. CRM is based on the effective use of all resources. Therefore, the knowledge and observations of the crew, which are part of these resources, can be accessed by the crew leader to help make decisions.

Teamwork

The fire and emergency services have long depended on teamwork to accomplish their assigned goals and objectives. Crews, which may also be referred to as *teams*, have specific tasks that they must perform together. Teamwork occurs when the crew effectively and efficiently performs its designated tasks. At an emergency incident, crews may be involved in suppression, rescue, overhaul, ventilation, or a myriad of other tasks (**Figure 3.9**).

The lack of teamwork, however, can result in loss of crew cohesion, failure to accomplish the objective or goal, and potential injury or death. Symbolic of a lack of teamwork is the individual or crew who *freelances* and performs the first task that comes to mind without coordinating with other crew members or the incident commander (IC).

Teamwork requires certain skills and actions on the part of crew members and the leader. Besides the obvious technical skills, a variety of social skills are required (some were introduced previously in the Communications section), including the following:

Figure 3.9 Emergency incident activities, such as vertical ventilation, require teamwork and effective communications. *Courtesy of McKinney (TX) Fire Department.*

- *Listening* — Effective listening on the part of both the leader and crew members is critical to the success of a team. The leader must be open to the ideas and concerns of the crew, and the crew must be ready to hear and act upon the leader's decisions.

- *Questioning* — Crew members must be able to ask questions in order to better understand the objectives of the team. The question must be concisely stated and focused on the objective and personnel safety.

- *Persuading* — Crew members must be able to freely provide and defend their ideas and suggestions without fear of criticism. If their viewpoints are based on fact, then they must have the right to persuade the crew that their opinions are valid. This action must be concisely stated and focused on the objective and personnel safety.

- *Respecting* — Basic to all interpersonal communication is treating and interacting with each other in a respectful manner.

- *Helping* — Crew members must be encouraged to help their coworkers. They also must be encouraged to ask for help when they need it. *Going it alone* is *not* a function of teamwork.

- *Sharing* — Crew members must be willing and able to share with the rest of the group. Sharing includes resources, workload, information, and personal concerns.

- *Participating* — Participation in a team's activities means being a part of the team. Crew members participate in the decision-making process, in work and leisure activities and in the success or failure of an operation.

Teams consist of two basic elements: leaders and followers. In the traditional approach to fire and emergency services organization, the leader had total and complete authority and sometimes (especially on the fireground) used an autocratic style of leadership. The followers traditionally carried out the leader's orders without question or comment. Autocratic leadership has lead to errors because the leader who was viewed as infallible actually erred.

The CRM Model redefines the relationship between leader and follower to create a new approach to teamwork. A description of each position is as follows:

- *Leader* — Leads others through formal authority or informal characteristics; therefore, the position may be assigned or assumed, depending on the circumstances.

- **Follower** — Accepts the formal authority of another person or subordinates himself/herself to the skills of a formal or informal leader but may, in some instances, become the informal leader of a group or unit.

Effective CRM teamwork depends on both the leader and follower possessing certain skills. These skills are briefly addressed in the sections that follow.

Leadership Skills

Books and publications on leadership in business, industry, and the fire and emergency services are many and varied. They can provide a leader with multiple models and examples for developing strong and effective leadership skills. The CRM Model provides the following four skills that are important to a leader:

1. *Authority* — A leader must have the legitimate authority to command and control the operation and resources to perform a mission. A leader has the responsibility for all decisions and crew safety and must be able to define and communicate mission goals to the crew. Formal authority, however, does not prevent the leader from soliciting input from crew members and altering command decisions accordingly if warranted.

2. *Mentoring* — A *mentor* is defined as a trusted and friendly guide or advisor. A leader teaches skills and transfers knowledge to crew members and advises them on all professional topics from career advancement to operational procedures. A leader also uses mentoring skills to motivate the crew and monitor its progress to ensure success **(Figure 3.10)**. In some situations, a crew member may mentor a leader.

3. *Conflict resolution* — Through the use of situational awareness and communication skills, a leader recognizes conflict within the group and resolves it to the best effect. At the same time, a leader must be open to the concerns of crew members that led to the conflict. These concerns may be varied, valid, and real, based on the experiences and observations of the crew. A leader must be ready to reassess the situation based on the concerns of the crew.

4. *Mission analysis* — This process is very similar to the size-up process used by emergency responders when they first arrive at an emergency incident. Mission analysis requires the leader to perform the following tasks:

   — Gather information on the situation.

   — Evaluate the level of risk against the potential benefit of an action.

Figure 3.10 Fire officers and senior firefighters can effectively mentor personnel as a way to improving skills and knowledge and providing direction for career advancement.

- Identify, define, and communicate objectives and goals. Create and implement an incident action plan (IAP). This step applies to the incident commander (IC).
- Evaluate the effectiveness of the plan.
- Develop contingency actions in the event the unexpected occurs.

Followership Skills

The CRM Model empowers followers to challenge the leader's decision or perception of the situation based on their situational awareness and ability to analyze the situation. Because the authority of the leader must be preserved, followers must have the knowledge and experience to recognize when the situation is not as the leader has defined it or that the leader's action plan will not accomplish the defined objectives.

To gain the right to question the leader's decisions, a follower must have the following traits:

- *Alert to the situation* — Focused on the mission and conditions that affect it
- *Physically fit* — Nourished, rested, and hydrated
- *Mentally fit* — Alerted to the surroundings, his/her personal condition, the condition of the leader and other crew members, and potential stress factors. *Mental aspects:*
  - Consider that people can generally only remember 10 items or tasks at one time.
  - Require the leader to provide objectives that are easily remembered by followers.

A follower must have a positive attitude that includes a *can-do* spirit, an optimistic and cooperative nature, a willingness to listen, and the ability to be open-minded. By developing each of these elements, the follower (and the leader) can create a strong team that can overcome most adverse situations.

Implementation Barriers

Barriers are any factors that prevent the implementation of the CRM model by the organization. These barriers may also prevent effective communication, situational awareness, effective decision-making, and good teamwork. Factors that may create barriers to implementation of the CRM Model include:

- *Resistance to Change* — Most people like the status quo, are comfortable with the condition or situation that they are used to, and fear any type of change. Any change generates actions or attitudes that can create barriers or obstacles that prevent change from occurring.
- *Fear of looking foolish* — People, and especially leaders, do not want to look stupid, feel foolish, or appear to be wrong.
- *Blame for misunderstanding* — Some people will place blame on others when there is a misunderstanding, rather than accepting the blame themselves. This situation can lead to conflict.

- *Intentionally withholding information* — Both leaders and followers can intentionally withhold information. This situation may occur for many reasons, including the fear of being wrong or losing control of the situation (information is power).

- *Complacency* — This condition occurs when a person is comfortable with the current situation or condition and does not want to risk the results that change might cause.

- *Physical, emotional, and mental fatigue* — When a crew member is unable to provide the maximum effort or is unable to concentrate on the situation, a barrier is created.

- *Halo effect* — This barrier occurs when we allow one personal trait or characteristic to influence our perception of a leader or crew member. By seeing a leader as infallible, we do not think we can question that person's decisions.

- *Being outside the group* — When crew members feel that they are not part of a group, they will not listen, make comments, or work effectively with the unit.

Disruptive Attitudes

Personal attitude may also create a barrier to a cohesive and safe crew relationship that can have a negative affect on the mission of the unit. The Federal Aviation Administration (FAA) has identified the following five negative or disruptive attitudes that can increase the risk to flight crew safety:

1. *Antiauthority* — Sign of independence and lack of teamwork. A symptom is the *Don't tell me what to do* response to an order or a crew member suggestion.

2. *Impulsivity* — Stress-induced desire to *Do something!* Symptomatic of a desire to act, even though all of the facts have not been gathered, the situation has not been analyzed, resources have not been gathered, or an action plan has not been developed. This attitude can lead to disastrous and sometimes fatal decisions.

3. *Invulnerability* — *It won't happen to me* syndrome leads to taking risks, ignoring safety policies, not using safe procedures, and not waiting for additional resources. Sometimes associated with teenagers but the syndrome certainly applies to members of the fire and emergency services.

4. *Machismo* — Belief that *I can do anything* . . . even if I lack the training, personal protective equipment (PPE), or resources required to tackle the hazard. Belief became part of the historical culture of the fire and emergency services described in Chapter 1, Risks of a High-Hazard Occupation. An example is the traditional view that respiratory protection was not needed for fire fighting. The mandatory use of self-contained breathing apparatus (SCBA) occurred as a result of fatalities and chronic respiratory diseases among firefighters.

5. *Resignation* — Belief that all action is useless and that the individual or group has no control over a situation. Strong motivated leadership is required to overcome this attitude.

The IAFC CRM publication includes two additional negative attitudes that can be present in firefighters. These attitudes are based on research done by U.S. Air Force Major Anthony T. Kern in his writings on flight safety and CRM. These last two attitudes are as follows:

- *Pressing* — Attitude that results from external pressures causing the leader and crew to feel the urgency to accomplish the mission simply for the sake of finishing quickly and returning to quarters. The pressure may be as simple as the desire to finish an interrupted meal, catch the last of a game on television, or get a few more hours of sleep. Generally, these external pressures are trivial. However, they can result in the leader making poor decisions and crew members taking shortcuts that compromise safety.

- *Air show syndrome* — Attitude that results from complacency with the belief that *we have always done it this way and never been hurt,* which causes a crew to *push the envelope* and attempt more dangerous activities. This sometimes fatal attitude can take a number of the following forms:

  — Desire to be the center of attention by performing a heroic (if not stupid) act for the benefit of onlookers, media, or peers.

  — Belief that *we have always done it this way,* which causes the leader or crew to apply one solution to every situation even though it might not be the best fit for that situation.

  — In training situations, this syndrome may be apparent in individuals or units attempting to outperform others to be the best.

Antidotes to Negative Attitudes

Negative Attitude	Antidote
Antiauthority Don't tell me what to do.	Follow the rules unless they are obviously wrong for the situation
Impulsivity Do something — Quickly!	Not so fast! Think first! Why am I doing this in this way!
Invulnerability It won't happen to me.	Consider the statistics: It CAN happen to me.
Machismo I can do it!	Is the possible gain worth the risk (risk analysis)?
Resignation What's the use?	I can make a difference in my world. Look at the individuals that have made a difference.
Pressing Let's hurry up and get this thing done so we can go home!	If a job is worth doing, it is worth doing right the first time. Let's have 0 tolerance for rekindles.
Air Show Syndrome I am going to look so good. Look at me!	Let's get the job done right and safely.

— Some company officers or unit leaders may also take the approach that the way the company or unit performs a task is superior to the way the task is taught at the training center.

— Some company officers or unit leaders may also take the approach that performing a task in a non-standard way will give them recognition.

The IAFC publication suggests that these negative attitudes can be prevented by using positive thoughts or statements to counteract them. The information box provides examples for antidotes to negative attitudes.

Workload Management

Most humans have limited capacities for remembering information and performing tasks. There are a few people in the population who can multitask; that is, they can perform multiple functions at the same time. Some people have Type A personalities that permit them to focus all of their energies and attention on getting a job done. For most of us, however, it is a challenge to keep all our responsibilities in order and remember which ones must be accomplished and in what order.

Workload management — The Crew Resource Management factor that attempts to reduce stress by reducing a job it its most basic tasks and assigning those tasks to individuals and groups.

We are constantly hearing of people who are overloaded, emotionally exhausted, and stressed due to heavy workloads. Workload management is a CRM factor that attempts to reduce stress and still meet the objectives of the unit. It can and should be applied to both emergency and nonemergency activities.

Basically, workload management involves reducing a job to its most simple tasks and assigning the tasks to individuals and groups. If a crew has trained together, the tasks will be easy to assign and accomplish when an emergency occurs. For instance, if a crew responds to a motor vehicle accident requiring extrication, each person will have a preassigned task such as patient stabilization, fire or vapor suppression, and power extrication equipment operation. The reality of life in both career and volunteer agencies is that crews are infrequently composed of the same personnel due to vacations, holidays, travel distances for volunteers, and the decreasing number of volunteers. Crews may be composed of personnel from other units, shifts, or simply the first to arrive at the scene.

If all personnel are trained to perform all the tasks in the same manner, then the officer can make specific task assignments as required by the situation with confidence that the personnel can perform the functions effectively and safely. An example of this situation could be the need to raise a long ground ladder (40 to 50 feet [12.2 to 15.2 m]) when no single company is staffed with sufficient personnel to do so safely. If all companies are trained in the same steps, then it is possible to safely call on personnel from multiple companies **(Figure 3.11)**.

Challenges occur when a situation changes and a crew is not prepared for it. For instance, in the scenario given earlier, if one member of the crew is absent, tasks may have to be reassigned and further prioritized. Another situation might involve the addition of a new or temporary crew member who has not practiced the tasks with other crew members and has to be told what to do and when. Or the new crew member might have a disruptive attitude that diminishes the crew's teamwork.

Figure 3.11 Cross training personnel, especially in volunteer departments, is important to effectively completing emergency scene tasks such as setting ladders that require many responders.

Workload management benefits from prioritizing tasks, assigning tasks to the appropriate crew member, and training for all contingencies. When it becomes apparent that a crew may be overwhelmed by the workload, additional resources must be requested. When considering command and control functions, the implementation of NIMS-ICS can significantly improve workload management.

Error Management

No matter how hard we try, mistakes still occur, errors are made, and accidents happen. While management theories such as Total Quality Management (TQM) may propose that *zero error* should be part of the manufacturing goal, it is not an attainable goal as long as humans are involved. And humans will always be involved in some part of the process. Therefore, it is necessary to manage or control errors in order to reduce the potential for their occurrence. Further, when considering emergency situations such as fire control, it is unlikely that the commanders will initially have adequate information to approach zero-error levels of decision making.

The CRM Model includes a process for error management that is intended to prevent, reduce, or repair errors. The process is based on the following three layers or defenses:

1. *Error avoidance* — Requires adhering to safety standard operating procedures (SOPs), training to a high level of proficiency, preincident planning, situational awareness, effective use of resources, and minimizing personal and crew distractions. This layer has the greatest potential for reducing accidents, injuries, and fatalities due to errors.

2. *Error trapping* — Requires that crew members recognize that an error has occurred and take steps to stop it and prevent it from developing into a disaster. You must be able to analyze what has happened and mentally

project it into the worst possible consequences. You must then take the actions necessary to reverse the process.

3. *Error mitigation* — Involves repairing the damage caused by an error, performing a postincident analysis and critique, and providing training, equipment, or SOPs to lessen the potential for the error to occur again. This layer is last and takes place once errors have occurred and could not be trapped in the second layer.

Application

The CRM Model presented in this chapter can be applied by an individual, a crew, or an organization. It can be applied to any situation whether emergency or nonemergency. While the material presented can give you an idea of how CRM works and the advantages of using it, it is best to participate in a formal training program. Training classes usually include role play, decision-making scenarios, and practical skills evolutions that will require participants to apply the CRM skills that they have learned. The sections that follow are examples of how CRM can be applied by an individual, a crew, and an organization.

CRM Training

To implement the CRM Model, the fire and emergency services organization must provide training in the CRM factors. Training should follow an accepted teaching format such as the one used by the commercial airline industry. The training model used by the commercial airline industry includes the following components:

(List elements of awareness addressed in CRM training.)

- *Awareness* — Learn about CRM. The awareness component includes the following activities:
  - Describe the current safety culture.
  - State why the current safety culture must be changed.
  - Describe the benefits of the new safety culture.
  - Describe crew resource management (CRM).
  - Explain how CRM applies to tasks and duties of fire and emergency responders.
  - Explain how CRM will help create and sustain the new safety culture.
  - Explain the limitations of CRM when applied to emergency situations.

(Explain how CRM factors are reinforced in CRM training.)

- *Reinforcement* — Apply CRM factors in controlled scenarios. Firefighters are required to solve problems, make decisions, and work as a team to accomplish a set of goals. These practical training evolutions may be based on any type of emergency or nonemergency situation. Positive reinforcement is important to the learning process because it fixes the problem but does *not* assign blame.

(Describe how CRM training is refreshed.)

- *Refreshment* — Provide, refresher training periodically for the organization's membership. This training, which may include a review of the CRM factors, role play, or practical training evolutions, emphasizes the importance of CRM and the organization's commitment to it.

Individual

You can apply one or all of the CRM factors in your daily life. The more often that you apply them, the more confident you will be in the outcome. In the sections that follow, each of the CRM factors is applied to a scenario in which an individual emergency responder might be involved.

In the scenario, you are a member of a small rural fire department that must depend on a mobile water supply apparatus to protect many parts of the response area. You are assigned the task of driving the water tender apparatus to the emergency scene by yourself **(Figure 3.12)**. The department has a SOP for the operation of this type of apparatus.

Communications

For the individual, communications can be both internal and external. Internally, you must listen to your inner voice that will tell you the condition of your mental, emotional, and physical states. The following examples would apply to a person operating any apparatus or motor vehicle:

- You must be mentally alert with your mind focused on your actions and surroundings.

- You must be emotionally stable and not allow the stress of the moment or other distractions to influence your actions.

- You must be fully aware of any physical conditions that could cause you to lose control of the apparatus. *Examples:*

  — Tightness in the chest is associated with high levels of stress that can result in cardiac arrest.

  — Alcohol and many types of medication can reduce reaction times and blur the senses. You should never operate any apparatus if you have been drinking alcohol or taking either prescription or nonprescription medications.

Figure 3.12 In volunteer, combination, and some career fire departments, individuals may be called upon to respond to an incident alone. This act will require the application of Crew Resource Management skills such as situational awareness.

External communications come from radio communications between the telecommunications center, the IC, on-scene crews, and you. Radio communications must be accurate and balanced. You must not allow these communications to cause you to commit unsafe acts such as failing to use both hands when maneuvering apparatus under difficult conditions, failing to stop at an uncontrolled intersection, or entering a slick curve at a high rate of speed.

Situational Awareness

Whether you are working alone or with a crew, you must always be aware of the situation you are in and any changes that are occurring. As the sole operator of the tanker apparatus in the scenario given earlier, you must be aware of the following factors:

- Your personal mental, emotional, and physical condition
- Apparatus condition

- Road and traffic conditions
- Weather conditions
- Emergency scene situation
- Radio communications

You must use the information that you receive from your senses to make the appropriate and correct decisions that guide your actions. At the same time, you must mentally prioritize this information and not allow yourself to react in an unsafe manner to unnecessary distracters.

Decision-Making

With the information gained from your senses, you will start making decisions before you leave the station. Most obviously, if you do not think that you can operate the apparatus, you will have to request someone else to drive it. If the apparatus itself is not ready due to the lack of water or a mechanical problem, you will have to decide on an alternative action such as informing the telecommunications center to send another unit.

During the response, you must stay focused on your immediate surroundings so that you can make decisions about route, speed, and travel time. Selection of the travel route will be based on available routes, your knowledge of weather, road, and traffic conditions and the maximum amount of time that can be allotted to travel. In all cases, your decisions must be based on the safest possible actions for handling the apparatus.

Teamwork

Even though in this scenario you are operating the apparatus by yourself, you are still part of the response team. It is your responsibility to ensure that you complete your task of delivering water to the scene to support the rest of the team's mission. You must communicate with the IC and telecommunications center and provide approximate travel and arrival times.

Workload Management

In this scenario, workload management should be rather simple: You drive the apparatus. Driving alone, however, will actually cause you to perform the following multiple tasks that involve almost all of your senses to help you manage the safe operation of the apparatus:

- You will use your eyes to watch the road, look for traffic, observe pedestrians or animals along the roadside, and look for indications of changes in the weather.
- You will listen to the radio communications and the apparatus engine sounds as well as audible warning devices from other emergency vehicles.
- You will use your sense of touch to determine how the apparatus is holding the road, to feel for any indications of sliding or tipping, and to feel any vibrations from the tires, steering mechanism, or brakes.

Error Management

Operating apparatus safely and following established policies and procedures are the best ways to avoid an error that could result in an accident. If an accident starts to occur, however, you will have to make the right decision to prevent the situation from worsening. For instance, if the right rear wheel slips off a hard road surface, you must be able to correct and prevent the apparatus from overturning. When an accident cannot be prevented, you must then know the procedures for reporting the accident and placing the unit back in service as soon as possible.

Crew

To successfully apply the CRM factors, a crew must be trained in their use and practice them in a variety of situations. Practice should be continuous and should involve changes in the conditions of each scenario. This practice will force a crew to develop and use contingency plans to fulfill its mission.

Communications

A crew leader is responsible for communicating the incident action plan (IAP) and tactical directions to members of the crew. The leader must also be able to listen to the orders provided by the IC and the suggestions and concerns of crew members. Openness to the comments of the crew is a key to the success of the CRM Model.

Crew members must also be able to listen to, evaluate, and act upon the orders given by the crew leader. They must also be empowered to question a leader's decisions based on their perceptions of the situation. Crew safety depends on this ability to ask questions and make suggestions. In emergency, time critical situations, good judgment must be exercised when asking questions. In emergency time critical situations, good judgment must be exercised when asking questions. Questions must be focused on the safe accomplishment of the crew's objective and personnel safety issues.

Situational Awareness

Members of a crew must apply their own situational awareness process and communicate any concerns that they may have about the situation. They must also communicate their personal status to the leader and other crew members. Physical or emotional stress, dehydration, fatigue, or fear can all affect the ability to function.

Decision-Making

One of the key factors of CRM is the ability of a leader to make decisions with the assistance of crew members. If a crew leader makes an error in judgment, the crew has the opportunity to trap the error before too much damage occurs. It is important that the crew leader not feel threatened by the crew's input in the decision-making process. Training and experience will help the leader to overcome any negative attitudes.

Great care must be exercised in countering a leader's decision to make sure the leader and other crew members are aware that the leader's direction is being modified. Lack of good judgment in this regard can lead to freelancing.

Teamwork

The unit applies teamwork by working together toward the same goal according to the IAP and the crew leader's tactical directions. Crew members perform their assigned tasks and assist others with theirs. The group works as a team and stays aware of each other's location and situation. When one person must leave the area, the team leader must consider the situation and determine if the entire team should withdraw. If the decision is made to withdraw, the team should leave together and inform the IC that they are withdrawing and why. When fire company personnel are geographically separated, e.g., an initially arriving ladder truck company is carrying out search and rescue and ventilation tasks concurrently, the company commander must continually track the location and condition of those personnel.

Workload Management

A crew leader assigns tasks based on a crew member's ability and training. Crew members should be assigned tasks that will not overload them mentally or physically. They should also assist other crew members and be ready to ask for assistance when needed.

Error Management

A crew must be able to recognize potential errors or mistakes and avoid making them such as ensuring that an apparatus is parked downwind and downhill from a hazardous liquids spill. When an error does occur, crew members must be prepared to either implement actions that will lessen the effects of the error or withdraw from the hazardous area.

Taking corrective actions to correct errors is much like a military retreat. It must be done with as much care as an advance. It must be disciplined. Preventing "freelancing" when pulling out of a position is just as important as it is when advancing.

Organization

The organization applies CRM at all levels in the management of emergency and nonemergency activities. The factors can be used to help the chief of a department negotiate the political waters of local government or implement change in the department as well as command a multiple unit or agency operation. NIMS-ICS should be used at all emergency incidents.

By adopting and actively using the CRM factors, the chief and administrative officers create a management culture that ensures safe, efficient, and effective operations. CRM will then gain acceptance and credibility with all members of the organization.

Communications

In nonemergency activities, the organization must have a free flow of information both from the top and from the bottom. Changes in SOPs should be explained, and comments from the membership of the organization should be accepted.

At an emergency incident scene involving multiple units or agencies, the IC must give clear and concise instructions, ensure that they are understood, and listen to reports and comments from personnel throughout the site. Company officers, Division and Group Supervisors, Branch Directors and Sections Chiefs must provide their immediate superiors (and where implemented, designated status collecting positions) with timely information regarding the situation and resource status. To ensure that objectives are met at specified times, the

IC may require that the telecommunications center periodically broadcast the current time (4 p.m., 8 p.m., etc.) in specified intervals such as 5, 10, or 15 minutes. This procedure is especially important if a structure has the potential to fail after a given amount of exposure to fire and heat.

Situational Awareness

The chief of a fire department should always be aware of the external and internal conditions of the organization. Customer feedback, media comments, and political agendas can generate external forces that may affect funding or political support for the department. Internally, labor leaders, officer groups, divisions (such as training or fire prevention), and informal leaders may all have needs and concerns that can affect the day-to-day operation of the organization.

At an emergency incident, the IC as well as all NIMS-ICS officers at the Company, Task Force, Division, Group, Branch and Section levels must constantly monitor the situation, including the weather, fire behavior, and the physical conditions of the operational crews. Contingency plans must be available for any changes in the situation. The IC must be prepared to change plans at a moment's notice.

Decision-Making

Current fire and emergency services leaders tend to make management decisions based on business models that involve other members of the organization, external customers, and stakeholders. Rarely are nonemergency decisions made autocratically and without input from others.

Decisions made at emergency incidents must be based on the most timely accurate information available, preplanning that follows SOPs accepted practice, and when the time permits the opinions of other personnel.

It can be argued that initially there is often little time for a group decision at an emergency; however, as an incident progresses, more input from personnel can be integrated into the decision-making process.

Teamwork

Management of the organization depends on teamwork. Each division (such as fire suppression, training, or fire prevention) works as a team causing the entire department to work together toward the common goal stated in the organization's mission statement.

Units and agencies that are operating at a major incident each have their assigned tasks based on their training, equipment, and role such as the following:

- Ladder/Truck company ventilation tasks must be coordinated with engine companies that are attacking the fire with hoselines.
- Law enforcement area evacuation tasks must be coordinated with incident control activities.
- Rapid intervention teams/crews (RITs/RICs) must be equipped and ready to respond the moment a call for help is made **(Figure 3.13, p. 110)**.
- Air attack must be coordinated with ground control activities.

Figure 3.13 Rapid intervention teams/crews must be equipped and prepared to enter a structure to rescue injured or trapped firefighters.

Workload Management

The proper application of teamwork should ensure that each division within an organization has the proper resources (personnel, funds, time, and equipment) to perform their tasks. When one division is overwhelmed by a task, other divisions can provide the necessary resources to accomplish it. For instance, if a state law mandates that all commercial buildings be inspected for fire hazards before a given date, fire companies may be required to assist fire-prevention officers in fulfilling this task.

The same approach can be taken at emergency incidents. Training officers and fire inspectors may be used for nonsuppression tasks such as scene control and staging to make suppression personnel available for fire fighting or other emergency activities. Civilian tractor operators may be used to cold trail wildland fires to make fire crews available to participate in fire attack.

Error Management

The leadership of a fire department should strive to ensure that errors and mistakes do not occur. Creating, implementing, monitoring, and altering the strategic plan as needed is one approach to avoiding errors. Another is developing and enforcing SOPs for all phases of a department's operation. Analyzing injury, fatality, loss, and fire statistics provides information for taking a proactive approach to safety and fire prevention.

When an emergency incident occurs, the organization can rely on the three-layer approach: avoid, trap (prevent), or mitigate (repair) errors. Adhering to SOPs and accepted strategy and tactics provides methods for avoiding errors at an emergency incident. However, remaining flexible and having the ability to implement a contingency plan help to trap (prevent) errors in command decisions when they occur. Finally, performing postincident analyses, including a review of all safety-related aspects and holding postincident critiques help to ensure that errors are not repeated at similar incidents.

CRMs Current Applications

CRM has proven itself in the aviation industry for over 25 years. Improved safety for commercial airlines and their passengers has resulted from the improved communications, situational awareness, decision-making, and teamwork of flight crews. In order to implement CRM, the commercial airlines had to overcome barriers within their organizations.

Today, the CRM Model is being applied in the fire and emergency services to help improve the safety and efficiency of fire crews. Individuals, crews, and organizations can apply each of the CRM factors, as well as the concepts of workload and error management, to ensure that *everyone goes home* safely. CRM is everyone's responsibility and has the potential for changing the culture of the fire and emergency services by reducing accidents, injuries, and fatalities.

High Reliability Organization

A High Reliability Organization (HRO) is defined as an organization that has succeeded in avoiding catastrophes in an environment where normal accidents can be expected due to risk factors and complexity. HRO is a total system that is engrained in every task performed by the agencies that have utilized it. The system promotes communication at all levels and encourages people of all ranks to participate in safety-related issues. The HRO process embraces CRM as part of the safety system.

The HRO concept was developed by Dr. Karlene Roberts of the University of California, Berkley, in the 1990s. It has been implemented by a wide variety of organizations including the U.S. Navy's carrier-based aviation program, commercial aviation, the nuclear power industry, the NASA, healthcare facilities, educational institutions, and a variety of fire and emergency services organizations.

Five characteristics are shared by HROs according to Dr. Roberts' research. They are as follows:

1. *Process Auditing* — System of checks intended to recognize expected and unexpected problems that can lead to an accident. This characteristic also includes safety drills and equipment testing as well as a reevaluation of solutions to previously recognized problems.

2. *Reward System* — Rewards are used to motivate individuals to perform in a safe manner and to discard unsafe behaviors. A reward system can also be applied to the organization as a whole.

3. *Quality Degradation Avoidance* — Avoiding a degradation or loss of quality is essential to maintaining a safety culture. A universally accepted standard is used to compare the individual or organization's safe behavior.

4. *Perception of Risk* — This characteristic consists of the following two elements:
   — The knowledge that a risk exists;
   — The extent to which the risk is recognized, minimized, or appropriately mitigated.

5. *Command and Control Elements* — This characteristic consists of the following five elements:

— Migrating (delegating) decisions to the person who has the technical expertise to make the decision.

For example, this can be accomplished at the incident scene by having a technical expert in the ICS-Planning Section able to give immediate advice to the Incident Commander or other command officers or initiating an ICS- Unified Command organization with one of the "joint ICs" having the necessary technical expertise.

— Creating redundancy by providing a backup system, made up of personnel or equipment.

For instance, ICS approaches this with the use of the Rapid Intervention Team or Crew (RIT/RIC) concept.

— Having senior managers who exercise situational awareness of the *big picture* and strategic goals. This type of manager does not micromanage but provides subordinates with the support necessary for them to make decisions and do their jobs.

For instance, the ICS is based on this concept, i.e., the IC does not interfere in the tactical decisions of subordinate officers as long as they function within the Incident Action Plan (IAP) and have not created an unacceptable personnel safety issue.

— Having formal policies and procedures that ensure predictability in the operation but also allow for flexibility in a fluid environment.

For instance, the fire service uses SOPs/SOGs to provide the element of predictability at incidents, but the ICS gives all levels of command the flexibility to deal with unexpected situations.

— Training that develops strong teams and creates redundancy through cross-training.

For instance, fire service training is primarily accomplished in team format (typically called "companies"). Progressive fire agencies train all members to utilize a wide array of tools, not just those assigned to the company on which they are assigned.

You should note some of the similarities between CRM and HRO. At the same time, you should also be aware that both systems are compatible. While the characteristics of the HRO apply to the organization, the characteristics of the CRM can be applied at all levels from the individual to the supervisor and finally to the administration.

Summary

To fulfill the 16 Safety Initiatives developed by the IAFC and NFFF, the fire and emergency services will have to adopt a behavioral-based safety culture. It does not matter which approach is taken just so long as a cultural shift is made. You have been introduced to three approaches to safety in the fire and emergency service: strategic safety planning, crew resource management, and high reliability organization . It is up to you, as an individual, a supervisor, or the head of the organization to determine the best model and adopt it for yourself, your unit, and your organization.

Review Questions

1. How does SWOT concept apply to the company level?

2. Describe the differences between flight crews and fire company crews and how that might affect Crew Resource Management functions

3. Discuss means of resolving conflict at incident scenes.

4. What does the DECIDE acronym stand for?

5. How can disruptive attitudes increase the risks to crew safety?

6. What does interference mean in a communication model?

7. What steps can a sender take to improve communication?

8. What steps can a listener take to improve communication?

Health and Fitness Program Requirements

Chapter Contents

Key Terms

Job Performance Requirements

This chapter provides information that addresses the following job performance requirements of NFPA® 1500, *Standard on Fire Department Occupational Safety and Health Program* (2007):

NFPA® 1500 References

5.2	10.1.1	10.2.3	10.3.4	10.5.1	10.6.4.1
5.2.1	10.1.2	10.2.4	10.4	10.5.2	10.6.5
5.2.2	10.1.3	10.2.5	10.4.1	10.6	10.7
5.2.3	10.1.4	10.3	10.4.2	10.6.1	10.7.1
5.2.4	10.1.5	10.3.1	10.4.3	10.6.2	10.7.2
5.2.5	10.2	10.3.2	10.4.4	10.6.3	10.7.3
5.2.6	10.2.1	10.3.3	10.5	10.6.4	10.7.4
10.1	10.2.2				

FESHE Learning Objectives

Occupational Safety and Health for Emergency Services

Describe the components and value of critical incident management programs.

Describe the components of a wellness/fitness plan.

Identify and analyze the major causes involved in line-of-duty firefighter deaths related to health, wellness, fitness, and vehicle operations.

Health and Fitness Program Requirements

Learning Objectives

After reading this chapter, students will be able to:

1. Describe 4 key elements to developing a wellness and fitness program.
2. Define job related injury.
3. Define job related fatality.
4. Describe importance of analysis of frequency and risk associated with firefighter duties.
5. Describe 4 key elements of a fitness for duty policy.
6. Describe 5 areas that are used to define fitness.
7. Describe importance of pre-employment medical requirements.
8. Describe the 8 Candidate Physical Abilities Test events.
9. Describe the types of work capacity tests.

Chapter 4
Health and Fitness Program Requirements

Case History

In 2002, Mark Noble, a nineteen-year veteran of the Olympia (WA) Fire Department, was diagnosed with brain cancer. After surgery to remove the tumor, Mark began a regimen of chemotherapy and radiation. During his treatment, Mark started researching the connection between firefighters and cancer. What he found was that firefighters are exposed to highly toxic substances in virtually every fire — especially during overhaul. These substances include asbestos, benzene, polycyclic aromatic hydrocarbons (PAH), and polychlorinated biphenyls (PCB), in addition to carbon monoxide (CO) and other well-known products of combustion. The toxic effects can accumulate in the body with repeated exposures.

In his research, Mark found that when compared to the general adult population, firefighters have a greater likelihood of developing a wide variety of cancers. His research indicated that firefighters are:

- 2 times as likely to develop intestinal cancer, liver cancer, prostate cancer, and non-Hodgkin's lymphoma;

- 2.25 times as likely to develop malignant melanoma and 3 times as likely to develop other skin cancers;

- 2.5 times as likely to develop testicular cancer;

- 3 times as likely to develop bladder cancer and leukemia;

- 3.5 times as likely to develop brain cancer;

- 4 times as likely to develop kidney cancer.

In 2005, at age 47, Mark Noble lost his battle with the brain cancer that he almost certainly developed because of the toxins he was exposed to as a firefighter. Mark loved being a firefighter, but he said that if he had it to do over, he would wear his self-contained breathing apparatus (SCBA) more and he would be more conscientious about connecting apparatus exhaust collection hoses.

Permission granted by Mrs. Rebecca Noble and ERGOMETRICS & Applied Personnel Research, Inc., who produced a video interview with Mark during his final months. The video is available at www.ergometrics.org.

Case History

Deputy Chief Suffers Unwitnessed Sudden Cardiac Death While Fighting a Wildland Fire in Steep Terrain - Massachusetts

On April 8, 1999, a 64-year-old male Deputy Chief of a paid call fire department was the first responder to a wildland fire in mountainous terrain. His initial assessment prompted him to request assistance from neighboring departments and from the State Air National Guard. The victim engaged in fire suppression activities for over 4 hours prior to his unwitnessed collapse. The victim was down for approximately 10 - 15 minutes when another firefighter found him unconscious, without a pulse or respirations, on the ground next to a fire truck. The victim died despite cardiopulmonary resuscitation (CPR) and basic life support (BLS) administered by fellow firefighters and emergency medical technicians (EMTs) at the scene, and advanced life support (ALS) administered by paramedics and hospital emergency department personnel. Autopsy findings included "coronary atherosclerosis, focally severe," "extensive fibrosis in the antero-septal region of the left ventricle," "left ventricular hypertrophy," and "chronic obstructive pulmonary disease." At autopsy the carboxyhemoglobin level was less than 5 percent, suggesting that carbon monoxide exposure was not the cause of his death.

The victim's last visit to a physician, approximately 5 years before his death, was for laryngitis and did not include any cardiology workup. In 1992, the victim was approved to wear a respirator for his regular job, at which time his resting electrocardiogram (EKG) and blood pressure were normal. Family members and colleagues reported that the victim did not exercise regularly but was not overweight. He had never reported symptoms suggestive of heart disease at home, during leisure activities, while performing his job as an electrician, or while performing Deputy Chief Officer duties. He had no family history of coronary artery disease (CAD). He had stopped cigarette smoking 22 years ago. The morning of the wildland fire, while at his regular job from 0730 to 1200 hours, the victim did not complain of chest pain, shortness of breath, dizziness, or any other symptom suggestive of cardiac problems.

The department had no voluntary or required physical fitness/wellness program, nor was periodic medical clearance for SCBA use required.

Recommendations relevant to this fire department include the following:

- ***Provide fire fighters with medical evaluations to determine their medical ability to perform duties without presenting a significant risk to the safety and health of themselves or others.***

- ***Initiate a wellness/fitness program for fire fighters to reduce risk factors for cardiovascular disease and improve cardiovascular capacity.***

Source: NIOSH F98-28 Investigation of line-of-duty-deaths (LODDs)

As the first case history indicates, not all firefighter fatalities occur during emergency operations. Many that go unreported as line-of-duty deaths (LODDs) occur months and even years after an injury, exposure to hazardous materials, or the onset of a long-term illness. As the second case history illustrates, cardiac arrest and strokes may occur on duty or during off duty hours due to stress or hypertension. The most effective means of preventing or reducing job-related injuries, illnesses, and fatalities is a proactive approach based on a comprehensive *health program*. Components of the health program include wellness and fitness programs.

Health program — Any program designed to assist and encourage employees to adopt and attain a healthier lifestyle.

A comprehensive health program consists of a number of elements. Examples exist in both the public and private sectors as organizations have realized the cost-saving benefits, and employees have taken advantage of these programs as benefits. Some labor unions have supported the creation of these programs and frequently include them in contract negotiations. Some or all of the following may be included in agency health and safety programs:

- Employee assistance programs
- Health and well-being survey data
- Wellness education
- Smoking or tobacco-use secession classes
- Fitness programs and annual testing
- Physical fitness facilities or centers
- Contract fitness center memberships
- Lifestyle change counseling
- Lifting education sessions
- Nutrition education
- Hydration education
- Supervised physical exercise programs
- Stress management programs
- Medical evaluations and examinations
- Health risk assessments
- On-site flu shots
- Health advice (phone) lines
- Online (Intranet or Internet) resources

The current trend of modern health programs is to move from maintaining general health and well-being to identifying, managing, or reducing specific and potentially costly health risks. New programs target high-risk behaviors such as disease management (diabetes, etc.), stress, tobacco use cessation, and weight-management. Financial incentives and consequences are sometimes used to motivate employees identified as being at high risk. Data is collected and used to track the success and effectiveness of these programs.

City of Los Angeles Fire Department (LAFD)
Example of a Successful Comprehensive Medical Program Funded by the Assistance to Firefighters Grant (AFG) Program

The LAFD provides fire fighting, emergency medical, terrorism response, fire prevention, hazardous materials mitigation, disaster response, and public education services to the 3.8 million residents of the City of Los Angeles from 103 fire stations strategically located across 471 square miles, in addition to providing automatic aid to the ten jurisdictions contiguous to the City of Los Angeles. With the volume of emergency incidents continuing to escalate annually (338,733 incidents in 2001/2002 and 344,546 incidents in 2002/2003), the continued health, fitness, and safety of emergency service members is crucial.

According to the USFA's *Firefighter Fatalities Report in the United States in 2005*, heart attacks are the most frequent cause of death for firefighters. After several unsuccessful attempts to fund the program through the city budget process, the AFG Program funded the LAFD's Comprehensive Medical Program. Previously, the LAFD was unaware of the health status of many of its members. By providing a mandated, biennial comprehensive medical assessment of each member's fitness to perform the rigors of fire fighting, individual health awareness and fitness has increased; potentially life-threatening conditions have been detected, brought to the member's attention, and addressed; and injuries and cardiovascular incidents have been prevented. A healthier cadre of firefighters is available to serve the residents of Los Angeles and those who live in contiguous communities.

The LAFD Comprehensive Medical Program is congruous with the Wellness-Fitness Initiative's positive, holistic, rehabilitative, and educational approach to firefighter wellness. Each comprehensive medical assessment encompasses a health questionnaire, a thorough physical examination, pulmonary function test, graded exercise treadmill test for members over 40 years of age and those under 40 years of age who have two or more cardiovascular risk factors, blood panel, chest X-ray, urinalysis, vision testing, measurement of body fat, audiometry, and skin-cancer evaluation. Follow-up testing is performed as indicated; for example, a Baseline Fast Heart Computed Tomography (CT) Scan as recommended by a cardiologist after an Exercise Stress Test. This grant award allowed the LAFD to perform 1,965 periodic medical examinations, approximately 55 percent of its workforce, and, through its findings, illustrated the importance of good health and physical fitness related to the duties firefighters perform on a daily basis and of developing a comprehensive wellness program.

Of the examinations performed, health care providers identified a sub-group of 116 members for follow-up evaluation based on the results of the initial medical examination and EKGs that presented rhythm irregularities. From the EKG reviews, 92 members received heart scans and 40 received noninvasive angiography. As of May, 2006, 91 members returned to duty, with one member undergoing a coronary artery bypass procedure. All 92 members received counseling on significant lifestyle modifications, encouraging a healthier life and reducing the risk of coronary heart disease. The Department considers discovery of the 92 members at risk for heart disease for firefighters as significant.

The medical examinations identified numerous other risk factors linked to heart disease and provided the Department with essential information regarding how best to help its members achieve a higher level of fitness and wellness. Because of these findings, the Department is furthering its efforts to build upon the success of the program by expanding the Comprehensive Wellness Program to include Behavioral Health Management and fitness and nutrition awareness and counseling.

The goal of most health programs is twofold. First, the goal is to ensure that employees are physically fit to perform in a physically intense occupation like the fire and emergency services. Many experts in the field believe that employees in good physical condition use less sick leave, require less medical care, and have a longer life expectancy. This helps to meet the second goal, reduced health- related costs for the employer in the form of less lost time due to illnesses or injuries, lower health and worker's compensation insurance premiums, and reduced overtime or substitute worker costs.

The health and wellness program is the responsibility of the organization's administration, supervisors, and the individual employee. The administration of the organization must develop, support, and implement a plan that has realistic objectives and meets the primary goals stated above. Supervisors must ensure that personnel assigned to them understand the program and participate in it. Finally, the emergency responders must embrace the program, participate in it, and provide accurate feedback to ensure that the program is providing what it is intended to provide.

Casualties Related to Health and Wellness

People who are unhealthy or physically unfit are less productive, more prone to injuries and illnesses, and contribute to loss-time costs for the organization. Therefore, a proactive health and wellness program, as discussed in Chapter 11, is the basis for improving the overall health of the organization's membership.

Because health- and wellness-related injuries and illnesses are not easily defined in the NFPA® and National Institute for Occupational Safety and Health (NIOSH) reports, this data is not consistently collected or evaluated. What can be determined is that cardiovascular-related incidents are the leading cause of death in the fire service. Between 1996 and 2001, 256 firefighters died from heart attacks and strokes according to a 2002 report by the USFA. A majority of these victims were over 40 years of age. Strains and sprains, common to emergency incidents, are more common in older firefighters and may be related to physical fitness, which is greater in younger personnel.

Some injuries and illnesses may not occur immediately. Cancer, cardiovascular, pulmonary, exposure-related, and skeletal conditions may not become apparent until well after a person's service career is over. While there is currently an attempt to track such occurrences, it can be difficult to directly relate them to specific incidents. In some cases, such as cancer, the exposure to apparatus fumes may have occurred over time. To prevent or reduce these injuries and illnesses, the department must begin early in a firefighter's career through education, awareness, fitness, and an aggressive health and wellness program.

Legal Requirements

The establishment of a health program is mandated by the federal government in regulations enforced by federal and state Occupational Safety and Health Administrations (OSHA). These requirements can be found in:

- 29 CFR 1910.134, *Respiratory Protection*
- 29 CFR 1910.120, *Hazardous Waste Operations and Emergency Response*
- 29 U.S.C. § 654, 5(a)1, the General Duty Clause may be cited in the lack of employee health protection.

State OSHA programs must meet the minimum requirements of the federal OSHA programs. State occupational safety and health programs will usually include more stringent requirements. Some of the requirements are based on the NFPA® standards, including 1500, 1581, 1582, 1583, and 1584, when they are adopted by the authority having jurisdiction (AHJ).

Additional laws that relate to health programs are the medical records acts. These require the retention of medical records for the duration of employment plus 30 years after separation from the fire and emergency service **(Figure 4.1, p. 122)**. These acts also require that the records are kept private with only the physician, the employee, and a specifically designated person authorized to gain access to them.

Figure 4.1 Medical records must be retained for 30 years after personnel are separated from the department.

Other laws related to employee health that have been enacted in the past quarter century are intended to provide benefits to the employee. They include the Family Medical Leave Act and Maternity Leave Act that allow an employee to take medical leave in the event that a family member is ill or incapacitated or when a child is born. In the case of each of these laws, the intent is to reduce the stress placed on the employee in times of medical crises.

Finally, laws that regulate wrongful termination have been established to protect employees. Some of these laws apply to termination for medical reasons. These laws specify the steps that the organization must take to terminate an employee on medical grounds. The laws protect the rights of the employee while establishing the protocol that must be followed. The employee is also given certain responsibilities in these cases plus the right to appeal a medical termination decision. Specific federal laws include the Americans with Disabilities Act (ADA), Title VII of the Civil Rights Act, the Family and Medical Leave Act, and various labor, health, and safety laws. It is good practice to add these specific laws to labor/management agreements and department rules and regulations.

Health Related Risks

Fire and emergency responders face multiple risks to their health during their careers. Accidents involving motor vehicles, structural collapses, or other physical injuries are covered in other chapters of this manual. Those health-related risks that can be mitigated or eliminated by a comprehensive health program are covered in this chapter. These risks can be divided into three broad categories:

- risks encountered in fire suppression and other emergency activities
- risks resulting from exposures to chemical and biological hazards
- risks created by lifestyle choices

Fire Suppression and Other Emergency Activities Risks

The primary cause of firefighter fatalities during emergency operations is stress-induced cardiac arrest. Stress can also cause strokes that can be fatal or result in long-term disabilities. Both cardiac arrests and strokes may not occur until after the emergency incident is over. Repeated exposures to stress can also weaken the heart muscles and cause repeated heart attacks. Age and lifestyle choices can also contribute to cardiac arrest and strokes. Statistics show that both volunteer and career firefighters who are over 50 years old are more susceptible to cardiac arrest than younger firefighters.

Age Comparison of Firefighter Death Rate to Cardiac Arrest

The death rate of firefighters due to cardiac arrest is double for those ages 50-59 compared to those ages 40-49. The rate for those ages 60 and over was four times the rate for those ages 40-49 according to a study published in the *New England Journal of Medicine* in 2007.

Source: *New England Journal of Medicine* study published March 27, 2007.

Stress, physical, physiological, and psychological, can be the result of emergency scene operations. Beginning with the shock caused by the alarm, often in the middle of the night, through the arrival at the incident and the activities at the incident to the final return to quarters, stress is a constant companion for fire and emergency responders.

As a result of studies done on the relationship between the use of alarm bells and stress, there has been a move to replace the bells with a gradual tone alarm in some departments.

Cardiovascular disease (CVD) is a family of diseases that includes hypertension (high blood pressure), atherosclerosis, coronary heart disease, and stroke. Besides causing cardiovascular disease and strokes, stress also contributes depression and emotional fatigue, physical fatigue, and professional burnout.

Pulmonary diseases are also the result of emergency scene operations. The exposure to smoke and the by-products of combustion can have very long-term effects on the body. Among these effects are reduced lung capacity, pneumonia, emphysema, cancer, and infections. Symptoms of some of these diseases may not appear until after responders retire.

An often-overlooked risk of emergency operations is hearing loss. Constant or repeated exposure to apparatus and equipment engine noise, audible warning devices, and on-scene noises can cause tinnitus. Tinnitus is the destruction of small nerve endings in the ear canal that normally carry sound into the hearing chambers. A symptom of tinnitus is a high-pitched ringing in the ears or simply silence and the loss of hearing at certain frequencies. While some hearing loss is the result of the aging process, the onset of tinnitus may occur at any age.

Exposure Risks

Besides exposure risks encountered at fire scenes, fire and emergency responders are increasingly exposed to biological hazards during medical responses. Both air and bloodborne pathogens can be present when dealing with victims and medical waste. Tuberculosis; acquired immune deficiency syndrome (AIDS), hepatitis A, B, and C; and other easily transmitted diseases must be protected against.

Exposure to toxic chemicals at spills, in the smoke from a burning building, during building inspections, and while working around the station may result in either short- or long-term illnesses. Asbestos, a known carcinogen, is still prevalent in older buildings and maritime vessels **(Figure 4.2)**. Inhaled, asbestos fibers will settle in the lungs causing cancer and pulmonary illnesses. Inhaling apparatus fumes, both from diesel and gasoline engines, particularly in a confined space, can expose personnel to carcinogens.

Other types of exposures occur during nonemergency activities. Personnel can be exposed to viruses, colds, flu, lice, and contaminated food in the confines of the fire station. Improper food preparation, lack of personal hygiene, unsanitary conditions, and sharing bedding and personal protective equipment (PPE) can all result in exposure to job-related illnesses.

Figure 4.2 Building materials that contain asbestos, such as siding, still exist and can cause a health hazard to firefighters.

Lifestyle Choice Risks

As Pogo, the alter ego of cartoonist Walt Kelly, once said: *We have met the enemy and he is us.* Our personal lifestyle choices can result in some of the greatest risks to our health. The following lifestyle choice risks can further increase the risks we must endure during emergency incident operations.

Use of tobacco products — The use of tobacco products can result in emphysema, pneumonia, cancer, and decreased lung capacity. Smokeless tobacco products have been linked to cancer of the mouth, throat, and stomach. As a result, medical costs to treat these diseases have increased, causing some organizations to prohibit smoking while on duty and to require recruits to sign a nonuse contract.

Misuse of narcotics — Narcotics are substances that affect the central nervous system and that may become addictive, including drugs and alcohol. Besides addiction or dependence, narcotics can impair judgment and lower reaction times. Career organizations generally provide employee assistance programs for anyone who is determined to be using narcotics.

Misuse of prescription and nonprescription drugs — Many prescription and nonprescription drugs can have harmful side effects including addition. Like narcotics, these drugs can induce sleep, reduce reaction times, and impair judgement. Many drugs even carry the warning to not operate heavy machinery while taking them. Another type of drug that can be misused or abused is the anabolic steroid. This drug has gained popularity, and recognition, in the field of sports as a means of increasing muscle strength. Misuse can have long-term effects on the body including increased chances of cardiovascular disease, aggressive behavior, liver disease, dehydration, and increased cholesterol levels.

Poor nutrition — Improper eating habits can result in obesity, diabetes, heart disease, and strokes. Currently, both obesity and diabetes are considered epidemic in the United States. Uncontrolled diabetes can result in blindness, kidney disease, cardiac arrest, and nerve damage in the extremities necessitating amputation. Some agencies, recognizing the connection between these diseases and nutrition, are providing education for proper eating habits.

Poor hydration — Approximately 60 percent of the body is composed of water. A reduction of this fluid level adversely impacts the body's ability to maintain its thermoregulatory mechanisms **(Table 4.1)**. This increases the firefighter's risk of heat-related injuries such as heat cramps, heat stress, and heatstroke.

In addition to traditionally recognized heat-related injuries, an overwhelmed thermoregulatory system adversely impacts the heart by reducing its cardiac output. An increase in body temperature will cause the intestinal and kidney vessels to dilate and further elevate the temperature while reducing the blood pressure. This in return, may contribute to swelling of the brain that produces elevated intracranial pressure that results in failure of the central nervous system.

Severe dehydration may also contribute to an imbalance of electrolytes that are essential to the proper operation of the heart. An imbalance of the electrolytes may produce dangerous disrythmias that could result in sudden cardiac death.

Lack of physical exercise — Inactivity and the lack of a structured physical fitness program increases the potential for obesity, physical stress, cardiac arrest, physical injury, and pulmonary exertion. Providing physical fitness programs, clothing, and equipment by organizations has been increasing in the past few decades.

Narcotic — Any class of substances or drugs that numbs, paralyses, or causes loss of feeling in the body. Narcotics may be addictive, alter moods or behavior, and induce sleep.

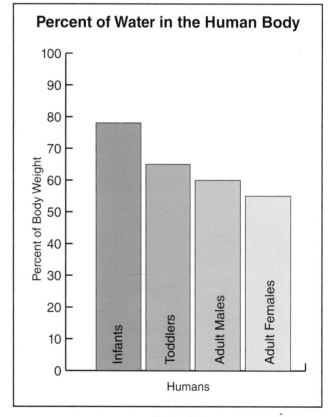

Table 4.1

Hiring Requirements and Procedures

Ensuring a healthy and physically fit fire and emergency services organization begins with hiring candidates that meet established medical and fitness criteria. The hiring criteria must be applied consistently to all applicants. Therefore, certified physical ability tests should be mandatory for all personnel hired by career or combination departments. Some jurisdictions require the use of a task-oriented, fitness selection test. Although in some states volunteer firefighters are considered employees and must meet these hiring criteria, not all states have this requirement. All volunteer organizations should seriously consider using the same criteria for entry-level personnel.

Any screening device employed during the hiring process by an agency must result from a valid hiring practice. That is, there must be a job analysis to determine the types of tasks that fire and emergency responders perform and the level of fitness and health that is required for those tasks. To ensure that the fitness requirements are valid, current employees should be able to perform the entry-level requirements. Based on that analysis, experts develop valid screening processes. Simply stated, you cannot require an applicant to demonstrate the physical ability to perform at a level that exceeds the requirements of the essential duties of the job. This requirement applies to volunteer as well as career departments. See **Appendix C** for additional information from the IAFF/IAFC on entry level testing.

Figure 4.3 The CPAT standardized physical ability test is based on job-related tasks. *Courtesy of Mesa (AZ) Fire Department.*

Hiring Requirements

NFPA® 1500 states that applicants shall receive a medical evaluation by the fire department physician and be certified as medically fit to perform the duties of a firefighter. Specific medical requirements are found in NFPA® 1582, *Standard on Comprehensive Occupational Medical Program for Fire Departments*. Departments that operate fixed or rotary wing aircraft must also require that pilots who work for the department maintain a commercial medical examination that conforms to Federal Aviation Agency (FAA) regulations for commercial pilots.

Applicants may be required to take third-party certified physical ability tests as part of the hiring procedures. These tests have been standardized and are based on work-related activities. Local agencies may develop their own tests and have them certified or acquire a standardized test such as the Candidate Physical Abilities Test (CPAT) or National Wildfire Coordinating Group (NWCG) Pack Test **(Figure 4.3)**. The key to these tests is how they conform to the types of activities normally performed by members of the department and the ability of current members of the department to successfully perform the tests.

Local political realities can influence hiring decisions. Pressure to hire an individual or give preference to a particular group can result in perceived bias or unfairness. To counteract the political pressures, relying on a set of job-related standards is essential. Those standards, which include medical, physical, mental, and emotional elements, must be strictly adhered to and applied evenly to all applicants. In the end, the organization will be viewed by the public and the politicians as a highly professional and ethical organization. At the same time, those standards will help ensure the safety and health of fire and emergency responders.

Hiring Procedures

Once the hiring criteria have been established, a consistent set of hiring procedures must be developed. Local policy will determine the order that the procedures are applied. The hiring procedures will in general, include:

- Announcing or advertising the vacancies to be filled
- Accepting applications during a specific time period
- Administering a written personality and/or mental aptitude (psychological) test
- Performing a medical evaluation
- Performing a background check
- Administering a physical ability test
- Selecting the applicants that are most qualified for the position

All of these steps are important to the process. However, only the physical ability test and the medical evaluation are pertinent to the primary theme of this manual. The following sections discuss two of the physical ability tests that are available and provide an overview of the medical evaluation.

Candidate Physical Abilities Test

Concern for the health and fitness of firefighters in order that they may adequately perform duties as assigned begins with minimum qualifications for entrance into the fire service. Most career fire departments have established some form of physical abilities testing as minimum qualifications for consideration for employment.

The International Association of Firefighters (IAFF) in cooperation with the International Association of Fire Chiefs (IAFC) developed a Fire Service Joint Labor-Management Wellness-Fitness Initiative in 1997. The approach was to develop a holistic and nonpunitive approach to both health and fitness. In the process, it was discovered that fire departments were hiring people that would not be physically capable of a successful career in the fire service. As a result, the task force developed and validated the labor management *CPAT*.

A fire and emergency services organization must have established minimum physical abilities standards for entry-level employees if there is any expectation that a health and fitness program is to be successful. The work required of a firefighter or emergency responder is demanding physically and mentally.

The *CPAT* consists of the following eight events performed while wearing a hardhat, gloves, and a 40 lb. vest to simulate the weight of an SCBA and protective equipment **(Figure 4.4)**:

Figure 4.4 Candidates wear a weight vest to simulate the weight of protective clothing and an SCBA during the test. *Courtesy of Mesa (AZ) Fire Department.*

- Stair Climb (climbing stairs while carrying an additional 25 lb. simulated hose pack)
- Ladder Raise and Extension (placing a ground ladder at the fire scene and extending the ladder to the roof or a window)
- Hose Drag (stretching uncharged hoselines, advancing lines)
- Equipment Carry (removing and carrying equipment from fire apparatus to fireground)
- Forcible Entry (penetrating a locked door, breaching a wall)
- Search (crawling through dark, unpredictable areas in search of victims)
- Rescue Drag (using a dummy to simulate removing a 175 lb. victim or partner from a fire building)
- Ceiling Pull (locating fire and checking for fire extension)

In accordance with recommended CPAT policy, provisions must be made to offer training and counseling to applicants wishing to take the test. The CPAT consists not only of physical test events but use of techniques and equipment unique to the fire service. To ensure fairness to all applicants, these pre test training and counseling programs should be made available.

Some agencies have found that possessing a CPAT certification is not a reliable indicator of an applicant's ability to perform all the functions required of a firefighter. It is not clear whether this is due to improper CPAT testing procedures, an elongated time span between passing a CPAT and undertaking agency testing, or some other unidentified factor(s).

NWCG Work Capacity Test

The *Work Capacity Test,* commonly referred as the Pack Test for Red Card Certification, is actually three tests designed for the work or task assigned to the employee. These tests were developed by the NWCG as minimum qualifications for Red Card Certification and for certification to complete other duties required in suppression of wildland fires or in support functions. Like the CPAT, the Work Capacity Test is designed to ensure that persons employed on the fireline as wildland firefighters are physically capable of performing the assigned tasks. The tests are also designed to ensure that even those employees assigned to typical office duties had attained a minimum fitness level.

The Work Capacity Test is designed for light, moderate, and arduous fitness requirements **(Figure 4.5)**. The tests are described as follows:

Fitness Requirement	Test	Description
Arduous	Pack Test	3-mile hike with 45-pound pack in 45 minutes
Moderate	Field Test	2-mile hike with 25-pound pack in 30 minutes
Light	Walk Test	1-mile hike in 16 minutes no pack

Arduous Work involves above-average endurance (aerobic fitness), lifting more than 50 pounds (muscular fitness), and occasional demands for extraordinary strenuous activities. All wildland firefighters perform arduous duty.

Moderate Work involves lifting 25 to 50 pounds, and the occasional demand for moderately strenuous activity. Safety officers and fire behavior officers perform moderate work.

Light Work involves mainly office-type work with occasional field activity.

Philosophically, the NWCG has established a model that demonstrates the intent to establish a correlation between a level of fitness and the work that employees perform. Fire service managers therefore have a model that may be reviewed in establishing a level of fitness for all employees irrespective of duties assigned.

Figure 4.5 Work Capacity Tests are designed to determine the ability of the candidate to perform wildland fire suppression or support activities.

Medical Evaluation/Examination

NFPA® 1582 provides an outline of medical evaluations that should be performed in examining the health and fitness of firefighter candidates. The recommended medical evaluations consist of a comprehensive examination of a number of functions and systems including:

- Head and neck
- Eyes and vision
- Ears and hearing
- Dental
- Nose, oropharynx, trachea, esophagus, and larynx
- Lungs and chest wall
- Heart and vascular system
- Abdominal organs and gastrointestinal system
- Reproductive system
- Urinary system
- Spine and axial skeleton
- Extremities
- Neurological disorders
- Skin
- Blood and blood-forming organs
- Systemic diseases and miscellaneous conditions
- Tumors and malignant diseases
- Psychiatric conditions
- Chemicals, drugs, and medications usage

NFPA® 1583 also includes a comprehensive list of 13 Essential Job Tasks and Descriptions. Prior to any medical evaluation, the physician performing the examination should have a thorough understanding of the nature of the work firefighters perform and of the demands that the tasks impose on the firefighter. The evaluation is intended to determine health and fitness as they relate to the demands of the work of a firefighter.

When the department has an assigned physician that conducts all medical evaluations, that physician should be experienced and familiar with NFPA® 1582 and 1583, as well as the demands of the essential tasks assigned to a firefighter. If the department does not have the resources to have an assigned physician, one that is selected to conduct the medical evaluations should not only be briefed as to the essential tasks, but should be provided an opportunity to observe and perhaps even participate in performing many of the tasks associated with the work that firefighters perform. The briefing should include but not be limited to documentation regarding the essential job and tasks and also the psychological as well as the physiological demands of the work.

The environmental conditions that a firefighter will be exposed to in performing essential tasks should also be well understood by the examining physician. Firefighters have a number of PPE requirements that must be considered in the medical evaluation by the physician. The full ensemble of boots, pants, coat, hood, and helmet as well as the requirement for SCBA in addition to tools carried and work performed place specific demands on the firefighter. Those firefighters assigned to enter and perform work in hazardous atmospheres may require full Level A suit protection, and the demands, especially from heat in those suits, must be considered in performing the medical evaluation. Finally, firefighters who will perform wildland fire suppression will face additional levels of physical stress. Physicians may not be fully aware of the extraordinary demands placed on the body with these layers of protective clothing's and safety devices. Firefighters are well aware of the heat buildup inside these PPE ensembles; the examining physician may not be that familiar with the amount of heat buildup and should be thoroughly briefed prior to performing any medical examinations.

Based on NFPA® 1582, any candidate that is determined to have any Category A preexisting condition shall not be certified for employment. Category B medical conditions must not prevent the candidate from fulfilling the duties of a firefighter.

The Category A conditions that may present a significant risk to the individual or other personnel include the following:

- Pulmonary hypertension
- Active tuberculosis (TB)
- Cerebral arteriosclerosis
- History of incapacitating hypoglycemia
- Angina pectoris
- Structural abnormality
- Limitations of motion in a joint

Applicants with Category B medical conditions that, based on the severity or degree of the condition, would prevent them from performing the duties of a firefighter in training or at an emergency incident shall not be certified as fit for duty. However, the AHJ can make reasonable accommodations to allow such an individual to perform other assigned functions. Some examples of Category B medical conditions include (but are not limited to) the following:

- Unequal hearing loss
- Allergic respiratory disorder
- Recurrent sinusitis
- Hernia
- Hypertension
- Diseases of the kidney, bladder, or prostate

As part of many hiring policies, the use of tobacco products of all types as well as addictive drugs may be cause for denial of employment. Applicants may be required to sign a statement that they do not currently use tobacco and will not use them throughout their employment. The justification for this prohibition on tobacco products is the established link between them and cancer. In 31 states, cancer is considered a job-related illness and is covered by worker's compensation.

Applicants may also be required to have certain vaccinations or immunizations prior to employment. Other vaccinations, such as Hepatitis A or B, may be provided to the new employee once hired **(Figure 4.6)**.

In addition to the medical evaluation, many fire departments require applicants to complete a psychological or mental aptitude test. This test consists of two parts: a written test and an oral interview. The written portion requires that the applicant answer many questions about himself or herself. Many of the questions repeat the same information in a variety of ways to determine

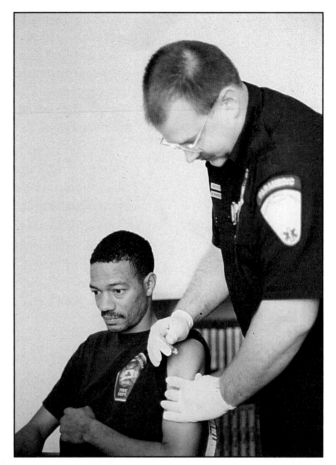

Figure 4.6 Entry-level personnel are given vaccinations to protect them from potentially infectious diseases.

consistency in the answers. The second portion of the test involves an oral interview with a trained psychologist. The purpose of the psychological test is to determine if the applicant can handle the emotional stress that fire and emergency responders are subjected to in the line of duty.

Health Program

The employee health program is intended to improve the health of the individual firefighter or emergency responder. Individual health improvement is accomplished through wellness education and physical fitness training. A holistic approach is proactive, attempting to address potentially hazardous situations before they occur. This section discusses the overall elements of the health program while the next presents the wellness and fitness components of the health program.

NFPA® standards require the establishment of committees and programs for managing programs designed to ensure the health of firefighters. Specifically, fire departments are required to have an established *risk management plan*. Elements of that plan guide the department in development of a comprehensive approach to management of risk.

An essential element of a risk management plan is the designation of a *safety and health committee.* It is recommended that the committee be inclusive of department management, individual members, and representatives of member (labor) organizations. Others can also be included, but the key partnership in such a committee is between the fire chief and the employees. Comprehensive management of a wellness program is dependent on the willing participation of employees; that participation is essential for the program to be successful. When administrative staff and members of the department work collaboratively to establish a workable and well-supported health and safety plan, success can be assured.

In order to lay the foundation for the development of a health program, a well-defined statement of health should be established. Managers of a departmental risk management program as well as participants must have clear communications and understanding of the mission of such programs.

Careful communication of the program defining health, providing clarification of the program mission as well as detailing process to ensure active participation and self-responsibility of department members are all elements of a comprehensive program **(Figure 4.7)**. After all, a health program must be accepted by every member of a department, and each member must willingly accept the notion of individual responsibility for one's own health in order for that program to be successful.

Figure 4.7 Health and wellness information can be distributed to members through notices posted in the workplace.

Selecting a health program model is the starting point for changing the organization's culture. Clearly, employees must have developed; an awareness of a problem before they will be willing to precede and be motivated to change their behavior. Ongoing open communications, discussion of problem issues, and agreement of a course of action opens the way to policy, procedure, equipment, and apparatus changes.

Essential to proceeding with the final development of a department's health program is a careful review by the safety and health committee of data relative to health and fitness in general. Such research should be completed with recommendations and included in a final report to the fire chief and the department.

A history of department members' injuries, illnesses, and medical issues, as well as job-related fatalities, while protecting the privacy of the members can be researched as an indicator of problem areas. Such research should be expanded, particularly in small or newer agencies as those agencies may not have sufficient experience to be meaningful. The NFPA® creates annual reports and multiyear summaries of firefighter injury and fatality data and analysis, which should prove useful to any agency initiating a health and safety program. As indicated earlier, line-of-duty death by cardiac problems is the number one cause of death. Obviously, a health program must include measures to screen for cardiac problems, improve members' nutrition, and fitness programs that support cardiovascular wellness.

Committee review should also include a best practices review and recommendations regarding medical requirements for a firefighter. NFPA® 1582 provides a basis against which the department's current practice can be compared. Any change and recommendations for improvement should be carefully documented and reported to the fire chief.

Any review of physical or medical requirements related to firefighter duties should also be well researched with recommendations made to the fire chief. It is simply not acceptable to assume that modern fire departments can require specific physical or medical requirements without demonstrating the relationship of those requirements to essential firefighter duties. Those requirements, if not proven to be job related, can and very likely will be challenged in court. Wearing the required PPE and carrying that the tools a firefighter must take to the incident scene impose significant physical demands on firefighters. That physical demand must be recognized in the design of any comprehensive wellness and fitness program. Generally, an evaluation by the department's physician is requested based on the consensus standard. The doctor's recommendation is given to the fire chief who makes the final decision on hiring the candidate.

Based on the work of the safety and health committee, the department should establish departmental policy regarding:

1. Medical requirements

2. Physical requirements

3. Infection control

4. Medical evaluation (annual physicals)

5. Fitness for duty

Of considerable concern to firefighters as employees is the confidentiality of their personal medical evaluation. While a committee may develop the evaluation criteria, they do not monitor each individual's performance. The department's administration must also be concerned about legal violations of the employee's privacy under the *Health Insurance Portability and Accountability Act (HIPAA)*. HIPAA incorporates a privacy rule that protects all individually identifiable health information. Of equal concern is the ability to maintain the health and wellness of each individual firefighter to ensure that an individual's medical condition does not expose that firefighter to additional risks or expose other firefighters to that risk. This inherent conflict of interest, it would seem, can only be surmounted by the employee's willingness to share that medical information, which may limit, or should limit, his/her ability to carry out the duties of his/her position.

Properly executed, a health program will allow employees to openly discuss with a physician their health and develop a treatment or wellness program to improve their physical fitness for the job. Benefits will be derived in reduced health care costs, improved employee longevity, reduced turnover, and reduced absenteeism.

Medical Requirements

Policies must be in place that not only define the medical criteria for being hired but also for continued employment and termination. These policies, usually associated with the labor/management agreement, protect the employee and establish the rights of management. Medical requirements should be the same for volunteer as well as career and combination departments. The medical requirements should include:

- Annual medical evaluations
- Return-to-duty criteria
- Medical-leave criteria
- Disability separation or termination criteria
- Limited-duty guidelines

Physical Requirements

Physical requirements are based on the task analysis used for hiring personnel. Incumbent members of the organization must be able to pass the same physical test that candidates must pass. Annual fitness tests, sometimes adjusted for age or medical conditions, are generally included in labor/management agreements. The test must be based on job-related tasks and mirror the test used for candidate selection. Physical fitness requirements should be the same for career, combination, and volunteer departments.

Infection Control

Infection control begins with a written plan that clearly explains its intent, benefits, and purposes. The plan must cover the standards of exposure control such as the following:

- Education and training requirements
- Vaccination requirements for potential threats such as anthrax or hepatitis B virus
- Documentation and record-keeping requirements
- Cleaning, decontamination, and disinfection of personnel and equipment **(Figure 4.8)**
- Exposure control and reporting protocols

The organization should establish the position of infection (exposure) control officer who ensures that an adequate infection control plan is developed and that all members are trained and supervised in the plan. The Health and Safety Officer (HSO) or other officer who has the knowledge, training, and skills to fulfill the required duties may fill this position.

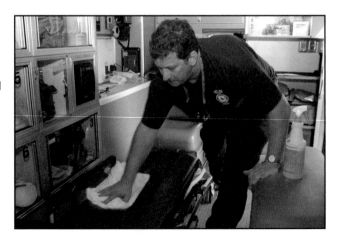

Figure 4.8 Infection control includes the cleaning and decontaminating of ambulance interiors.

Exposure reports must be maintained in a record-keeping system. Because some symptoms may not become apparent for many years following exposure, these records must be maintained for 30 years following termination of an employee. These records are confidential and may only be released to the member or a designated representative. Records are also maintained for training that involves the proper use of PPE, exposure protection, postexposure protocols, and disease modes of transmission as they relate to infectious diseases. The HSO and safety and health committee members are required to annually review the infectious disease plan, updates, protocols, and equipment. Policies should also be developed that specify the types of protection that must be used to prevent an exposure such as that in the following example:

> *Members will wear a particulate respirator when entering areas occupied by individuals with suspected or confirmed tuberculosis (TB), performing high-risk procedures on such individuals, or transporting individuals with suspected or confirmed TB in a closed vehicle. A NIOSH-approved 95-percent efficient particulate air respirator is the minimum acceptable level of respiratory protection.*

Annual Medical Evaluation

Any established physical evaluation of employees and the corresponding medical counsel to the individual by a provider must be confidential. The physical exam, the results of that exam, and the recommended follow-up are between the firefighter and the physician. The fire chief need only be concerned with the medical opinion that either the firefighter is fit for duty or not; any other information is privileged and confidential.

Fitness for Duty

A firefighter's performance should be constantly under review by immediate and secondary supervisory officers with the expectation that the performance will meet or exceed standards. Any observed failure to perform can then be evaluated to determine the cause. Should the initial evaluation reveal that there is a possible physical, emotional, or psychological problem relative to performance, a *fitness-for-duty evaluation* can be required. The procedures for requiring a *fitness-for-duty evaluation* must clearly indicate that the decision is based on observed and documented observation of the firefighter's inability to perform the assigned duties. Drug and alcohol screenings also fit into the category of fitness for duty.

The medical authority trusted with the evaluation should be thoroughly familiar with the essential duties of a firefighter and the associated physical and physiological demands of the work. The referral for the evaluation must be handled with the utmost confidentiality. In addition, should there be a finding of injury or illness, a determination must be made as to whether the issue is job related or not.

Injury or illness that is not job related must be managed by the employee and their chosen physician. The organization's primary concern is in the fitness for duty of the firefighter when they are medically released. Policies and procedure relevant to managing the return to full duty should be in place. A significant element to that policy is to ensure that the attending physician fully understands the demands of the work.

Figure 4.9 Limited or light duty assignments can be used for personnel on injury leave who can still perform some duties for the department.

A return to duty policy could also incorporate some form of a restricted or *limited (light) duty* assignment that will return the firefighter to some capacity of work while they continue to rehabilitate from an injury or illness. That policy must be well-defined, coordinated and equitably applied with relevant sick leave, injury leave, and family leave policies. Work assignments need to be carefully evaluated and compared to the restrictions established by the fire department physician to ensure that the activities will not aggravate existing injury or illness or contribute to new injuries or illnesses **(Figure 4.9)**. Fire departments have a substantial investment in the training and experience of firefighters, developing a light duty policy allows for the transition of that valuable employee back into the work force while at the same time the organization can benefit from the work they complete **(Figure 4.10)**.

If a firefighter is unable to perform the essential duties of their position due to a line-of-duty injury or illness, the department should also have a well-defined policy to manage worker's compensation issues. In these cases the medical treatment and rehabilitation may be the same; however, the department is, in these cases, obligated to support the treatment and rehabilitation of that firefighter. Return to duty and light duty may be treated in much the same manner for either on or off duty illness or injury; however, the use of sick/injury leave may not be required of the employee for on duty illness or injury. State laws differ regarding worker's compensation issues. The organization must seek consultation from an experienced risk manager or legal advisor to coordinate the agency's policy with relevant state law.

Fire Department Physician

In order to develop and implement an effective health and fitness program, proper management and direction must be used. The selection and appointment of a fire department physician is a crucial element. The requirements for the fire department physician are found in 29 CFR 1910.134, as well as NFPA® 1500 and NFPA® 1582. Appendix B of NFPA® 1582 contains a comprehensive list of these requirements. Briefly, these requirements include the following:

- The fire department shall officially designate a physician who is responsible for advising, steering, and counseling the members regarding their health, fitness, and suitability for various duties.

- The fire department physician shall provide medical guidance regarding the management of the occupational safety and health program.

- The fire department physician shall be a licensed medical doctor or osteopathic physician who is qualified to provide professional expertise in occupational safety and health relating to emergency services.

- The fire department physician shall be available for consultation and for providing expert services on an emergency basis.

In order to provide an NFPA®-compliant medical certification program, the fire department physician must have a clear understanding of the services that the department provides. Fire department work frequency combined with task

MESA FIRE DEPARTMENT
Return to Duty Progress Report – Sworn

Revised 06/14/07

Physician Name:_____ Physician Phone Number: _____

Employee Name: _____

Contact: Wellness Captain (480) 644-3072 Fire Personnel & Wellness (480) 644-4694

Instructions: Please complete the section that applies. Fax (480) 644-5160

❑ MODIFIED WORK

The City of Mesa has a Transitional Work Program. This program allows an employee to return to work in a modified duty capacity when he/she cannot perform normal job junctions. If this employee can perform modified work, please specify the physical limitation.

Maximum number of hours released to work: Per Day _____ Per Week _____

Physical Tolerance:

❑ Sit – (Hours/Minutes) _____

❑ Stand – (Hours/Minutes) _____

❑ Walk – (Hours/Minutes) _____

❑ Lift – (Maximum # of Pounds) _____

❑ Climb Stairs _____

❑ Driving _____

	Yes	No
Bend		
Crouch		
Squat		
Kneel		

Physician Comments:_____

_____

_____

_____

May Begin Modified Duty Date : _____

Expected Date to Full Duty : _____

❑ FULL DUTY

I have read and reviewed the attached list of essential firefighter duties. I have assessed the firefighter listed above and believe he/she can perform these essential duties safely and without compromise.

Physician Signature: _____

May Return to Full Duty on (date):_____

❑ NO WORK STATUS

Employee cannot perform or function in any type of work activity.

Expected date employee can return to modified work: _____

Next Appointment: _____

Physician Signature: _____ Date: _____

Figure 4.10 Sample Return to Duty Form. *Courtesy of Mesa (AZ) Fire Department.*

analysis helps in providing this understanding. It may be necessary to provide training for the fire department physician in the tasks, operating procedures, and evolutions inherent in fire fighting. Participation on the health and safety committee also helps in his assimilation into the fire service. In the case of smaller departments, joining together to hire a physician to provide services for all of them can be a cost-saving approach.

Health and Safety Officer as Liaison

While the fire department physician is responsible for the medical aspects of the program, the HSO provides the liaison between the administration and the fire department physician. The HSO is also responsible for ensuring that exposure report information is collected by department officers or members and that the appropriate forms are completed.

Incumbent Physical Evaluation

Following employment with the fire and emergency services organization, a periodic medical evaluation shall be performed on each employee involved in emergency response operations. This evaluation can be in conjunction with the annual physical ability test or be separate. A medical evaluation is required following a lost-time injury or illness and the associated rehabilitation program. The employee must also receive a medical evaluation before returning to duty. Under the mandates of worker's compensation, strict procedures must be followed in the event of a member's occupational injury or illness and return to duty. An organization should also have procedures for reporting nonoccupational injuries and illnesses and return to work from those incidents.

The periodic medical evaluation shall consist of the following items:

- Interval medical history
- Interval occupational history, including significant exposures
- Height and weight
- Blood pressure
- Heart rate and rhythm

A medical examination shall be required periodically depending on the age of the employee. The schedule required by NFPA® 1582 is as follows:

- Age 29 and under, every 3 years
- Age 30 to 39, every 2 years
- Age 40 and over, every year

Additional medical examinations are required for those personnel assigned to hazardous materials teams including blood tests to establish a baseline. These examinations are annual in many departments.

In addition to complete medical physicals, firefighters may be required to participate in annual hearing and pulmonary function testing. Audiometric tests or hearing tests are usually required during the hiring or admittance phase to determine a baseline and annually if the firefighter is repeatedly exposed to a given level of sound to determine occupational hearing loss. The same is required for spirometry or pulmonary function testing for all firefighters who are required to wear a respirator including an SCBA.

Wellness and Fitness Components

Two of the essential components of a health program are wellness and fitness. Taken together they provide both the behavioral and physical aspects needed for good health. Wellness provides the knowledge required to recognize the need for change in a person's behavior such as improved dietary habits. The fitness component provides the required level of structured physical activity that will help maintain strength, stamina, and flexibility needed to safely perform assigned tasks. It will also help to strengthen heart muscles, increase stamina and lung capacity, and reduce weight.

Wellness Component

A good wellness plan is a necessary component in the holistic approach to health and safety. Wellness involves educating the membership in the cause and effect of lifestyle choices on the human body. Employees have the opportunity to see the benefits of making the changes in their lifestyles through:

- thorough nutrition
- proper hydration
- back care
- heart and lung diseases training sessions
- psychological stress recognition and control sessions

Training helps to prevent the possibility of psychological stress; however, should stress occur, the HSO must be able to recognize the symptoms and provide professional care through critical incident stress management. The HSO can also assist members in efforts to stop using tobacco in any form by referring them to a member assistance program for tobacco-cessation plans.

Nutrition

The importance of a good, balanced diet cannot be emphasized enough. Recent studies have linked poor diet to heart disease, cancer, diabetes, high blood pressure, high cholesterol, and other chronic diseases. It is estimated that $^2/_3$ of the U.S. population is overweight and 1/3 of those are considered obese **(Figure 4.11, p. 140)**. As part of the employee wellness and fitness program, the HSO can provide members of the organization with training regarding the importance of good nutrition. This training can include learning the negative effect of certain foods, guidelines for a balanced diet, results of good nutrition, and safe and effective weight-control diets. Some progressive organizations provide a recommended daily meal plan for each station via e-mail.

An effective educational plan on nutrition can have the added bonus of altering the off-duty lifestyles of employees and their families. Information on nutritional issues can be obtained from the U.S. Center for Disease Control and Prevention (CDC), state/territorial/provincial or local health department, or the American Heart Association (AHA).

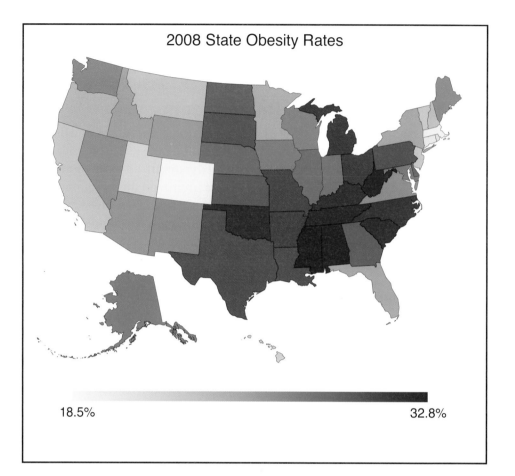

Figure 4.11 2008 State Obesity Rates

Health Effects of Overweight and Obesity

Overweight and obesity are known risk factors for:

- diabetes
- coronary heart disease
- high blood cholesterol
- stroke
- hypertension
- gallbladder disease
- osteoarthritis (degeneration of cartilage and bone of joints)
- sleep apnea and other breathing problems
- some forms of cancer (breast, colorectal, endometrial, and kidney)

Obesity is also associated with:

- complications of pregnancy
- menstrual irregularities
- hirsutism (presence of excess body and facial hair)
- stress incontinence (urine leakage caused by weak pelvic floor muscles)
- psychological disorders such as depression
- increased surgical risk
- increased mortality

Hydration

Frequently overlooked, hydration is important before, during, and following strenuous activities both on and off duty. The body has the ability to lose in excess of 2 liters per hour of fluid through sweat, urine, stool and exhaled air than what it can absorb though ingestion, approximately 1 to 1.5 liters/hour, which may result in serious injury or death **(Figure 4.12)**. To ensure proper bodily functions, firefighters must continuously monitor and maintain their levels of hydration both on and off duty.

Firefighters should monitor their intakes of both food and water as well as the frequency, color, and smell of their urine. Infrequent urination, dark colored urine, or urine that has a strong odor may suggest an inadequate level of hydration **(Figure 4.13, p. 142)**.

The process of hydration should begin no less than 2 hours prior to scheduled events where the firefighter will knowingly participate in activities requiring a large amount of heavy exertion. Fluids should continue to be taken in small sips throughout the various activities with larger quantities taken during scheduled times of rehabilitation. Finally, hydration should continue to be consciously monitored for a minimum of 2 hours following the completion of the event.

Water is generally sufficient for the purpose of maintaining hydration prior to and during events that last for less than two hours. However, research suggests that a commercially available sports drink containing both electrolytes and carbohydrates, diluted to half strength, should be provided for events

Effects of Dehydration on the Human Body

Water Loss (Percent of body weight)	Effect
0.5 to 2.0%	Increased body core temperature Increased strain on heart muscle Reduced aerobic endurance
2.0 to 3.0%	Reduced muscular endurance Significant increase in body core temperature
3.0 to 4.0%	Reduced muscle strength Reduced motor skills Heat cramps
4.0 to 5.0%	Heat exhaustion Fatigue Increased cramping Reduced mental capacity Decrease in strength by 20 to 30% Continued increase in body core temperature
5.0 to 6.0%	Physical exhaustion Heatstroke Coma Muscle spasms
6.0 to 10% or greater	Excessively high body core temperature Heat injury and circulatory collapse

Figure 4.12 Effects of Dehydration on the Human Body.

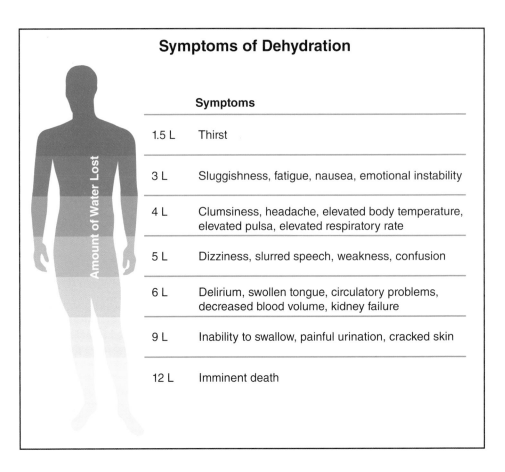

Figure 4.13 Symptoms of Dehydration

Symptoms of Dehydration

Amount of Water Lost	Symptoms
1.5 L	Thirst
3 L	Sluggishness, fatigue, nausea, emotional instability
4 L	Clumsiness, headache, elevated body temperature, elevated pulsa, elevated respiratory rate
5 L	Dizziness, slurred speech, weakness, confusion
6 L	Delirium, swollen tongue, circulatory problems, decreased blood volume, kidney failure
9 L	Inability to swallow, painful urination, cracked skin
12 L	Imminent death

extending beyond two hours. It is also recommended that cold drinks should be served cool and hot drinks be served warm to allow the fluids to facilitate absorption into the body after being ingested. Alcoholic drinks or those containing caffeine such as coffee, tea, or soda pop contribute to dehydration and should be avoided.

Back Care

Back injuries are common for both Americans and Canadians between the ages of 25 and 50. These injuries are most common in people who do not adhere to a regular physical fitness plan. Due to the nature of fire-fighting and emergency services operations, back injuries rank just below heart disease as a cause of firefighter and emergency responder injury. Therefore, the organization should provide back-care training sessions for all employees. This training would provide the following information:

- Techniques for proper lifting and carrying both equipment and victims **(Figure 4.14)**
- Use of tools and equipment that assist in lifting or carrying of equipment or victims
- Proper methods for dragging hoselines
- Exercises for strengthening the back muscles

Heart and Lung Diseases

As part of the overall cardiovascular system, the heart and lungs are susceptible to diseases that can be addressed jointly through education. The organization must implement a policy for the use of SCBA or a supplied-air respirator

(SAR) in all toxic atmospheres and in suspected or unknown atmospheres. All facilities must provide a smoke-free environment or at least provide segregated designated smoking areas that have proper ventilation. Training includes providing the following information:

- Importance of unimpaired lung capacity and unobstructed blood circulation
- Causes of heart and lung diseases
- Methods for reducing the potential for these diseases

Stress Recognition and Control

In the past half century, stress has become a common part of the lives of most people. Surveys indicate that one in five people experience stress daily. In the fire and emergency services, stress is a continuing fact of life from the moment the alarm sounds until they return to quarters **(Figure 4.15)**. For that reason, a stress management plan is an essential part of the wellness portion of the employee physical fitness and wellness program. Stress in the fire and emergency services can take two forms: physiological stress (work environment experiences) and psychological stress (total life experiences).

Physiological stress. Physiological stress is caused by the emergency responder's work environment and can be created by the following situations:

- Sound of the alert tone in quarters
- Shift work involving irregular schedules
- Lack of sleep or interrupted sleep cycles
- Loud noises (sirens, air horns, station radios, explosions, etc.)
- Station overcrowding
- Hazardous or toxic environments
- Exposure to extremes of heat and cold
- Long hours of physical exertion

Figure 4.14 The two-person carry is an effective way to move some victims.

Figure 4.15 Emergency operations generate physical and psychological stress on firefighters.

- Carrying heavy PPE
- Fatigue

Environmental stress can be reduced by policies, education, and prevention. Policies should be established that control the length of time a fire or emergency responder works during an incident and how often rehabilitation must occur. Personnel must be educated to recognize their own physical limitations while using respiratory protection equipment and not exceed them.

At emergency incident scenes or training evolutions, the incident safety officer (ISO) or HSO on site notifies the incident commander when members or crews have reached their limits of physical effectiveness. In many cases, rest time is used as the only measurement as to when a person is able to return to work. Personnel are considered fit for duty when their body core temperatures and vital signs have returned to acceptable levels. The rehab unit provides the following services:

- Emergency medical technicians (EMTs)/paramedics make an assessment of a person's physical condition and general appearance.

- EMTs make an assessment of a person's physical limitations by establishing a baseline and continued monitoring of vital signs.

- Mental status and level of orientation are established. Personnel who may be confused, disorientated, or agitated without cause may be symptomatic of a serious life-threatening condition and should immediately receive advanced medical care.

- A medical rehab unit is established. Transfer those who are injured, severely dehydrated, severely exhausted, or those who show other significant chief complaints where they may receive additional medical monitoring, definitive care, and/or transportation to a receiving hospital.

- Vital-sign assessment data is provided to the incident safety officer (ISO).

- Services such as nutritious food, replacement fluids, and a rest area allow personnel to return to acceptable physiological levels.

- When core temperatures are elevated, provide methods of active cooling such as placement of hands and arms into five-gallon buckets of tap water.

Fire or emergency units/companies are not returned to service until the ISO releases them as fit for duty. In some cases, it may be necessary for the emergency unit/company to return to quarters for a change of uniform and dry protective clothing before returning to service. Firefighters have received steam burns and been scalded while wearing damp or wet clothing under their protective clothing after working a succession of incidents.

Psychological stress. Psychological stress is more individual in nature and is created not only by the work environment but also by the total life experience of the individual. A short list of psychological stressors that emergency responders experience in their daily lives and take to work with them are as follows:

- Personality conflicts with coworkers
- Perceived lack of respect from managers, administration, or the public
- Boredom

- Lack of job satisfaction
- Concerns over promotion, layoff, or retirement
- Feelings of inadequacy
- Fear of failure
- Personal injury to self or coworkers
- Death of coworker or family member
- Personal family problems

As part of the holistic employee physical fitness and wellness program, the HSO includes activities that help members deal with stress. The National Fire Academy (NFA) has developed a stress management model program that can assist in the training of members. It defines stress, outlines the symptoms, lists the physical and psychological reactions, and provides activities for stress management. Four basic strategies for managing stress are as follows:

- Minimize the stress-causing factors by avoiding, eliminating, or reducing them.
- Change how the individual perceives or views problems.
- Learn methods for relaxing both the mind and body.
- Build the body's physical resistance to stress.

The HSO (in conjunction with the training division) develops training sessions that assist in meeting these four strategies. Recommended topics for training include the following:

- Time management
- Role management
- Conflict resolution
- Assertiveness training
- Relaxation methods
- Diet and nutrition guidelines

Critical Incident Stress Management

Stress has always been a part of the emergency responder's life due to the high level of uncertainty, limited control over the work environment, and the psychological effect of repeated emergency calls. Add major events that exceed the normal level of stress and the ability of the body to cope, and critical stress develops. For this reason, a critical incident stress management plan must be part of the organization's employee physical fitness and wellness program **(Figure 4.16)**. This plan is intended to manage the stress experienced by fire and emergency service personnel after incidents involving the following situations:

- Mass casualties
- Fatalities involving children
- Serious injuries or fatalities involving members of the organization

Figure 4.16 Department members who are trained in critical stress debriefing work with other members who have been exposed to situations that can create high levels of stress.

- Suicides

- Serious injuries or fatalities involving close friends, relatives, or colleagues

- Violence directed toward firefighters or other emergency responders

- Deaths of civilians as a result of emergency operations

- Excessive media attention generated after incidents

Signs and symptoms. When the body undergoes normal levels of stress, it responds with increases in heart rate and blood pressure, oxygen consumption, muscle tension and strength, and dilation of the pupils among others. Excessive stress, however, results in further emotional and cognitive responses. Symptoms associated with excessive stress are as follows:

- Difficulty concentrating or staying focused

- Temporary loss of short-term memory

- Obsessive thoughts

- Loss of mental flexibility

- Tendency to withdraw or become isolated

- Invulnerable feelings

- Fantasy or wishful thinking experiences

- Autopilot mind focus

- Abuse of alcohol and drugs

In order to address these symptoms, a critical incident stress debriefing team is established and trained. The team consists of emergency services personnel, health-care professionals, and clergy who have specialized training in dealing with critical incident stress. This team provides the necessary support for those members experiencing excessive stress. It also provides training/support for emergency personnel about the causes and results of critical incident stress, stress reduction methods, and sources of professional assistance. At major incidents, team members provide guidance to the incident commander and the command staff as well as support for individuals at the scene.

Following a prolonged incident, team members help in the transition back to normal working conditions through demobilization services. Debriefing the participants begins in this stage and continues into the defusing stage where team members discuss with the participants their reactions and feelings. The final phase, if necessary, is the formal debriefing meeting that occurs from 1 to 3 days following the event. Follow-up services, consisting of additional debriefing meetings, individual sessions, and referrals, can occur over a period of several months following the incident. Throughout the process, team members maintain strict confidentiality.

Training. Although participation in the critical incident stress plan is not mandatory, pre- and postincident training can be used to prepare personnel for the possibility of a critical stress incident. The first step in reducing the effects of critical incident stress is to manage it by training personnel about the psychological hazards of the job. This training should include information about critical incident stress, including its causes, symptoms, and some effective coping techniques. This type of training is different from normal fire-fighting training and is not exclusive to entry-level firefighters or emergency responders. All fire and emergency services personnel, including chief officers, should be included in critical incident stress management training. It is strongly advisable to have personnel who have experienced critical incident stress as members of the training team. This makes their personal experience and observations available and creates credibility.

Member Assistance Program

The levels of stress normally encountered by fire and emergency personnel can result in dependence on tobacco products, abuse of alcohol or drugs, domestic violence, excessive gambling, and financial difficulties. In recognition of this fact, NFPA® 1500 mandates the establishment of a member assistance program within the fire and emergency services organization. A member assistance program is essential in the holistic approach to personnel health and wellness. Because it is impossible for the organization to provide the professional counseling services to meet these needs and ensure complete confidentiality, a member assistance program often includes outside contract counselors to provide the services. The HSO is responsible for prevention, education, and referral of employees to program counselors. All officers must also be aware of the symptoms of alcoholism, drug abuse, and other types of abuse in order to provide direction and care as soon as possible.

The effectiveness of the member assistance programs cannot be denied. Studies indicate that as many as 80 percent of those individuals who receive counseling return to full productive status within the workforce. Member assistance programs can also reduce the cost of prolonged medical care and lost-time benefits. In addition to fire and emergency service members, NFPA® 1500 provides for assistance to immediate family members through the program. Finally, the use of member assistance programs can result in improved employee morale.

Counseling. The member assistance program outlined in the organization's policy should have the ability to provide a wide range of counseling services for both the employee and the family. Besides smoking-cessation and substance-abuse assistance, domestic violence, child abuse, and family and financial counseling services should be available. Success of the program depends on the quality of the services offered, support of the administration, involvement of the members, and confidentiality of all services.

Smoking or tobacco-use cessation. Of all the addictions of the American public, smoking and the use of tobacco products have received the greatest attention since the 1970s. It is estimated that over a quarter of the population members over 25 years of age were smoking in 1990. This figure was a decrease from 33.5 percent in 1979. Smoking results in the loss of an estimated 400,000 lives in the U.S. every year **(Figure 4.17, p. 148)**. These deaths result from lung cancer, cardiovascular disease, and chronic obstructive pulmonary disease. According to the American Lung Association (ALA), smoking costs the nation approximately $65 billion per year in health care and lost productivity. Surveys indicate that Canadians have experienced the same rates of disease due to smoking. Add the results of using smokeless tobacco products (mouth cancer, gum infections, tooth decay, and lowered sense of taste and smell), and the problem takes on greater magnitude.

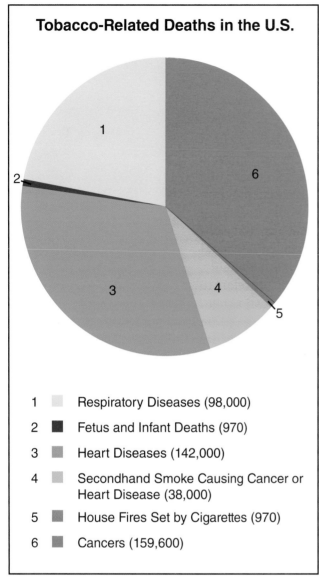

Tobacco-Related Deaths in the U.S.

1　　Respiratory Diseases (98,000)

2 ■　Fetus and Infant Deaths (970)

3 ■　Heart Diseases (142,000)

4 ■　Secondhand Smoke Causing Cancer or
　　　Heart Disease (38,000)

5 ■　House Fires Set by Cigarettes (970)

6 ■　Cancers (159,600)

Figure 4.17 Tobacco-Related Deaths in the U.S.

Based on the national mortality rate due to smoking, the fire and emergency services lose approximately 1,800 members per year. Because firefighters are exposed to the unburned products of combustion resulting in chronic lung disease and loss of lung function, they are already at risk. Smoking, according to the ALA, increases these hazards. Studies indicate that firefighters who smoke have a higher level of risk from heart and lung diseases than do firefighters who do not smoke. Smoking also results in lowered lung capacity and shortness of breath. These results can impair the stamina of firefighters during physically strenuous emergency operations such as wildland or high-rise incidents.

In order to create a healthy and smoke-free work environment, the fire and emergency service organization can approach the smoking problem from two directions. First, mandate that candidates or applicants be nonsmokers, which can be justified through existing documentation on the effects of smoking. This documentation includes the cost of long-term health care for lung and esophageal cancer and emphysema. New members can be required to be smoke free throughout their employment. Many organizations have implemented this employment criterion. Second, the organization can provide smoking-cessation training for current members. These programs can be part of the training cycle for all members or specific cessation classes for individuals or groups. The basis is an education program that points out the hazards and costs of smoking, the reasons for quitting, and the methods available for quitting. Information on smoking-cessation training is available from the CDC, American Cancer Society, ALA, state/territorial/provincial health departments, and tobacco industry.

Substance abuse. Smoking can have a negative effect on the health of the individual fire and emergency responder; drug and alcohol abuse has an even greater widespread effect. Because drugs and alcohol impair judgment and slow reaction times, their effects are not only on the individual but also on those who work and live with the individual as well as the public. The USFA estimates that as many as 10 percent of the 1.1 million firefighters in the U.S. may be abusing drugs. The IAFF estimates that 75 percent of the total firefighter population in the U.S. has used either drugs or alcohol. Similar figures are available for the rest of the emergency response services.

To offset the potential danger of this type of abuse, the fire and emergency services organization must establish a written policy as part of the member assistance program in the employee physical fitness and wellness program. The HSO's duties in this area involve developing and presenting education sessions on the negative effects of alcohol and drug abuse, assisting the administration in developing a drug and alcohol policy, directing members to the member assistance program when necessary, and ensuring a liaison between the member assistance program and the administration. The policy must include control mechanisms that provide rehabilitation procedures for employees as well as reporting and privacy procedures. Some organizations have established periodic random drug testing for all employees. The medical department of the jurisdiction usually performs this testing by contracted health-care facilities. It is essential that the results of this testing (like the medical records of the members) are confidential. A process should be established within the program whereby members who refuse to participate or who become repeat offenders following rehabilitation may have their employment terminated.

Sleep Deprivation

In 2007, the IAFC completed a study on sleep deprivation and its affect on firefighter safety and efficiency. The study was prompted by research into the effects of sleep deprivation in other professions, primarily transportation and health care. The primary result of the study was that additional data must be collected on how sleep deprivation affects fire and emergency responder performance and health.

The need for additional data fell into five categories:

1. identifying the affect of fire fighting and EMS duties on responder families

2. gathering and analyzing data on motor vehicle accidents that occur while commuting to and from work

3. determining the benefits of strategic napping during work shifts

4. determining the prevalence of sleep disorders among emergency responders

5. determining the relationship between cardiovascular risks and work hours

Some of the information contained in the report provides an indication of the need to control and eliminate sleep deprivation among emergency responders. That information includes:

- The average adult requires between 6 and 10 hours of sleep each day.

- More than one-third of Americans get less than the required daily average.

- One-half of all Americans report being significantly fatigued at least one day a week.

- There is a link between sleep loss and weight gain based on medical studies.

- Work shifts lasting more than 10 to 18 hours have been clearly linked to time-dependent errors in tasks requiring vigilance and focused alertness.

- Chronic sleep loss results in decreased ability to think clearly.

Because the results of the study are new and additional data is needed, approaches to correcting sleep deprivation in the fire and emergency services is inconsistent. Some career organizations are reducing the work hours to shorter shifts while others are increasing the length of the shifts to 48 hours with 96 hours in between shifts. The latter approach is intended to reduce the amount of time employees spend commuting. Still other organizations are sanctioning naps during the work shift to increase productivity and reduce fatigue.

Physical Fitness Component

According to the CDC, the overall results of a good physical fitness plan include the following:

- Reduces the risk of dying prematurely from all causes
- Reduces the risk of dying prematurely from heart disease
- Reduces the risk of developing adult-onset diabetes
- Reduces the risk of developing high blood pressure
- Reduces high blood pressure in people who already have it
- Reduces the risk of developing colon cancer
- Reduces depression and anxiety
- Controls weight
- Builds healthy bones, muscles, and joints
- Promotes psychological well-being
- Improves sleep habits

In order to develop a comprehensive, holistic physical fitness plan, the HSO must prepare an analysis of the tasks performed by the organization's members. This analysis is used as the basis for developing the physical fitness plan and for establishing hiring criteria for new personnel.

The physical fitness component of the employee wellness and fitness program must address flexibility, cardiovascular fitness, muscular fitness, and body composition. The HSO lists specific exercises using appropriate equipment to improve an individual's deficiencies in each of these areas. Criteria necessary for determining the effectiveness of the plan must be established. Effectiveness is determined through periodic monitoring of the physical fitness component and the organization's illness and injury reports collected before and after implementation of the component.

Task Analysis

Preparing the task analysis and developing the employee physical fitness component are the responsibilities of the HSO, the occupational safety and health committee, or a physical fitness subcommittee. Members of the subcommittee would include the HSO, members of the organization's administration, representatives of the member organization, emergency response personnel, the organization's physician, and a qualified/certified exercise physiologist. The exercise physiologist provides the professional knowledge necessary to analyze the tasks in terms of physical exertion and can recommend the appropriate test criteria.

If the subcommittee is given the development responsibility, the first step is for committee members to perform the task analysis and develop a list of the basic services that the organization provides. These services may include (but are not limited to) structural fire-fighting operations, wildland fire-fighting operations, emergency medical services, light and heavy rescue operations, hazardous materials responses, training functions, and building inspections and surveys. Next, committee members determine the types of tools and equipment that each service activity requires. The tools and equipment (such as hoselines and nozzles, pike poles, ground ladders, and axes) indicate the types of physical tasks that each activity requires. The tasks must be performed while wearing the appropriate PPE and respiratory protection. These physical tasks may include the following:

- Lifting an inert weight such as an unconscious victim
- Pulling a hoseline, both charged and uncharged (empty)
- Operating a hoseline and nozzle
- Climbing a ladder
- Climbing a flight of stairs in full protective clothing with SCBA
- Entering and operating in a confined space with hand tools
- Pulling a ceiling with a pike pole **(Figure 4.18)**
- Using an axe to cut through a substantial timber

Once the list of tasks is complete, it can be categorized into general groups of similar activities such as lifting, pulling, climbing, etc. A suitable physical fitness plan can be devised to meet the needs of each task. The physical fitness plan is not only used to improve the physical abilities and stamina of the current members but also provide preemployment testing criteria for candidates/applicants for the organization.

NFPA® 1500 requires the fire and emergency services organization to establish and provide a physical fitness plan that meets the requirements of NFPA® 1583. This plan is designed to enable members to develop and maintain an appropriate level of fitness to safely perform their assigned functions. The maintenance of fitness levels shall be based on fitness standards determined by the organization's physician. These levels must reflect the member's assigned functions and activities and the severity of occupational injuries and illnesses associated with these activities. Results are compiled in a personnel file for each employee and maintained for analysis purposes.

Figure 4.18 Overhaul activities require firefighters to open ceilings and walls to locate any hidden or remaining fire, sometimes resulting in muscle strains to the arms and back.

All members must be required to participate in the physical fitness plan. They should be evaluated annually and certified to perform their assigned duties in emergency operations. Members who cannot meet the physical performance requirements must be required to participate in a physical rehabilitation program to assist them in meeting their designated levels. The HSO, physician, or physical fitness officer shall devise a set of exercises and schedules to assist employees in meeting their goals.

A physical fitness plan is holistic, positive, rehabilitating, and educational. It is *not* punitive or used as punishment. The goal of physical fitness is to improve the quality of life for all emergency service personnel and help them live a long, healthy life. Good physical fitness also improves the quality of services provided to the community and professional image of the organization. Adhering to a complete physical fitness plan helps to reduce the potential for fatigue, heart disease, stroke, shortness of breath, and stress.

Before exercises can begin, a physical performance assessment (series of exercises) must be made of each individual's level of fitness to determine the correct exercise plan to meet the individual's needs. It also allows the officer or coordinator to establish a baseline for evaluating progress, set a realistic range of expectations or goals, and help the participant remain motivated. The individual's performance is scored and compared to a predetermined scale. Once the assessment is complete, a physical fitness plan can be developed for each individual based on age, need, and gender. The USFA has developed the *Physical Fitness Coordinator's Manual for Fire Departments* (FA-95, 1990) that contains a detailed description of the assessment process and recommended exercises. It is important to recognize that purchasing workout station equipment and placing it in a workout room in your station does not make a wellness program complete. There must be an analysis, training, and follow up components in any well structured program.

Exercise Plans and Equipment

The exercise plan is designed to provide exercises that will improve body flexibility, cardiovascular fitness, muscular fitness, and body composition. **Table 4.2** provides a quick reference comparison between each of these areas and lists the appropriate exercises. In all types of exercise, it is always important to warm up before and cool down after each workout. These activities reduce the potential of injuries occurring during the exercise session. The four exercise plan areas are described as follows:

- *Flexibility* — Flexibility (or stretching) exercises are intended to improve the body's mobility and range of motion. Divide the flexibility training plan into attainable objectives that ultimately reach the goal established in the assessment. Attainable objectives may include a reasonable degree of improvement on a sit-and-reach test. Increase these objectives periodically (such as weekly) until the maintenance-level goal is attained.

- *Cardiovascular fitness* — Cardiovascular fitness exercise will improve the ability of the body to supply oxygen through the circulation system to the muscles while performing vigorous activities over an extended period of time. Once the assessment is complete, develop an exercise program to maintain or improve the individual's ability. Playing sports is not a recommended substitute for any of the recommended exercises, although it may be used to supplement them and make the fitness plan more interesting. However, the USFA has determined that many injuries to emergency personnel occur during these types of activities.

- *Muscular fitness* — Good muscular fitness allows personnel to perform assigned tasks more effectively and efficiently and reduces the potential of personal injury. To increase muscular fitness, the plan needs to address the following items:

  — *Strength:* Maximum amount of force a muscle can generate

  — *Power:* Ability to exert strength quickly

  — *Endurance:* Ability to perform the activity repeatedly

As the individual's strength increases through exercise, power and endurance also increase. Exercises are directed at the six muscle groups (midsection, legs, arms, chest, shoulders, and back).

- *Body composition* — Body composition is the relationship of lean-to-fat tissues in the body; it can be roughly estimated by comparing height to weight. More accurate measures can be achieved by using special equipment such as skinfold calipers. The resulting number is converted to points that are added to the points achieved during the exercise portion of the assessment. Exercise helps to improve the muscle fitness of the individual. However, a healthy diet reduces the quantity of fat tissues in the body

Table 4.2
Exercise Plan Comparison Table

Exercise Plans	Desired Result	Exercise Type
Flexibility	Improved mobility and range of motion in back and legs	• Posterior Thigh • Calf Stretch • Anterior Thigh • Inner Thigh • Iliotibial Band and Lateral Thigh • Soleus
Cardiovascular fitness	Strengthened heart muscles	• Weight-bearing exercises: — running — stair climbing — rope jumping • Non-weight-bearing exercises: — bicycling — rowing — swimming
Muscular fitness	Increased strength and endurance	• Weight training: — bar bells — leg presses — bench presses — dead lifts • Endurance training: — push-ups — dips — curls — squats — sit-ups — pull-ups
Body composition	Weight loss and reduction in body fat or mass	• Jogging • Swimming • Dancing • Cycling • Brisk walking • Aerobics

The equipment necessary for a physical fitness plan varies depending on the finances available to the organization. The HSO and safety and health committee provide the administration with a list of recommended equipment necessary to implement the plan. The organization must determine what equipment will be provided and how to fund the plan. At a minimum, equipment should consist of workout uniforms, T-shirts, jogging shorts, warm-up suits, and athletic shoes to be worn by participating members. In addition, workout mats need to be available in each facility or at a central location where the exercising takes place. Free weights, exercise machines, powered stair-climbing equipment, and stationary bicycles may also be provided. The organization might find it more economical to enter initially into a contract with a local health-care facility or gymnasium for the use of its weights and machines. A physical fitness plan can either cost the organization a small amount of money or a large expenditure depending on the approach taken. Either way, the cost is justified in the health and well-being of the members and the reduced cost of lost-time injuries.

Figure 4.19 The organization's physical fitness plan must have a means of tracking the exercises performed by the members on a daily or weekly schedule.

Physical Fitness Plan Effectiveness

The physical fitness plan can be monitored through three methods: (1) using daily or weekly exercise logs that list the participation of members **(Figure 4.19)**, (2) comparing lost-time injury data before and after the start of the program, and (3) conducting annual physical fitness tests based on the original testing criteria to determine effectiveness. If the plan is mandatory, then the collection of participation data is permitted and can be used as a monitoring tool. However, the program must be flexible enough so that nonparticipation for legitimate reasons is not punished. Rather, alternative programs should be available to the membership. For instance, some members may belong to a health club and prefer to perform their exercise activities there. Others may prefer bicycle riding to jogging. In any case, the alternative to participation in the organization's plan should be noted in the log.

If the program is effective, then job-related injuries and occupational illnesses related to physical fitness should decrease over time. It is important to remember that USFA statistics show that injuries related to the physical fitness plan itself are leading causes of job-related lost-time injuries. This factor obviously alters the data collected on post-implementation injuries. Consider two solutions: In the first, the injuries that are incurred in the physical fitness plan should not be included in the comparison data. Only injuries that occur at incident scenes or in non-exercise work are compared to preplan data. The second is to take the data on physical fitness related injuries and use it to determine the need for change in the exercise plan. The lack of warm-up and cool-down exercises is a contributing factor to exercise-related injuries. However, other factors such as poor nutrition, smoking, or personal health problems can also contribute to such injuries.

Any alterations in the physical fitness exercise plan should be made with the knowledge and assistance of the organization's physician and exercise physiologist. Changes may also be indicated in the wellness portion of the employee physical fitness and wellness program.

A fitness program deals with proactive efforts to increase strength, endurance, flexibility, and other physical attributes to not only increase performance but to reduce injury due to muscle strain, sprains, and other related physical injuries. While there are a number of definitions regarding fitness, certainly a definition that is most appropriate for firefighters is one that acknowledges the established fire service standards for *fitness* and is consistent with entrance standards. The American College of Sports Medicine defines fitness in five areas:

1. Muscular strength

2. Muscular endurance

3. Flexibility or range of motion

4. Cardiorespiratory efficiency

5. Body composition

NFPA® standard on health-related fitness has a very similar list when discussing fitness assessment:

1. Aerobic capacity

2. Body composition

3. Muscular strength

4. Muscular endurance

5. Flexibility

Program components developed to ensure fitness of firefighters is also well defined in NFPA® 1583. The recommended components provide a clear model for development of a fire department fitness program.

Summary

To improve the lives of fire and emergency responders, the organization must develop, implement, and manage a health program. That program must provide both education, in the form of the wellness component, and physical fitness, in the form of a consistent exercise program. The health program lays the groundwork for all employees during their tenure in the fire and emergency services. It begins with the basic medical and fitness criteria used to select candidates and ends with the criteria for medical or disability termination.

Review Questions

1. How does analyzing the risks associated with firefighter duties help improve job-related safety?

2. What are the four elements used in a fitness-for-duty policy?

3. What is the purpose for preemployment medical requirements?

4. What areas are used to define fitness?

5. What areas does the Candidate Physical Abilities Test (CPAT) assess?

Training Safety

Chapter Contents

chapter 5

Key Terms

Job Performance Requirements

This chapter provides information that addresses the following job performance requirements of NFPA® 1500, *Standard on Fire Department Occupational Safety and Health Program:*

NFPA® References

5.1.1	5.2.1	5.3.3	5.3.9	5.4.3
5.1.5	5.2.2	5.3.4	5.3.10	5.4.4
5.1.8	5.2.3	5.3.5	5.3.11	5.5.1
5.1.9	5.2.4	5.3.6	5.3.12	5.5.2
5.1.10	5.2.5	5.3.7	5.4.1	5.5.3
5.1.11	5.2.6	5.3.8	5.4.2	

FESHE Learning Objectives

Occupational Safety and Health for Emergency Services
Apply the knowledge of an effective safety plan to pre-incident planning, response, and training activities.

Learning Objectives

After reading this chapter, students will be able to:

1. Summarize the common types of training-related injuries.

2. Summarize the reasons for increases in training-related casualties.

3. Describe the training requirements in entry-level training, specialized training, recertification training and in-service training.

4. Explain the administration's role in creating a safe training environment.

5. Explain the instructor's role in creating and delivering safe training.

6. Describe the participant's role in creating a safe training environment.

7. Describe the operational concepts used during emergencies which should be applied during training evolutions.

Chapter 5
Training Safety

Case History

On January 25, 2009, two male career fire fighters, age 28 (Victim #1) and age 45 (Victim #2), died after falling from an elevated aerial platform during a training exercise in Texas. The fire fighters were participating in the exercise to familiarize fire department personnel with a newly purchased 95-foot mid-mount aerial platform truck. A group of four fire fighters were standing in the aerial platform, which was raised to the roof of an eight-story dormitory building at a local college. The platform became stuck on the concrete parapet wall at the top of the building. During attempts to free the platform, the top edge of the parapet wall gave way and the aerial ladder sprung back from the top of the building, then began to whip violently back and forth. Two of the four fire fighters standing in the platform were ejected from the platform by the motion. They fell approximately 83 feet to the ground and died from their injuries.

Key contributing factors identified in this investigation include the fire fighters being unfamiliar with the controls on the newly purchased aerial platform truck, training in a "high-risk" scenario before becoming familiar with new equipment, failure to use fall restraints, the design of the platform railing and integrated doors, and the location of the lifting eyes underneath the platform which contributed to the platform snagging on the building's parapet wall.

National Institute for Occupational Safety and Health (NIOSH) investigators concluded that, to minimize the risk of similar occurrences, fire departments should:

- *Ensure fire fighters are fully familiar with new equipment before training under "high-risk scenarios."*

- *Ensure fall protection is used whenever fire fighters and other personnel are working in elevated aerial platforms.*

- *Follow standard operating procedures (SOPs) for training, including the designation of a safety officer.*

- *Ensure SOPs covering the operation and use of fire apparatus (including aerial platform apparatus) are developed and followed during training exercises as well as in fire-suppression activities.*

  Fire apparatus manufacturers should:

- *Provide fall-protection belts with all aerial ladder and platform apparatus and ensure that fall protection is used during manufacturer-provided training.*

- *Ensure that aerial platforms and other aerial devices are designed to reduce or eliminate the potential for snagging on buildings or other elevated surfaces.*

- *Ensure aerial platform doors or gates are designed to prevent opening in the outward direction.*

Source: NIOSH Firefighter Fatality Investigation Report F2009.06

There is a popular saying in the fire and emergency services that you should train the way you work and work the way you train. The concept is that if you learn and repeatedly practice a skill the "right way" in training, then you will be able to recall and apply the "right way" at an emergency incident. The logic of this concept should result in training exercises becoming as realistic as possible.

As the number of fire-related emergency responses has decreased, the need for realistic training evolutions has increased. Opportunities for *on-the-job training* have decreased. Training scenarios therefore must be provided that prepare you for any type of emergency even though you may rarely experience that emergency.

Unfortunately, the more realistic the practical training evolutions are, the more dangerous they tend to be. As would be expected, the more dangerous the training becomes, the more likely that people will be injured or killed during training. NFPA® reported in 2006 that 13 percent of all firefighter fatalities the previous year occurred during training. During 2006, over 7,000 reported injuries occurred during training evolutions. Sprains and muscular pain, the same type of injury that is most prevalent on the fireground, were the most frequently reported training-related injury.

The fact of the matter is that all training injuries and fatalities are preventable and unnecessary. Training evolutions are controlled situations that provide a chance to apply knowledge, practice skills, and improve abilities. These evolutions must not increase the risk to students or instructors in the name of realism.

Safety Training and Safe Training

Safety in training takes two forms. The first is safety training that involves learning the importance of safety while performing your duties. This training may be provided by the health and safety officer (HSO), an instructor, or your company officer. You may also have had a safety training module during entry-level or recruit training. Safety training topics include proper lifting techniques, smoking cessation, nutrition, the organization's employee assistance program, and similar information.

The second form of safety in training is the application of safety to training. That is *safe training*. If you take the concept of *training the way you work and work the way you train* and apply it to safety, then safe training will result in safety at the emergency incident.

Training-Related Injuries and Fatalities

According to the United State Fire Administration *Special Report: Trends and Hazards in Firefighter Training* (USFA-TR-100/May 2003), the leading cause of training-related fatalities is stress/overexertion resulting in cardiac arrest. The report, which collected and analyzed training-related firefighter injuries from 1987 to 2001, determined that these injuries increased by 21 percent. Because the data was based on reported injuries, it is very likely that the increase was much greater if unreported injuries were included. **Figure 5.1** provides a comparison of the percentage of injuries by type based on the data collected in the report.

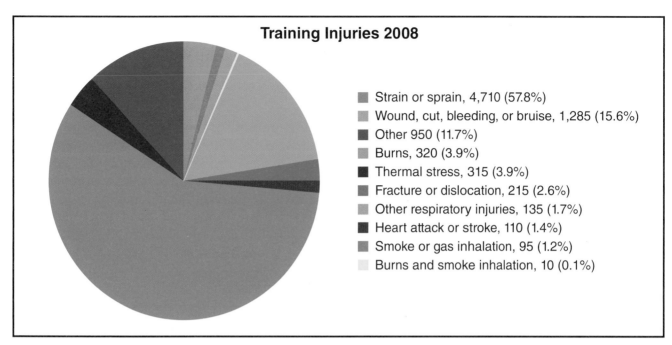

Training Injuries 2008

- ■ Strain or sprain, 4,710 (57.8%)
- ■ Wound, cut, bleeding, or bruise, 1,285 (15.6%)
- ■ Other 950 (11.7%)
- ■ Burns, 320 (3.9%)
- ■ Thermal stress, 315 (3.9%)
- ■ Fracture or dislocation, 215 (2.6%)
- ■ Other respiratory injuries, 135 (1.7%)
- ■ Heart attack or stroke, 110 (1.4%)
- ■ Smoke or gas inhalation, 95 (1.2%)
- ■ Burns and smoke inhalation, 10 (0.1%)

Figure 5.1 Common Types of Training Injuries.

While the leading type of injury was strains and sprains, heart attacks or strokes contributed to the highest number of fatalities. These results are similar to the numbers of injuries and fatalities caused by emergency-related activities. The report indicates that the cause for these fatalities and injuries is the lack of physical fitness, occurring primarily during equipment or apparatus training drills and live-fire training exercises.

The report concludes that there are several reasons for the increase in training-related casualties. The reasons include the following:

- Wider range of services provided by the fire and emergency services requiring greater time to be spent on a variety of training topics and less time on basic fire suppression skills

- Decreasing number of fires that formerly provided experience.

- Departments must now rely on live-fire training in burn buildings and acquired structures

- Improved protection provided by better personal protective equipment (PPE) leading to overconfidence that causes firefighters to go farther into buildings and stay longer than is safe

Training-related injuries and fatalities are not just associated with structural fire-fighting activities. Injuries and fatalities also result from emergency medical training, wildland and urban-interface fire training, and high-hazard training. The types of injuries continue to be the same as those occurring during actual incidents and lack of physical fitness is still the primary cause for cardiac arrest.

Types of Training Required

Training requirements can be collected into three categories. They include entry-level training, specialized training, and recertification training. The initial training in each category will be provided by the organization's training division or a state, regional, provincial, or national training agency. Company- or unit-level training in many of the topics will be continuous and will occur on a schedule established by the organization to fulfill state/provincial or national mandates.

Entry-Level Training

Entry-level training can take a multitude of forms. In states/provinces that conform to the NFPA® professional qualifications standards, entry-level or recruit training will provide the knowledge and skills required to certify for Firefighter I criteria. Some municipalities may include the requirements for Firefighter II in the training curriculum resulting in a longer recruit school. All basic firefighters must also certify to the Awareness and Operations Levels of hazardous materials (haz mat) training as described in NFPA® 472. Many organizations also include either a basic first responder course or Emergency Medical Technician-Basic (EMT-B) training as part of the entry-level course. Training must progress from basic to advanced skills in a logical manner.

Figure 5.2 Recruit training is structured to provide the new firefighter with the skills necessary to perform a variety of tasks safely, such as the use of respiratory protection equipment.

Recruit school is the first time that fire-suppression training takes place and live-fire evolutions are critical to this training. Considered high-hazard training, live-fire training has claimed the lives of a number of recruits as well as veteran instructors. Prior to any live-fire training, all Fire Fighter I candidates must receive respiratory protection training. This training includes all types of respiratory protection equipment that they will be expected to use but especially self-contained breathing apparatus (SCBA) **(Figure 5.2)**.

Essential to all entry-level training are safety-related topics. These are usually taught by the organization's health and safety officer (HSO) or a qualified training officer. Safety training should be continuous throughout the recruit course, and safety policies and procedures must be applied to all training activities.

Other types of training provided in the entry-level course include the selection, donning, use, and care of personal protective equipment (PPE) and firefighter self-rescue or survival training. Practical training evolutions are used to reinforce this training and provide an opportunity to practice the skills. Personnel who may engage in wildland fire fighting must be trained and tested annually in the proper deployment of the fire shelter used by the organization.

Large industrial complexes, such as refineries or chemical plants, usually provide training for personnel who are expected to extinguish incipient fires or respond as part of the permanent industrial fire brigade. Similar to the skills and training provided to career and volunteer firefighters, requirements for this type of training is included in the IFSTA **Industrial Emergency Service Training: Incipient Level** and **Industrial Exterior and Structural Fire Brigades** manuals and in NFPA® 600, *Standard on Industrial Fire Brigades*, and NFPA®1081, *Standard for Industrial Fire Brigade Member Professional Qualifications.*

Public safety officers are personnel who are responsible for both law enforcement and fire protection functions. While their primary function may be law enforcement, they must also be fully certified as a minimum of Fire Fighter I.

Specialized Training

Specialized training is usually provided for personnel who have completed a specified time as a Fire Fighter I or II depending on state/provincial or local policies. Most specialized training will be provided in order to create a team of specialists to meet a specific need or hazard. Specialized training includes:

- *Driver/operator training* — Apparatus driver/operator training may be provided to personnel who will be assigned or promoted to the position responsible for the operation of an ambulance or fire-fighting apparatus **(Figure 5.3)**

- *EMT-Advanced* — Higher levels of medical training provided to meet local requirements

- *Extrication* — Basic vehicle or machinery extrication is taught in the entry-level course. More advanced training is required if the firefighter or paramedic is assigned to a vehicle that deploys specialized power extrication tools

- *Technical Rescue* — Technical rescue training may include any or all of the following types of rescue:

  - *Rope rescue*
  - *Confined Space*
  - *Trench rescue*
  - *Structural collapse*
  - *Water or dive rescue*
  - *Ice rescue*

- *Marine* — Shipboard fire fighting training is provided for personnel assigned to marine vessels, fire department fireboats, and land-based units that protect harbor facilities and respond to shipboard emergencies

- *Aircraft Rescue and Fire Fighting (ARFF)* — ARFF training is required by the U.S. Federal Aviation Administration (FAA) for all personnel assigned to certain airports and other personnel who may respond to aircraft-related incidents on those airports' property **(Figure 5.4)**

- *Hazardous Materials* — Training required for units specifically designated to respond to hazardous materials incidents. While fire department personnel must meet the Awareness or Operations Levels training, haz mat team personnel must be trained and certified to Technician Level.

Figure 5.3 Driver/operator training is intended to provide firefighters with the skills needed to drive the apparatus, respond safely to emergency incidents, and operate the pumps and aerial devices mounted on the vehicle.

Figure 5.4 Aircraft rescue and fire fighting training for personnel who respond to or are stationed at airports in the U.S. is mandated by the Federal Aviation Administration.

Recertification Training

Fire and emergency services organizations are required by the federal and state/provincial government regulations to provide refresher/recertification training in a variety of topics. These include respiratory protection (SCBA), haz mat, and medical, among others. The number of training hours for each topic is also mandated.

Another recent requirement is to provide training to meet national credentialing criteria. The purpose of national credentialing is to ensure that personnel trained in one area of the country can respond to and provide assistance at an emergency anywhere in the country.

In-Service Training

Both career and volunteer departments require continuous in-service training. In volunteer and career departments, there is usually an agency-set schedule, using agency-approved lesson plans and periodic testing. In volunteer organizations, this training typically takes place on a weekly basis at evening or weekend drills **(Figure 5.5)**. Career departments train during work shifts at the fire station, training facility, or within the response area. Training may be conducted by the company officer or a department training officer.

Training should also occur when any type of new equipment is purchased and issued or when there is a change in protocol, procedures, or policies. This type of training is provided by training officers, manufacturer's representatives, the HSO, or company officers or crew members who have received initial training (sometimes referred to as *train the trainer*).

Figure 5.5 Volunteer and combination fire departments provide weekly training for their members to maintain efficiency and learn new skills.

Train the trainer — Training program that relies on training key members of the organization, such as company officers, to train other members, such as company personnel, in a specific topic, procedure, or policy. An effective train-the-trainer program will teach trainers such skills as how to facilitate a meeting, how to deal with students who are having trouble, and how to teach effective note-taking.

Creating a Safe Training Environment

Your organization is obligated to provide a safe training environment while making the training as realistic as possible. Administrators, supervisors/instructors, and participants must realize that safety takes precedence over realism in all training. The role that each plays in creating a safe training environment is described in this section.

Administration's Role

It is the administration's responsibility to provide the resources needed for a safe training environment. The administration must know what is required legally, for example, the national haz mat regulations. The resources include the following:

- Leadership and commitment
- Risk management program
- Qualified and certified instructors
- Props, equipment, and acquired structures

Leadership and Commitment

It cannot be overemphasized that safety depends on the commitment of the organization's leaders. From the chief of the department throughout the senior staff to the training officer(s), these officers must understand their responsibility for the safety of their employees. They must be willing to change the current culture, establish a new culture based on the 16 Safety Initiatives, and nurture that culture until it spreads throughout the organization. They cannot permit unsafe acts to occur during training, and they must enforce safety policies and procedures.

Risk Management Program

To reduce the potential risks to personnel, all fire and emergency services organizations regulated by legally adopted NFPA® standards are required to have a risk management plan. After plan implementation, it must be monitored for effectiveness. This plan is designed to accomplish the following objectives:

- Identify risks.
- Evaluate the potential based on frequency and severity of the risk.
- Establish appropriate controls to minimize or eliminate the risk.

The risk management plan includes all job-related activities in which fire and emergency services personnel (emergency and nonemergency) normally participate, including emergency, nonemergency, training, and support activities. The risk management plan process can be found in Chapter 11.

Instructors play a significant role in preventing injuries and fatalities during training evolutions. However, it is imperative that the organization's administration support and enforce safety, fitness, and health and wellness programs in all aspects of the organization's operations. The administration should commit to taking the following responsibilities:

- Provide adequate personal protective equipment (PPE).
- Ensure that all apparatus and equipment are appropriately maintained.
- Ensure that all safety equipment is properly installed and operating.
- Provide policies and procedures for the safe use of the apparatus and equipment.
- Address the fitness, health, and wellness of personnel through the following elements:
  — Job-related physical fitness testing
  — Annual medical evaluations and periodic examinations
  — Health- and wellness-related information and training
  — Employee assistance programs

Some of the recommendations provided in the USFA Special Report, *Trends and Hazards in Firefighter Training*, May 2003 (TR-100) include the following:

- Follow established guidelines and currently accepted procedures and standards for training.
- Ensure that training safety standards and procedures are followed at all times.

- Include the use of live-burn evolutions in a variety of structure types to provide realistic fire-fighting experiences for students.

- Train firefighters or emergency responders to recognize the visual and physical clues to impending danger (such as changes in smoke conditions) and anticipate fire behavior in a variety of building types.

Based on investigations of training-related fatalities, the National Institute for Occupational Safety and Health (NIOSH) investigators have concluded that fire departments should take the following actions to minimize the risk of similar occurrences:

- Ensure that the fuels used in live-fire training have known burning characteristics and the structure is inspected for possible hazards before the training exercise.

- Ensure that ventilation is closely coordinated with interior operations.

- Ensure that fires are *not* located in designated exit paths.

- Ensure that a method of fireground communication is established to enable coordination among the incident commander (IC) and all participants.

- Ensure that standard operating guidelines or procedures (SOGs/SOPs) specific to live-fire training are developed and followed.

- Consider using a thermal imaging camera and other equipment used on the actual fireground during live-fire training exercises **(Figure 5.6)**.

Additionally, NIOSH recommends that states consider taking the following actions:

- Develop a permitting procedure for live-fire training to be conducted at acquired structures.

- Ensure that all the requirements of NFPA® 1403, *Standard on Live Fire Training Evolutions,* have been met before issuing a permit.

Figure 5.6 Thermal imaging cameras are increasingly used in live-fire training.

Instructor Preparation

The administration is responsible for the preparing instructors who teach the organization's training courses. A lack of preparation can result in casualties and a legal liability placed upon the instructor and the organization. Unqualified instructors must not be assigned to teach courses they are not trained to teach. Instructor preparation must include:

- **Instructor certification** — Instructors must be trained, certified, and authorized to teach the specific courses they are assigned.

- **Instructor-student ratio** — For high-hazard training the instructor-to-student ratio should be no less than 1 to 5.

- **Safety officers and technical safety officers** — Incident safety officers must be provided for all high-hazard training. Technical safety officers, certified in the topic being taught such as high-angle rescue, must be present for technical-type training.

- **Instruction materials and equipment** — Instructors must be provided with the appropriate tools and equipment to teach the course.

An essential element for preventing training-related casualties during live-fire training is application of NFPA® 1403. Instructors must be familiar with the requirements of this standard for all live-fire training in purpose-built burn buildings, acquired structures, and burn props **(Figure 5.7)**. Safety requirements must be enforced by the instructor in charge of the training evolution, the designated incident safety officer (ISO), and the organization's administration.

Figure 5.7 Burn buildings and props can be very complex and include safety shutoffs and temperature sensors that must be inspected regularly and prior to use.

Props, Equipment, and Acquired Structures

The administration must budget for and provide the props, tools, equipment, and safety materials used in training. They must also provide live-burn buildings or acquired structures that meet the requirements of NFPA® 1403.

- **Props** — Props are specifically designed to simulate some type of situation or equipment that may be encountered during an emergency incident. Props range from gas-fired live-burn structures to wrecked vehicles. Some props are designed and purchased from companies that specialize in engineering fire behavior props. Others are created locally, such as mazes used to simulate rescue situations.

  NOTE: You must remember that constructing props or altering commercially purchased props places a legal liability on you and the organization. Constructing confined-space props that do not have the proper safety requirements will place those who use it at increased risk. Altering a commercial fire behavior prop without approval from the manufacturer will not only create liability, it will also void the warranty.

- **Maintenance** — All training props, tools, and equipment must be inspected before use, after use, and on a periodic basis to ensure that they are working properly. Preventive maintenance as well as corrective maintenance must be provided for the equipment.

- *PPE* — Personal protective equipment used by participants and instructors must be inspected and cleaned after each use in accordance with the manufacturer's recommendations. The administration and supervisors must not tolerate the use of contaminated PPE during training. This condition provides a negative message to participants, especially new recruits **(Figure 5.8)**.

- *Safety information* — The HSO is generally responsible for safety training. This position must be given the resources to provide safety training.

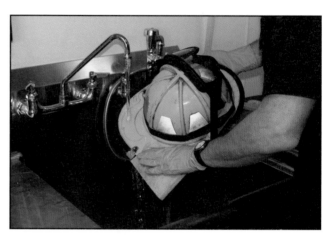

Figure 5.8 Training officers and instructors must set a good example for their students by maintaining their personal protective equipment. Equipment must be clean and in good condition.

Instructor's Role

It should be apparent that fire and emergency services personnel work in high-risk environments that are inherently unsafe. Although it is impossible to remove the risks completely to create a totally safe work environment, it is possible to reduce the risk and increase the safety level. Through the efforts of fire and emergency services organizations and instructors, agencies can reduce risks through a variety of the following methods:

- Provide and require the use of approved PPE.

- Provide training in safety topics and practices.

- Teach the safest methods for accomplishing skills and tasks.

- Adhere to safety regulations and standards both in training and at emergency incidents.

- Establish and adhere to the National Incident Management System (NIMS) Incident Command System (ICS).

- Use a personnel accountability system.

- Investigate, analyze, and correct unsafe activities and behaviors and near misses.

- Establish safety policies and procedures for use in all work-related activities.

- Establish and use appropriate rehabilitation based on NFPA® 1584.

- Conduct a brief safety critique at the end of every training drill.

The instructor is actively involved with each of these methods. The instructor's role in creating and ensuring a safe work environment can be divided into four broad training-related categories:

- Provide a safe training environment.
- Teach safety-related topics.
- Fulfill the duties of the incident safety officer (ISO) when qualified and assigned.
- Provide students with a positive, proactive safety role model.

The most important thing to remember is that safety must be taught and practiced during training. By requiring safety training and applying the use of safe practices, fire and emergency services personnel will automatically apply the safe practices during emergency incidents. Safety must be a permanent behavioral trait that becomes automatic to the firefighter or emergency responder during daily activities.

Participant's Role

Your safety depends on you. You must consciously perform your role in the training process in a safe manner. Some of the things that you can do to ensure safety in training are as follows:

- Follow the organization's SOPs
- Follow all directions and orders
- Provide feedback to the instructor if something seems unsafe
- Report unsafe acts
- Do not participate in horseplay
- Inspect, clean, and care for PPE according to manufacturer's recommendations
- Report all injuries no matter how small
- Actively participate in post training critiques

Practical Training Evolutions

Fire and emergency responder training is divided between cognitive and psychomotor training. Cognitive training may occur in the classroom through computer-based programs or as distance-learning courses broadcast on television. Cognitive training provides the knowledge necessary to make sound decisions and respond quickly to changing conditions.

Psychomotor training involves physical skills. It occurs during practical training evolutions at the training center, fire station, or in the field. These evolutions are intended to be as realistic as possible while still being as safe as possible. When these skills are learned, they must be reinforced by frequent practice so you will be able to carry them out by second nature.

To be safe and effective, practical training evolutions must include training regarding the application of concepts that will be applied to actual emergency incidents. Among these concepts are the following:

- *NIMS/ICS* — In all high-hazard practical training evolutions, the locally adopted version of the National Incident Management System/Incident

Command System (NIMS/ICS) must be established. The use of the ICS not only ensures greater control of the training evolution but also reinforces the need for it at emergency incidents.

- *Accountability system* — An essential element of the ICS is the establishment and use of the personnel accountability system. Using an accountability system ensures that the location and actions of all personnel involved in the practical training evolution is known **(Figure 5.9)**.

- *Incident Safety Officer (ISO)* — An ISO must be assigned to all high-hazard training evolutions. If the evolution is large and involves multiple units spread over a wide area, assistant ISOs must be assigned.

- *Communications protocol* — The locally adopted communications protocol must be used during all practical training evolutions. Usually, a separate training radio frequency is provided for all units involved in the evolution. Proper communications protocol includes requirements for communications between emergency medical technicians (EMTs) and emergency room personnel during firefighter Mayday or rescue procedures.

- *Personal Alert Safety System (PASS)* — Many fire agencies purchase and issue PASS devices to all personnel. PASS devices may be individual units or integrated into SCBA units. Unfortunately, injuries and fatalities have occurred because personnel have not activated the PASS devices prior to entering hostile atmospheres. The repeated use of the devices in training can help to make their use during emergencies more likely.

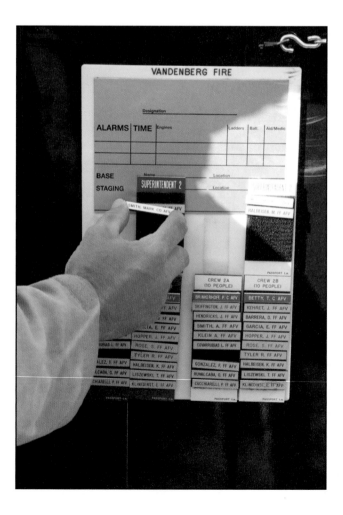

Figure 5.9 An accountability system helps the incident commander keep track of personnel at the incident. Systems must be used in training to reinforce their use at actual incidents.

- **Proper respiratory protection** — Training in the selection and use of respiratory protection equipment reinforces the need to sample unknown atmospheres and determine the correct type of protection needed.

- **Firefighter rescue training** — Because firefighter rescue conditions are rare, training must be provided as often as possible. Self-escape or survival techniques must be developed and training must be provided so the actions will be instantaneous. Terms generally used to describe conditions that require rescue include *Mayday, Firefighter down,* or *Firefighter needs assistance.*

- **Crew resource management (CRM)** — As discussed in Chapter 3, crew resource management or some form of it, must be applied to all high-hazard training evolutions. Repeated use of CRM will help improve the ability to make decisions both by individuals and by the unit.

Besides these concepts, other safety-related issues must also be addressed when performing high-hazard practical training evolutions. Some of these have been mentioned previously but are worth reinforcing by including here. They include the following:

- **Live-fire certified training officer** — Many states/provinces now require that a live-fire trained and certified training officer be present at all live-fire training evolutions.

- **Training officer certified for skill** — Training officers at high-hazard training evolutions should be trained and certified or qualified in the specific skill they are teaching.

- **Adherence to NFPA® 1403 and 1500 standards** — Both NFPA® 1403 and 1500 were developed to ensure fire and emergency responder safety. They will only provide this safety if organizations, supervisors, and participants adhere to these standards during training evolutions.

- **Adherence to Occupational Safety and Heallth Administration (OSHA) requirements for respiratory protection** — Prior to any training evolution that involves immediately dangerous to life or health (IDLH) atmospheres, including those that may have low oxygen content, all participants must be trained and certified to use respiratory protection equipment. This certification includes a required medical examination and facepiece fit testing.

- **Adherence to OSHA confined-space training requirements** — Confined-space training evolutions may require permits from the facility owner prior to some types of training.

- **Adherence to OSHA infection-control requirements** — Medical training must meet infection-control requirements.

- **Adherence to NFPA® medical fitness requirements** — All participants, including instructors, must meet the adopted physical fitness and medical requirements prior to participating in any high-hazard training evolutions.

- **Adherence to 29 CFR 1910.147 lock-out/tag-out requirements** — When training in a working facility around stored-energy or hazardous equipment, the facility's policies must be followed.

- **Adherence to the organization's hearing conservation policies** — Hearing protection must be used during all training activities that produce excessive levels of noise (greater than 90 dB).

Summary

Applying safety to all training activities provides multiple benefits. It reinforces the importance of safety in the organization. It provides safety training in the context of actual training evolutions. It gives the participants the opportunity to develop safe skills. In the end, safety training and safe training will provide the foundation for changing the culture and enforcing the importance of safe practices within the organization.

Review Questions

1. What is one reason for training-related injuries?

2. Describe two of the most common types of training-related injuries.

3. Explain the difference between entry-level training and specialized training.

4. What role does administration play in creating a safe training environment?

5. What four broad areas make up the instructor's role in creating and delivering safe training?

6. Explain how each participant can help create a safe training environment.

7. Describe three operational concepts used during emergencies that should also be used during training evolutions.

8. Why is it important to employ the same operational concepts used during emergencies when executing a training evolution?

Fire Apparatus Safety

Chapter Contents

Key Terms

Job Performance Requirements

This chapter provides information that addresses the following job performance requirements of NFPA® 1500, *Standard on Fire Department Occupational Safety and Health Program:*

NFPA® References

6.1.1	6.1.8	6.2.5	6.2.11	6.3.4	6.4.3
6.1.1.1	6.2.1	6.2.6	6.2.12	6.3.5	6.4.4
6.1.2	6.2.2	6.2.7	6.2.13	6.3.6	6.4.4.1
6.1.3	6.2.2.1	6.2.7.1	6.3.1	6.3.7	6.4.5
6.1.4	6.2.3	6.2.8	6.3.2	6.3.8	6.4.6
6.1.5	6.2.3.1	6.2.9	6.3.3	6.4.1	6.4.7
6.1.6	6.2.4	6.2.10	6.3.3.1	6.4.2	6.4.8
6.1.7	6.2.4.1				

This chapter provides information that addresses the following job performance requirements of NFPA® 1001, *Standard for Fire Fighter Professional Qualifications:*

NFPA® References

5.3.2	A5.3.2	5.3.3

This chapter provides information that addresses the following job performance requirements of NFPA® 1002, *Standard for Fire Apparatus Driver/Operator Professional Qualifications:*

NFPA® References

4.1	5.1	6.2	10.1
4.2	5.2	8.1	10.2
4.3	6.1	8.2	

FESHE Learning Objectives

Occupational Safety and Health for Emergency Services

Apply the knowledge of an effective safety plan to pre-incident planning, response, and training activities.

Fire Apparatus Safety

Learning Objectives

After reading this chapter, students will be able to:

1. Summarize the five common causes of fire apparatus crashes.
2. Summarize the legal requirements impacting the design of fire apparatus.
3. Describe the safety components found on fire apparatus.
4. Explain the difference between preventive and corrective maintenance.
5. Describe the difference between an acceptance and a service test.
6. Describe the common requirements for fire apparatus driver operators.
7. Summarize the basic concepts of defensive driving.
8. Define visual lead time.
9. Summarize the company officer's responsibility for safe operation of a fire apparatus.
10. Summarize the operational concerns for a driver/operator during an emergency response.
11. Explain the difference between total stopping distance, reaction distance, and braking distance.
12. Describe the considerations for parking fire apparatus at various types of incidents.
13. Describe the casual factors responsible for firefighters coming in contact with other vehicles at roadway incidents.

Chapter 6
Fire Apparatus Safety

Case History

On October 5, 1999, a captain (fire officer), a driver/operator, and a firefighter assigned to an engine company responded to a medical call that had been dispatched as a patient with shortness of breath. Traveling north, the engine approached a four-way intersection that was crossed on the north side by an overpass supported by concrete columns. The intersection was controlled by electronic traffic lights. The traffic signal was red for the engine's direction of travel. The driver, initially reducing engine speed, checked that traffic had cleared and resumed speed through the intersection.

At the same time, a civilian operating an automobile traveled through the intersection into the engine's path. The driver of the engine was unable to avoid the automobile, and the two vehicles collided. The driver lost control of the engine, which then struck one of the concrete columns supporting the overpass. The engine struck the column on the driver's side, and the officer, who was not wearing a seat belt, was ejected through the windshield, landing in a lane for oncoming traffic. The engine continued past the column and came to a stop in the same lane next to the officer.

The driver was knocked unconscious, and the firefighter riding in the rear crew compartment received minor injuries as a result of flying objects within the cab. The officer was flown by life flight helicopter to a nearby hospital where he was pronounced dead upon arrival. The two injured firefighters were transported by ambulance to the hospital, where the driver was admitted in critical condition, and the firefighter was treated for his injuries and released. The civilian driver of the automobile was not injured.

It should be apparent from this example that the urgency of an emergency response must never compromise personal safety. Driver/operators and fire officers must adhere to safety policies by wearing seat belts. They must practice defensive-driving techniques and above all maintain situational awareness. Finally, the department must provide a safe environment within the apparatus crew compartment by providing seat belts and securing or removing equipment that can injure occupants in a collision.

Source: NIOSH Firefighter Fatality Investigation Report 99-F36

As emergency service workers, we are in the business of saving lives, property, and protecting the environment. We should be part of the solution to problems, not create new problems by being involved in a traffic accident during response. In 2006 alone, we responded to over 24 million emergency and nonemergency incidents that required us to apply our knowledge and skills in the safest manner possible. That may seem easy, but in reality, we need to realize that unsafe acts by responders can and do occur. Whether those unsafe acts are intentional or not, it is the responsibility of everyone involved at an incident to be responsible for their own safety and those around them.

Likewise, we cannot perform our duties at the emergency incident if we are involved in an accident while en route. Statistics indicate that fatalities resulting from apparatus accidents exceed fatalities that occur at the emergency incident. En route accidents not only eliminate the unit involved but they also require additional resources to assist at the accident scene. Emotionally, the en route accident will also place an additional stress on other emergency responders who are concerned with the safety of their friends and coworkers.

Apparatus-Related Fatalities and Injuries 2008

Fatalities	24
Injuries	4,965

Source: United States Fire Administration, *U.S. Firefighter Injuries – 2008; Fire Fighter Fatalities in the United States in 2008.*

Apparatus safety depends on a number of controllable factors such as the following:

- The driving skill of the apparatus driver/operator may be the most significant factor. The driver/operator must be trained in the safe operation of the vehicle during emergency and nonemergency operations and in all types of weather. The driver/operator must be trained to operate the vehicle safely in a variety of weather and traffic conditions.

- Personal factors including situational awareness, internal awareness, and physical fitness all contribute to the driver/operator's ability to safely operate the vehicle. When and where available, simulation technology should be employed for driver/operator training. This new technology provides the ability to present unusual and dangerous situations to drivers without the possibility of actual accident. However, there is no substitute for physically operating the type of apparatus that the driver/operator is going to be assigned to drive.

- The mechanical condition of the apparatus contributes to the safe operation of the vehicle. The organization must provide a preventive maintenance program that ensures that the vehicle is in a safe condition at all times **(Figure 6.1)**.

- Apparatus must meet the nationally and locally established design criteria for emergency and nonemergency vehicles. When the fire and emergency services organization creates purchase specifications for new vehicles, the specifications must, at a minimum, meet National Fire Protection Association® (NFPA®) and U.S. Department of Transportation (DOT) or Canadian requirements.

- Finally, a standard operating policy for all vehicles must be established, implemented, and enforced by the organization. The policy must cover all aspects of operation of all types of department-owned vehicles. It must take into consideration both how the vehicle is driven and how passengers ride in the vehicle (including privately owned vehicles [POVs]) during emergency responses.

This chapter will provide general information and guidance that will allow members of the organization's administration to develop a preventive maintenance program and to write specifications for new apparatus. It will also provide supervisors with information for enforcing safe apparatus operation during all types of situations. Finally, it will provide members of the department with information on how they can operate and maintain the vehicle using safety-conscious behaviors. The end result will be increased safety through the control of many factors that will ensure the safe arrival of fire-fighting units at their destination.

Figure 6.1 Driver/operators must be trained in preventive maintenance and inspection of their vehicle.

Apparatus Accidents

Lawsuits against fire departments arising from accidents involving fire apparatus are a significant legal and financial liabilities that a department may face. A single occurrence of such an incident can have a devastating and lasting effect on the personnel involved, on their families, on any civilians involved, and on the financial condition of the fire and emergency services organization. The public expects that fire apparatus driver/operators will efficiently and safely drive their vehicles through the community in a controlled manner. When that fails to happen and an accident occurs, the loss of citizen confidence will result. In addition, a lawsuit seeking monetary compensation for injuries or property damages suffered may be filed. In some instances criminal charges may also occur.

According to the NFPA®, during the 30 years between 1977 and 2007, a total of 406 firefighters died in vehicle-related accidents **(Figure 6.2)**. This figure constituted the second highest number of fatalities during that period. The majority of the personnel who were killed were volunteer firefighters (75 percent), 14 percent were career personnel, and the rest were contract employees. Responding to emergency calls in personal vehicles resulted in 37 percent of the deaths.

The NFPA® study determined that over three-quarters of the victims were not wearing a seat belt or other type of restraint. The responsibility for these types of fatalities must be shared by the administration, the supervisor, and the firefighter. The administration is responsible for providing seat belts or restraints in the vehicles and for

Figure 6.2 Accidents involving apparatus traveling to or from an emergency incident account for the second highest number of firefighter deaths. *Courtesy of Ron Jeffers.*

having a consistent and enforceable policy on the wearing of seat belts. The supervisors are responsible for ensuring that all personnel in the apparatus secure their seat belts before the apparatus responds. And all firefighters are responsible for using the seat belts that are provided in the apparatus and in their personal vehicles. In the majority of states, seat belt use laws exist and must be complied with.

Beyond the firefighter fatalities and injuries are those sustained by the civilian population in collisions with fire apparatus. According to the U.S. National Highway Safety Administration (NHSA), 170 people died and 18,722 were injured as a result of a collision with emergency vehicles in 2004. Part of the reason for these fatalities and injuries is the difference in size and weight of emergency vehicles and passenger vehicles. In 2004, 9 drivers of ambulances or passengers died while 24 civilians were killed in accidents involving ambulances.

A final consideration for the effect of emergency vehicle accidents is the cost of the accident. Accident-associated costs may include the following:

- Vehicle repairs
- Vehicle replacement
- Insurance claims
- Medical benefits
- Court costs
- Replacement personnel

The majority of apparatus accidents are preventable. Yet the number of apparatus accidents continues to remain high. So critical is the need to reduce the number of these incidents that in 2004, the U.S. Federal Emergency Management Agency (FEMA) issued its *Emergency Vehicle Safety Initiative* (FA-272). This document is the culmination of a study of 12 fire and emergency services organizations and their approaches to apparatus safety. The results, ranging from the design of apparatus to their operations during responses, provide the reader with numerous ways for reducing the possibility of apparatus accidents. This FEMA document is available on its web site.

In the **Pumping Apparatus Driver/Operator Handbook**, IFSTA suggests that the causes of all fire-apparatus crashes can be grouped into one of the following five categories:

1. *Improper backing of the apparatus* — Backing crashes are among the most frequent of all types of fire-apparatus crashes **(Figure 6.3)**. While they are seldom serious in terms of injury or death, they do account for a significant portion of overall damage costs.

2. *Reckless driving by the public* — This category includes a variety of reckless actions, including failure to obey traffic signals, excessive speed, failure to yield to emergency vehicles, and other common civilian driving behaviors.

3. *Excessive speed by the fire apparatus driver/operator* — Excessive speed may result in the driver/operator losing control of the vehicle or being unable to stop the vehicle before hitting another vehicle or object or leaving the roadway.

4. Lack of driving skill and experience by the fire apparatus driver/operator — This may be due to insufficient training of the driver/operator or unfamiliarity with the vehicle being driven.

5. Poor apparatus design and/or maintenance — While poor design may not be the case with custom-built fire apparatus, it is a more significant problem in departments that use retrofitted or home-built apparatus. However, a lack of adequate maintenance is universal to all types of apparatus.

Of these causes, only the second is completely out of the control of the fire and emergency services, the supervisor, and the firefighter. However, good defensive driving skills on the part of the driver/operator and the application of situational awareness on the part of the supervisor and crew members can reduce the potential caused by reckless driving of the public.

Figure 6.3 Spotters must be used when backing fire apparatus.

Analyzing a Fatality Accident

In the case history presented at the beginning of this chapter, you learned that a fire officer was killed and two members of the company were injured in a collision during an emergency response. The ensuing investigation determined that a variety of safety policies were either ignored or not in place, leading to the fatality. The recommendations of the NIOSH investigation include the following:

- *Fire departments should ensure that drivers of emergency fire apparatus adequately reduce their speed to maintain vehicle control and proceed with caution through intersections*

Intersections are extremely dangerous places for responding firefighters and civilian motorists. When a responding emergency apparatus approaches an intersection against a traffic signal, even with the apparatus' red lights and sirens activated as required by law, the driver must check that the right-of-way has been yielded by all oncoming motorists before proceeding. Department protocols may require all drivers of fire apparatus and other department vehicles equipped with emergency warning devices to proceed cautiously when responding to emergencies.

Proceeding cautiously may include slowing to a speed "prepared to stop" when entering a traffic-controlled intersection or "coming to a complete stop" when necessitated by observable circumstances. According to *NFPA® 1500*, during emergency responses, driver/operators shall bring vehicles to a complete stop before proceeding when safe to do so.

A red traffic light is one of the observable circumstances. The fire apparatus may, once it is determined safe to do so, proceed against the red light. Fire departments should consider exploring the technology that exists for enhancing traffic control, such as intersection control devices that allow emergency vehicles to control traffic lights at intersections.

Continued on page 182

Analyzing a Fatality Accident (Concluded)

- *Fire departments must consistently enforce standard operating procedures (SOPs) for safely driving fire department vehicles during emergency response*

SOPs for safe driving fire department vehicles during nonemergency travel and emergency response may include specific criteria for maintaining appropriate vehicle speed, following distances, crossing intersections, traversing railroad grade crossings, use of spotters when backing, and using emergency warning devices. Such procedures for emergency response emphasize that the safe arrival of fire department vehicles at emergency scenes is the first priority.

- *Firefighters should ensure that all equipment within the fire apparatus driving and crew compartments is properly mounted or stowed.*

One of the most significant hazards to firefighters involved in vehicle collisions is the equipment carried inside the apparatus crew cab. Equipment such as self-contained breathing apparatus (SCBAs), tools, PPE, and portable radios can become projectiles in the event of vehicle acceleration, deceleration, or impact. The storage of these items in enclosed compartments or in brackets with mechanical means of securing the equipment is very important.

In this incident, equipment and other items, including SCBAs, a water cooler, and PPE, became airborne upon impact. The occupants of the engine may have been struck and injured by these airborne objects.

- *Fire departments should ensure that all firefighters riding in emergency fire apparatus are wearing seat belts and belted securely.*

Firefighters make many life-and-death decisions during a tour of duty, and one of the most important is snapping on a seat belt after climbing aboard an emergency apparatus that has been called to respond. The fire department involved in this incident had a directive from the fire chief and a memo from a city official that required the use of seat belts by all employees while operating city vehicles.

The captain (the victim) was not wearing his seat belt and was ejected from the engine. Both the driver and the injured firefighter were wearing their seat belts and survived the accident.

Discuss the following questions based on this incident:

1. If the D/O found that traffic had "cleared," how could the civilian vehicle travel into the engine's path "at the same time"?
2. Is there reason to believe that the captain would have survived if he had been wearing his seat belt?
3. Is there technology available to advise the D/O that not all personnel's seat belts are in use?
4. Does a preemptive control device guarantee that the intersection is safe to enter?
5. Should the D/O have initially moved the apparatus knowing that the captain was not belted?

Apparatus Safety Design Requirements

Fire apparatus safety begins with the vehicle's design. The vehicle, whether a custom-made ambulance or aerial device or a commercially designed pickup, van, or car, must meet nationally established design requirements. Several organizations or government agencies stipulate the minimum requirements for the design of motor vehicles. Each of these will be discussed in the following section.

Even though vehicles are constructed to meet required specifications, continual service can cause them to become unsafe. The department must have a maintenance program that will appropriately maintain all safety-related equipment throughout the service life of the vehicle. The maintenance program should consist of both a preventive element and a corrective element.

Warning devices used by emergency vehicles decrease the likelihood of an accident by making the presence of the vehicles obvious **(Figures 6.4 a-d)**. In the past decade, many new requirements for audible and optical warning devices have been developed. At the same time, new technology in the design of warning devices has improved the level of warning they provide. Unfortunately, many modern automobiles have become all but soundproof, and the availability of numerous gadgets, electronic and otherwise, lead to distracted drivers. It can be argued that drivers currently have less awareness of their driving environment than was the case 50 years ago.

Legal Requirements

The design of all motor vehicles is regulated to some extent by the U.S. or Canadian governments. Following an alarming rise in vehicle-related fatalities in the mid-1960s, the U.S. federal government enacted the National Traffic and Motor Vehicle Safety Act. This act gave the government the authority to mandate certain safety features on cars and trucks. Included in these features were headrests, seat belts, and laminated safety glass in windshields. As time passed, and technology improved, impact-absorbing bumpers and driver and passenger airbags were required. Many of these requirements have found their way into all types of emergency vehicles.

NFPA® 1901

Emergency vehicle design requirements date from the early years of the 20th century. As inventors and manufacturers began to develop and produce the early motor cars, attempts were made to replace hand and horse-drawn fire apparatus with motorized drive systems. The earliest requirements for the design of fire apparatus were created by the NFPA® in 1906. This was followed by the adoption of NFPA® 1901, *Automotive Fire Apparatus, Suggested Specifications for Combination Pumping*

Figures 6.4 a-d A variety of visual and audible warning devices are required on emergency apparatus including: a) roof-mounted light bars; b) directional arrows on the rear end; c) air horns mounted in or under the front bumper; and d) sirens mounted in or on top of the front bumper.

Engine and Hose Wagon, in 1914. Over the years, the NFPA® standards have expanded to include all types of fire apparatus to include pumpers, aerial devices, wildland apparatus, and aircraft rescue and fire fighting apparatus (ARFF).

As a result of firefighter fatalities and injuries in apparatus-related accidents, the NFPA® increased the safety requirements for apparatus. Personnel were prohibited from standing on the tailboard or running board of moving apparatus and fully enclosed crew compartments were added. The adoption of the 1991 edition of NFPA® 1901, *Standard for Automotive Fire Apparatus,* saw the greatest advances in apparatus safety.

Requirements for additional visual and audible warning devices, reflective trim, and improved braking systems have improved safety. A maximum noise level was established for the cab and crew compartment to reduce firefighter hearing loss. This requirement resulted in the removal of air horns and siren speakers from the cab roof, relocating them to the front bumper.

The 1991 edition of NFPA® 1901 has been recognized as the current benchmark from which the new, improved apparatus have evolved. It is recommended that only apparatus that meet the 1991 or later editions of NFPA® apparatus standards or that has been refurbished in accordance with NFPA® 1912, be allowed to operate in emergency service. This will ensure that the latest improvements and upgrades are available for the safety of the emergency responders.

Contained in NFPA® 1901, 2009 edition, were also requirements for such items as the following:

- Intersection lights for increased visibility
- A flashing light in the cab to indicating an open cab or body door
- An audible back up alarm
- An automatic transmission to allow the driver to focus on safety issues (unless the purchaser had a specific reason for a manual transmission)
- Seat belts and seats for all crew members on the apparatus
- Fail safe door handles to prevent coat sleeves from catching on the handle and opening the door
- Signs requiring everyone to be seated and belted

On apparatus containing pumps, the standard included changes to the pump panel to improve driver/operator safety. Included in these changes were specifications that 3-inch (77 mm) or larger valves be *slow close valves.* The caps were tested to 500 psi (3 447 kPa), an intake relief valve was provided to help manage incoming pressure, and all 3-inch (77 mm) and larger discharges were eliminated from the pump panel to reduce the possibility of injuries to the pump operator.

In the apparatus body area, the minimum step surface size and load-carrying capabilities were increased, handrails were required to be slip-resistant, and reflective striping was required on all four sides of the apparatus. Electrical system requirements for line voltage were upgraded to require the use of Underwriters Laboratories Inc. (UL) listed components that are grounded.

Many requirements were added to increase the operating capabilities of all aerial devices including water towers equipped with ladders. For aerial ladders, the requirements were as follows **(Figure 6.5)**:

- Minimum design strength of the rungs was increased

- Height requirement for the handrails was specified

- Minimum load-carrying requirement for folding steps was specified

- Minimum carrying capacity was 250 pounds (113.40 kg) at the tip at 0 degrees elevation at maximum extension

The carrying capacity was elevating platforms at 0 degrees full extension was raised to 750 pounds (340.19 kg). Elevating platforms were also required to have handrails, an escape ladder from the platform, and a water curtain cooling system under the platform.

All aerial devices must be capable of supporting a static load of 1½ times their rated capacity in any position. A requirement for a stabilizer movement alarm and reflective striping with warning lights was added. Interlocks to prevent inadvertent movement to an unsupported side and to prevent raising the aerial device prior to the stabilizers being deployed were specified. 100 percent nondestructive tests became a requirement with increased safety and strength of materials being required.

Figure 6.5 Warning labels are applied to aerial devices by the manufacturer. *Courtesy of Pat McAuliff.*

Any apparatus, whether in first-line or reserve service, should have the following features as a minimum:

- Fully enclosed seating is provided for all members riding on the fire apparatus

- Warning lights meet the current standard

- Reflective striping meets the current standard

- Slip resistance of walking surfaces and handrails meets the current standard

- A low-voltage electrical system load manager is installed if the total continuous load exceeds the alternator output

- Where the gross vehicle weight rating (GVWR) is 36,000 pounds (16 329.33 kg) or more, an auxiliary braking system is installed and operated correctly

- Ground and step lights meet the current standard

- Noise levels in the driving and crew compartment(s) meet the current standard or ear protection is provided

- Brakes, brake lines and wheel seals have been serviced in accordance with the manufacturers' maintenance schedule

- Tires and suspension are serviceable

- All horns and sirens are relocated from the roof to a position as low and far forward as possible

- Seat belts are available for every seat and are in serviceable condition **(Figure 6.6, p. 186)**

- Sign plates are present stating no riding on open areas

- A complete weight analysis has been completed assuring that the fire apparatus is not over individual axle or total GVW ratings

- The alternator output meets its rating

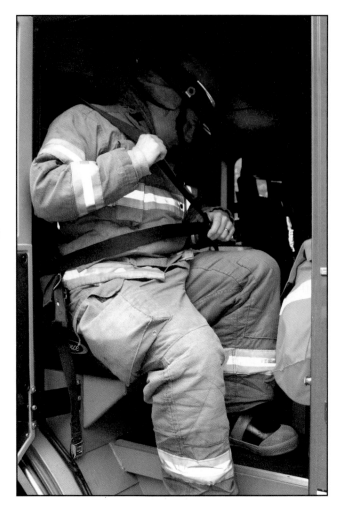

Figure 6.6 The installation and use of seatbelts in emergency apparatus is mandatory by law.

- Water tank and baffles are not corroded or distorted
- A transmission shift pump interlock is present and working properly on vehicles equipped with an automatic transmission
- All loose equipment in the driving and crew areas is securely mounted to prevent its movement in case of an accident, abrupt braking, or change of direction.
- If so equipped, the generator and line voltage accessories are tested and meet current standards.
- If equipped with an aerial device, a complete test to original specifications has been conducted and certified by a certified testing laboratory

These standards have been adopted into law by the U.S. Department of Transportation (DOT) making them mandatory for the design of all fire apparatus manufactured in the U.S. The U.S. Code of Federal Regulations (CFR) contains regulations for the design of all types of motor vehicles in 49 CFR 393, *Federal Motor Carrier Safety Regulations* (FMCSR), and 49 CFR 571, *Federal Motor Vehicle Safety Standards* (FMVSS). These federal government requirements are enforced by the DOT. The regulations govern such design components as the gross vehicle weight (GVW), braking power, warning devices or lights, and number of axles, to name a few.

The Federal Aviation Administration (FAA) enforces regulations that describe the minimum requirements for aircraft fire rescue apparatus based on NFPA® 414, *Standard for Aircraft Rescue and Fire-Fighting Vehicles*. In addition, NFPA® 1500, 1901, and 1912, among others, were adopted by the U.S. Department of Homeland Security (DHS) in September 2007.

State departments of transportation also contain laws that affect apparatus design. Some of these, such as those enacted in California, are stricter than federal laws. For instance, states may require additional axles to carry the increasing weight of fire apparatus. Other laws may regulate the use of engine retarding brake systems or the color of warning lights.

KKK-A-1822F Ambulance Specifications

Ambulances have their own design standards developed by the federal government. They are found in the Federal Specifications, KKK-A-1822F for the Star-of-Life Ambulance. This document establishes five classifications of ambulances **(Figures 6.7 a-e, p. 188)** which include the following:

- *Type I Ambulance (10,001 TO 14,000 GVWR)* — Type I vehicle shall be a cab chassis furnished with a modular ambulance body.

- *Type I-AD (Additional Duty) Ambulance (14,001 GVWR or more)* — Type I-AD shall be a Cab-Chassis with modular ambulance body, increased GVWR, storage, and payload.

- *Type II Ambulance (9,201 to 10,000 GVWR)* — Type II ambulance shall be a long wheelbase Van, with Integral Cab-Body.

- *Type III Ambulance (10,001 to 14,000 GVWR)* — Type III shall be a Cutaway Van with integrated modular ambulance body.

- *Type III-D (Additional Duty) Ambulance (14,001 GVWR or more)* — Type III-AD shall be a Cutaway Van with integrated modular body and increased GVWR, storage, and payload.

Canadian ambulances are built to the Star-of-Life standard although no nationally adopted requirement exists. Provinces have adopted various requirements such as the Star-of-Life standard.

Refurbishment

While all of these standards and laws regulate the manufacture of new emergency apparatus, the NFPA® has also created standards on the refurbishment of older vehicles. NFPA® 1912, describes the minimum requirements for refurbishing apparatus. Recognizing that small career and volunteer fire departments may not be able to afford a complete overhaul of an aging apparatus, NFPA® created two options for refurbishment. The first option, termed a Level I refurbishment, describes the requirements for the assembly of a new apparatus using the following new components:

- Chassis frame

- Driving and crew compartments

- Front axle

- Steering and suspension components

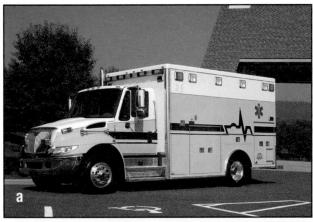
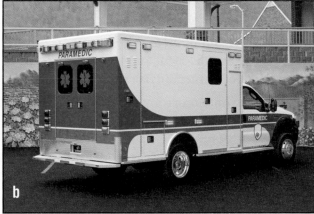
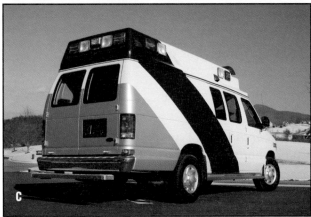
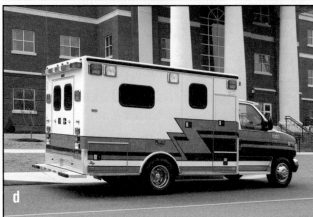
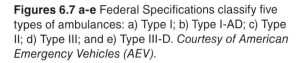
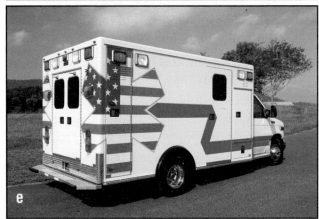

Figures 6.7 a-e Federal Specifications classify five types of ambulances: a) Type I; b) Type I-AD; c) Type II; d) Type III; and e) Type III-D. *Courtesy of American Emergency Vehicles (AEV).*

The remainder of the apparatus may be constructed from new or existing components, such as a pump, aerial device, hose bed, tool compartments, and rear axle, to name a few.

The Level II refurbishment consists of new components added to the existing apparatus chassis. The new components, such as front axle, engine, or crew compartment, must meet or exceed the requirements for the original manufacture of the apparatus.

The standard also requires that all current safety requirements of NFPA®1901 and 1906, *Standard for Wildland Fire Apparatus* must be included on either Level I or II refurbishments.

Fire and emergency services managers must remember that attempting to build apparatus without professional design help can lead to illegal and unsafe results. Fatalities and injuries have occurred as a result of vehicle rollovers of homemade water tenders (tankers), which were often former flammable liquid carriers. Consult an apparatus design professional before attempting to modify any vehicle for fire fighting operations.

Safety Components

When writing specifications for a new apparatus or for the refurbishment of an existing apparatus, it is essential to consider the types of safety components that are required by law. At the same time, some safety equipment not mandated by law may be worth the additional funds necessary to install them. Safety components should be included on all vehicles and apparatus that respond to emergencies. In addition, they may prove to be beneficial to some support, staff, and other nonemergency vehicles. Safety components may include the following:

- Audible warning devices
- Optical warning devices
- Passive visibility markings
- Rearward- and forward-facing video cameras
- Electronic data logging devices
- Position tracking devices
- Traffic preemptive control devices
- Auxiliary braking systems
- Antilock braking systems

In addition, *ergonomic* factors must be taken into consideration in the apparatus design. As storage compartments are located higher on the apparatus and tools and equipment become heavier and larger, it is essential to plan for the safest location of this equipment. Skeletal and muscular injuries often result when personnel are removing tools and equipment from apparatus compartments. Apparatus manufacturers are designing apparatus that address ergonomic factors by lowering hose beds, providing steps for cab and body access, and pullout compartments for improved access. This issue becomes even greater significance with the hiring of people who are of smaller stature by fire and emergency service organizations.

Ergonomics — The applied science of workplace equipment design intended to maximize productivity and minimize potential injuries by reducing operator fatigue and discomfort.

Audible Warning Devices

Audible warning devices have been used on ambulances and fire apparatus for close to 100 years. Audible warning devices have included bells, exhaust or steam whistles, public address systems, air horns, and sirens. Bells and whistles are rare on modern apparatus though they may be present on restored vintage equipment that is no longer used for emergency operations. A public address system can be used to alert pedestrians in congested areas. Air horns, while effective, have the disadvantage of depleting the air pressure required for the air brakes. Current apparatus designs include auxiliary air tanks for the air horns. Sirens may be electronic, electric, or manual and designed to emit a variety of tones.

When considering the installation of audible warning devices, two safety considerations are important **(Figure 6.8)**. First, the device must have the ability to alert the public to the approach of the emergency vehicle. That means the devices must be loud enough to be heard over other ambient noises as well as the barriers caused by vehicles with closed windows. The federal standard for ambulances states that the minimum sound level at 10 feet (3 m) must be 123 dB. The device must also project its tone at an angle that is wide enough to be heard by drivers in cars that are not directly in front of the emergency vehicle; that is, the warning must be audible to vehicles that are approaching the emergency vehicle's line of travel from a side road. Finally, the tones used by the audible warning device must be easily associated with an emergency vehicle.

The second safety consideration is for the driver/operator and passengers of the emergency vehicle. The tones emitted by the audible warning device must not be so loud within the cab that they cause temporary or permanent hearing loss. The NFPA® has established a maximum sound level of 90 dB inside the cab of fire apparatus. The federal ambulance standard states that the maximum sound level in the cab must conform to 29 CFR 1910.95 unless the state or municipality sets the level. In the patient area, the maximum is 80 dB.

NOTE: In some jurisdictions, the use of air horns and sirens may be prohibited or restricted based on the time of day or location.

To meet these sound levels and to protect the hearing of crew members, NFPA® 1901 prohibits mounting audible warning devices on the roof of the apparatus cab. Air horns and siren speakers may be located on the frame beneath the bumper, in the bumper when openings are provided, on the bumper, or on the front fenders.

Studies have shown that civilian drivers respond better to sounds that frequently change pitch. Short bursts with the air horns and the constant up-and-down oscillation of a mechanical or electronic siren are the surest ways to catch a driver's attention. At speeds above 50 mph (80 km/h), an emergency

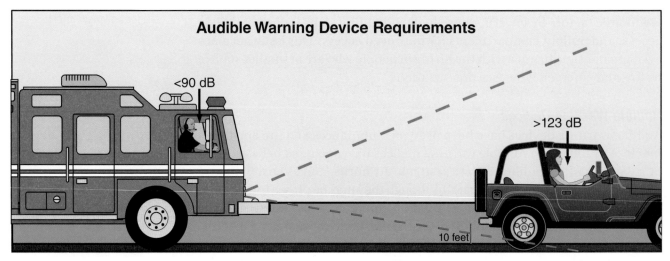

Figure 6.8 Audible warning devices must generate a minimum of 123 dB at 10 feet while projecting the sound at an angle to be heard by vehicles approaching from the side and create a noise level not to exceed 90 dB in the cab of the apparatus.

vehicle may over take the effective range of its audible warning device. One study indicated that a siren operating on an emergency vehicle moving at 40 mph (64 km/h) can project 300 feet (90 m) in front of the vehicle. At a speed of 60 mph (97 km/h), however, the siren is only audible 12 feet (3.7 m) or less in front of the vehicle. Driver/operators must drive within the effective range of their audible warning devices.

Optical Warning Devices

Optical warning devices used on emergency vehicles have two primary purposes: to alert other drivers and pedestrians of the approach of an emergency vehicle and to indicate the location of the vehicle when it is stopped at the incident scene. Optical warning devices consist of flashing lights mounted on the body of the vehicle. NFPA® 1901 divides each vehicle into four zones **(Figure 6.9)**. The standard specifies both the color of the lights and the flashing rate for each zone. In addition to the optical warning devices, the headlights, parking lights, and taillights can be connected to a device that causes them to flash when the primary optical warning devices are activated. State law should be consulted prior to specifying optical warning devices.

Passive Visibility Markings

To improve the visibility of emergency vehicles when they are at an incident, NFPA® also requires the application of retroreflective trim on the vehicle body to meet ASTM D4956, *Standard Specification for Retroreflective Sheeting for Traffic Control*, Type I, Class 1 or Class 3. The retroreflective trim must extend across a minimum of 25 percent of the front of the vehicle and 50 percent of the sides and rear. The trim must be a minimum of 4 inches (100 mm) in width. The retroreflective trim should also be applied to the inside of cab doors so that they will be visible when open at night **(Figure 6.10, p. 192)**.

The choice of vehicle or apparatus colors has been discussed for many years. The traditional *Fire Engine Red* continues in use throughout North America although some departments use other colors. During the 1970s, studies on apparatus visibility at night indicated that red tended to disappear into the darkness. As a result, lime green and white were chosen to replace red. By the 1990s, red was in use again and lime green had lost favor. Ambulances tend to be red or white depending on local preferences. There is no national

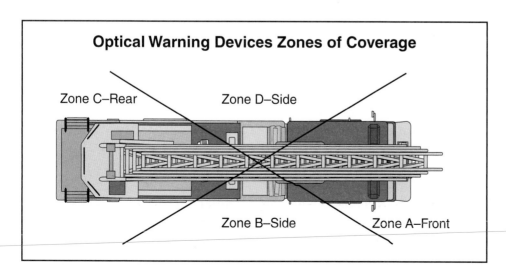

Optical Warning Devices Zones of Coverage

Zone C–Rear   Zone D–Side

Zone B–Side   Zone A–Front

Figure 6.9 NFPA® 1901 has designated four zones around the apparatus that have specific visual warning light overage.

Figure 6.10 Recent studies have indicated the effectiveness of retroreflective stripes on apparatus.

Figure 6.11 Aircraft rescue and fire fighting apparatus are generally painted a distinctive tone of yellow.

requirement regarding color of community fire apparatus, however, the FAA does include color of aircraft rescue and fire fighting apparatus (ARFF) in its requirements **(Figure 6.11)**.

Rearward- and Forward-Facing Video Cameras

The increase in digital video technology has resulted in the development of apparatus-mounted video cameras. The cameras provide a number of benefits including the ability of the driver/operator to see any obstructions or people directly behind the apparatus. The rearward-facing camera, however, does not eliminate the need for a radio-equipped member of the crew from directing the rearward movement of the vehicle. These cameras should be installed so that the screen is not visible while the vehicle is in forward motion. Having a monitor display in the cab while moving is another unnecessary distraction.

The forward-facing camera can be used to provide a view of the emergency scene to other responding units or to a central command center. In the event of an accident, the video record of the incident can help to determine the cause of the accident.

Electronic Data Logging Devices

Electronic data logging devices have been mandatory on passenger busses for many years. These devices are used to determine the number of hours that a driver has been working (restricted by the DOT) and the speeds the vehicle has been traveling. Information can also be gathered to diagnose the operation of the vehicle's engine that can be used to determine the need for maintenance. Additional information that can be gathered on the use of seat belts, wheel speeds, foot pressure on accelerator or brake, and other data can be crucial for accident investigations.

Vehicle Tracking Devices

Vehicle tracking systems can provide essential information for the telecommunications or dispatch center. By knowing the location of all apparatus, the center can dispatch the unit that is closest to the emergency incident. At the same time, the ability to track responding units can provide an additional level of safety. Responding units can be tracked and their relative positions reported to reduce the chance of a collision. First-arriving units can also be informed of the estimated arrival time and identity of other units. When their data is recorded, these devices can also be used to closely estimate a vehicle's speed prior to an accident.

Traffic Preemptive Control Devices

Some jurisdictions use traffic control devices to assist emergency vehicles during their response. Driver/operators must be aware of the traffic control devices used in their jurisdiction and how they operate. One of the simplest involves placing a traffic signal in front of the fire station to stop the flow of traffic so that the apparatus can exit safely. This signal may be controlled by a button in the station, by the telecommunications center, or by the alert tone or signal used to dispatch the unit. Some jurisdictions have systems that control one or more traffic lights in the normal route of travel for fire apparatus **(Figures 6.12 a and b)**. These may be controlled from the fire station, remote controls on the fire apparatus, or the telecommunications center.

Another common system for controlling traffic signals for fire apparatus is the preemption device. These systems may use strobe lights (emitters) mounted on the fire apparatus to activate sensors in the traffic lights. The emitter generates an optical signal that is received by the sensor on the traffic light as the apparatus approaches. The sensor converts this signal to an electronic impulse that is routed to the phase selector in the traffic light control cabinet. The phase selector then provides a green light for the direction that the apparatus is traveling and red signals in all other directions.

In some jurisdictions, the traffic light standard may be equipped with a white light that indicates to the driver that the signal has been received and a green light is forthcoming. On some apparatus, the emitter is wired into the parking brake system. When the parking brake is set, the emitter will be turned off. On apparatus that do not have this feature, the driver/operator should remember to turn off the emitter when the apparatus is parked on the scene of an emergency. Otherwise, the emitter could affect any traffic signals that are within reach and disrupt the normal flow of traffic.

Another type of traffic control system is activated by the emergency vehicle's siren as it approaches an intersection. A microphone on the traffic signal hears the siren and sends a signal to the traffic signal controller, ordering a preemption of the current traffic signal. The microphone may be adjusted to order the preemption from distances of anywhere from a few hundred feet (meters) to about ½ mile (1 km). Intersections equipped with this system will have 3 inch (76.2 mm) white and blue lights in each direction of travel, somewhere

Figures 6.12 a and b Traffic preemptive control devices allow approaching a) emergency vehicles to control and b) intersection traffic lights, reducing the chance of a collision.

to the side of the regular traffic signals. As soon as the microphone sends the preemption signal to the signal controller, the direction of travel for the emergency vehicle gets a white light indicating that the signal was received and that a green traffic light is forthcoming. All other directions of travel get a blue light that indicates an emergency vehicle coming from one of the other directions has gained control of the signal first. This is extremely important when emergency vehicles are approaching the intersection from more than one direction. Vehicles getting the blue light know that they will have to come to a stop because a green signal is not immediately forthcoming in their direction of travel.

Regardless of the type of traffic control devices used in a jurisdiction, they are no substitutes for using proper defensive driving techniques. When entering an intersection with a green light, driver/operators must maintain a speed that will allow for evasive actions in the event another vehicle enters the intersection. If the emergency vehicle does not have a green light, the driver/operator should bring the vehicle to a complete stop at a red light before proceeding after determining it is safe to do so. Keep in mind that if two emergency vehicles equipped with preemption devices approach the same intersection from different directions, only the apparatus whose sensor activates the system first will get a green light. The later-approaching apparatus gets a red light. Do not assume that just because you did not get a green light that the system is not working. Approach the intersection with caution and come to a complete stop. When it is determined that it is safe to enter the intersection, regardless of the color of the light, proceed with caution.

Auxiliary Braking Systems

Because of the weight and increased stopping distances, fire apparatus that weigh in excess of 36,000 lbs (16 329.32 kg) are required to have auxiliary braking systems. These systems may be engine, transmission, exhaust, or driveline retarders. Because some of these retarders create loud noises, they are prohibited by local ordinances in some jurisdictions. One version of the driveline retarder is the electromagnetic braking system that augments and works in conjunction with the vehicle's conventional service brakes. These frictionless braking systems are connected to either the driveshaft or the rear axle of the vehicle. These systems can be programmed to activate when driver/operators remove their foot from the accelerator, steps on the brake pedal, or uses a manual selector lever mounted on the steering column. Under normal conditions, this system does not activate at speeds under 2 mph (3.2 kmh).

Antilock Braking System (ABS)

Most new fire apparatus are equipped with an all-wheel, antilock braking system (ABS). These systems are effective because they minimize the chance of the vehicle being put into a skid when the brakes are applied forcefully. ABS works using digital technology in an onboard computer that monitors each wheel and controls pressure to the brakes, providing straight-line braking, even under maximum application. A sensing device monitors the speed of each wheel. The wheel speed is converted into a digital signal that is sent to the onboard computer. When a wheel begins to lock, the sensing device sends a signal to the computer that the wheel is not turning. The computer compares this signal to the signals from the other wheels to determine if this particular wheel should still be turning. If it is determined that it should be turning, a signal is sent to that wheel, reducing the brake pressure and allowing the wheel to turn. Once the wheel turns, it is braked again. The computer makes these decisions more than twenty times per second, until the vehicle is brought to a halt. Thus, when driving a vehicle equipped with an ABS, maintain a steady pressure on the brake pedal (rather than pumping the pedal) until the apparatus is brought to a complete halt.

Some of the newer vehicles are using the ABS as the foundation for additional systems such as Traction Control and Stability Control. As implied, traction control assists the driver of the vehicle stuck in the mud or on slick roadways by applying the brake to the spinning wheel. Using some of the same basic principles and equipment, Stability Control allows the vehicle to measure wheel speed and other forces applied to the vehicle while operated in high-speed turns. The computer-operated system automatically applies specific brakes and may temporarily disable the accelerator to assist the driver/operator in maintaining control of the vehicle. Combined with the skills of the driver/operator, the Stability Control system reduces the likelihood of a rollover.

Maintenance and Inspections

A proactive emergency vehicle maintenance program has the benefits of ensuring that the vehicles are safe to operate and that the jurisdiction will have a good return on its investment. NFPA® 1500 mandates a preventive maintenance program supported by accurate record keeping. The program elements are based on NFPA® 1915, *Standard for Fire Apparatus Preventive Maintenance Program.* The success of the program depends on a commitment by all levels of the fire and emergency services organization: the administration that develops and implements it, the supervisors who enforce it, and the members who actively participate in it.

Besides the preventive maintenance, the organization will also need to have a corrective maintenance capacity. Corrective maintenance is reactive in nature, responding to wear, tear, and damage that has already occurred. A key to corrective maintenance is determining if the cost of the repair can be justified or if the vehicle should be replaced. If a vehicle is nearing the end of its service life, then it may be more cost-effective to replace the vehicle than to repair it. Further, the reliability of fire apparatus should be a key factor in this decision.

Preventive Maintenance

It is extremely important that all emergency and non-emergency vehicles in the organization function safely at all times. It is the responsibility of everyone within your organization to ensure all equipment performs well when needed.

The apparatus driver/operator is primarily responsible for ensuring that assigned department vehicles perform effectively **(Figure 6.13)**. A daily apparatus inspection checklist may be designed to assist the driver/operator to accomplish the task of filling out the necessary information needed to make competent repairs. Check sheets

Figure 6.13 Apparatus driver/operators perform daily inspections of their equipment at the beginning of the work shift.

may not need to be designed to include each and every item that is necessary to inspect, but to act as a guide to ensure that the inspector checks the most important items. As a result, timely inspections and repairs can be made.

Corrective Maintenance

An effective preventive maintenance program will ensure that most but not all repairs are made before a serious breakdown occurs. Unforeseen damage can occur at anytime, resulting from metal fatigue, collisions, or operator error. When these incidents occur, the organization must have a procedure for notifying administrative officers (including the safety officer), mechanics, and tow truck operators. Depending on the severity of the damage, it may be necessary to remove the unit from service and replace it with a reserve vehicle.

Inspections and Tests

Both career and volunteer fire and emergency services organizations should have SOPs that establish the frequency and level of apparatus inspections. NFPA® 1500 requires a minimum of once a week inspections. Career organizations usually require inspections at shift change or daily. A more thorough inspection may be scheduled weekly when the driver/operator performs major cleaning of the vehicle. Some departments schedule tool and equipment inspection on different days than apparatus inspection, especially on aerial ladder and rescue vehicles. In keeping with NFPA® requirements, apparatus are also inspected within 24 hours after any use and immediately after major repair and prior to being placed into service. Semiannual or annual apparatus inspections are generally performed by maintenance personnel when the vehicle has its scheduled preventive maintenance.

Tests for all types of fire apparatus are specified in NFPA® 1911, *Standard for the Inspection, Maintenance, Testing, and Retirement of In-Service Automotive Fire Apparatus.* Two types of tests are specified: acceptance and in-service tests.

Acceptance tests occur when a new or refurbished apparatus is delivered by the manufacturer or maintenance center. The tests include measuring the mechanical, pneumatic, hydraulic, or electrical systems' characteristics and comparing them to the equipment specifications and the required design standards.

Service tests are performed by a third party or the organization annually or after any major repairs or alterations to the apparatus or component. Annual service tests are required for the following items:

- Fire pumps **(Figure 6.14)**
- Aerial devices
- Compressed air foam systems
- Foam proportioners
- Road tests of the chassis components
- Breathing air compressors

In addition to the annual tests, a verification that the weight of the vehicle has not changed since it was constructed is also required. The standard defines how apparatus is to be loaded and requires that each axle be weighed separately and then the entire apparatus be weighed as a whole. If the units exceeds the corresponding weight limit, changes must be made to the vehicle or the unit is taken out of service.

In addition to the required annual test, aerial devices are required to be tested following any use that may have placed it under severe strain or if it is believed that the manufacturer's operational recommendations have been surpassed. The tests are called nondestructive tests and must be performed by a certified technician.

Apparatus Service Life

In the fire service, some fire apparatus with 8-10 years service are worn past the point of cost-effective repair/maintenance. There are also fire apparatus that were built with quality components that had excellent maintenance and have responded to a minimum number of runs that are still serviceable after 20 or more years. It is generally accepted that the quality and timeliness of preventive and corrective maintenance are perhaps as important as the use factor in determining how well an emergency vehicle ages.

Even with an aggressive preventive maintenance program, apparatus will reach a point where continued operation is unwise from the standpoint of safety, reliability, and cost-effectiveness. NFPA® 1911 does not define a specific age at which apparatus must be replaced, stating only that safety is the primary consideration. Determining when apparatus should be replaced is based on many factors including the following:

Figure 6.14 Apparatus fire pumps are tested annually to ensure proper operation.

- Safety
- Mileage
- Age
- Aging design or technology
- Government-mandated safety or design changes
- Cost-effectiveness of refurbishment
- Quality of the preventative maintenance program
- Quality of the original builder and components
- Availability of parts

Record Keeping

Justifying apparatus refurbishment or replacement is based on accurate and thorough record keeping. NFPA® requires organizations to maintain records on all fire and emergency apparatus as part of the records management system **(Figure 6.15, p. 198)**. The department should have data containing all information collected on each emergency and nonemergency vehicle owned, leased, or assigned to it. Information maintained in the file should include, but is not limited to, the following:

APPARATUS MAINTENANCE CHECKLIST

For use of this form, see AR 420-90; the proponent agency is ACSIM

APPARATUS	MONTH	CALENDAR YEAR

SECTION A - DAILY INSPECTION STATUS *(see legend below)*

L I N E	ITEM INSPECTED	DAYS OF THE MONTH 1-31
1	FUEL *(fill at 3/4)*	
2	ENGINE OIL	
3	RADIATOR	
4	TIRE AIR PRESSURE	
5	LIGHTS	
6	BATTERY	
7	FAN BELT	
8	BRAKES	
9	AIR PRESSURE *(drain tanks)*	
10	CLUTCH	
11	WINDSHIELD WIPERS	
12	START ENGINE	
13	WARNING DEVICES	
14	ENGAGE PUMP	
15	INSTRUMENT PANEL	
16	FIREFIGHTING AGENTS	
17	RELIEF VALVE SET *(150 psig)*	
18	PRIMER OIL	
19	FLUSH PUMP *(monthly)*	
20	VACUUM TEST *(monthly)*	
21	DRAFTING OPERATION *(annual)*	
22	GRID MAPS	
23	JAWS OF LIFE *(weekly operation)*	
24	HOSE CHANGE *(pumper - monthly)*	
25	K-1 2 SAW *(weekly operation)*	
26	RADIO CHECK	
27	APPEARANCE	
28	OPERATOR'S INITIALS	
29	SUPERVISOR'S INITIALS	

Column headers across: 1 2 3 4 5 6 7 8 9 10 11 12 13 14 15 16 17 18 19 20 21 22 23 24 25 26 27 28 29 30 31

LEGEND: Daily Inspection Status
(Enter status of item inspected in columns above)

1 = OK	5 = NEEDS ADJUSTMENT	8 = INCOMPLETE
2 = LOW	6 = NEEDS REPLACING	9 = MISSING
3 = DAMAGED	7 = MALFUNCTION	10 = FIRE DEPARTMENT RESPONSIBILITY
4 = DIRTY		11 = MAINTENANCE RESPONSIBILITY

DA FORM 5379-R, JAN 85

USAPA V1.01

Figure 6.15 Sample apparatus maintenance form.

198　Chapter 6 • Fire Apparatus Safety

- Safety-related issues
- Accident involvement
- Reliability
- Original specifications
- Acceptance test results
- Vehicle assignment
- Unit inspection records
- Preventive maintenance records
- Major overhauls, refurbishments, or repairs
- Purchase and maintenance costs

Driver/Operator Requirements

Operating fire and emergency services vehicles requires the driver/operator to be in good health and physically fit and alert. The driver must be mentally alert and able to respond to changing traffic conditions. Using defensive-driving skills can reduce the likelihood of an accident.

During emergency responses in apparatus, the driver/operator must exercise even greater care. The weight and speed of the apparatus may dictate longer stopping distances and make control on rain or ice covered road surfaces more difficult.

Responding to emergency incidents in POVs is common in volunteer and combination type departments. The use of POVs presents its own special concerns, including awareness issues with the public, ability to audit vehicle condition, driver training, and liability.

Health and Physical Fitness Testing

The public image of the fire and emergency services places the firefighter and medical technician in the awkward position of having to meet sometimes unrealistic expectations. We are expected to perform heroic feats under conditions of extreme danger and stress. In theory, we should be in better physical condition and be far better trained than the majority of the community we serve.

There is no reason why the community's expectation of a high-performance level should not be reflective of efforts to achieve and maintain good health and physical fitness. There are many health and physical fitness programs that are sponsored by organizations like the International Association of Fire Fighters (IAFF) and International Association of Fire Chiefs (IAFC) and used by emergency service organizations to help personnel stay fit. For driver/operators, many states have requirements when driving certain vehicles in which a medical exam is required **(Figure 6.16)**.

Regardless, it is the responsibility of each member of the organization to monitor their personal health and fitness. If your organization does not promote health and welfare of its employees, take the lead and bring in a program that is equitable to the organization.

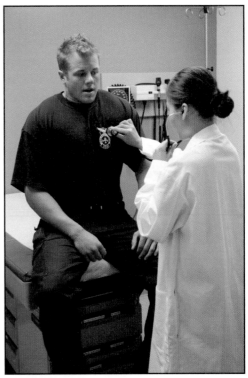

Figure 6.16 Many states require fire and emergency services driver/operators to undergo periodic medical examinations.

Certification

Each state/province has its own requirements for the licensing of emergency vehicle driver/operators. A commercial driver's license (CDL) is a special license issued by the state/province that allows a person to operate certain types of vehicles. For instance, a CDL is required in order to drive the following:

- A single vehicle with a gross vehicle weight rating (GVWR) of more than 26,000 pounds (11 793.4 kg)

- A trailer with a GVWR of more than 10,000 pounds (4 535.9 kg) if the combined weight of the vehicle and trailer is more than 26,000 pounds (11 793.4 kg)

- Any size vehicle that is required to display hazardous materials placards.

At the same time, the majority of states/provinces exempt personnel assigned to operate ambulances and fire apparatus from possessing a CDL. A regular driver's license is all that is required for this activity. This exemption does not permit a fire department driver/operator to also drive a support vehicle that requires a state/province CDL. Therefore, a member of the fire department who is assigned to operate the fuel truck or supply truck may be required to have a CDL.

Many states/provinces and emergency services organizations are adopting a *zero-tolerance* policy toward driver/operators regarding the use of alcohol or drugs. The zero-tolerance concept was originated by the IAFC in 2004 in response to an increasing number of vehicle accidents that resulted from the use of alcohol. The basic policy states that if a person has consumed alcohol or still appears impaired by the effects of alcohol consumed within the previous 8 hours, that person may not engage in fire-fighting activities, including the operation of an emergency vehicle.

Volunteer, combination, and career fire and emergency services organizations are encouraged to adopt a zero-tolerance policy and enforce it. In addition, arrests and convictions for driving under the influence (DUI) or driving while intoxicated (DWI) should be considered when assigning an individual as an emergency vehicle driver/operator. Loss of a regular driver's license is valid reason to remove a person from the position of driver/operator.

NOTE: The state/province issues the appropriate driver's license to an individual based on the type of vehicle to be driven. It is up to the fire and emergency services organization to confirm that the license is appropriate for operation of an emergency vehicle and to ensure that the license is still in affect.

Training

The driver/operator of the unit is responsible for the safe operation and transportation of occupants riding in the unit. The company officer shares responsibility for the safe and efficient operation of the unit. This responsibility includes the training and consistent application of rules and regulations with regards to vehicle safety, apparatus placement, vehicle inspections, and record keeping. The administration of the organization is responsible for providing, either internally or through other agencies, training for all driver/operators. In some cases, company officers may be tasked with verifying the ability of personnel to operate the vehicles within their command.

Minimum qualifications for driver/operators can be found in NFPA® 1002, *Standard for Fire Apparatus Driver/Operator Professional Qualifications.* It is common in fire service organizations to provide a driver/operator program, which may be a template of a local, regional, state/provincial, or federal program designed to prepare personnel to operate emergency vehicles in both in emergency and nonemergency situations. The program will detail the minimum requirements needed to certify or qualify personnel with enough knowledge, skills, and ability to operate emergency vehicles.

Driver/operator training courses generally consist of classroom presentations, driving exercises, and a final driving examination. NFPA® 1002 specifically defines four exercises including alley dock (4.3.2), serpentine (4.3.3), three-point turn (4.3.4), and diminishing clearance (4.3.5). There are additional exercises that can be used to teach emergency vehicle operation and to test the student. They apply to all types of vehicles, not just fire apparatus. Other exercises specific to tractor-trailer, tiller-operated, aircraft rescue and fire fighting, and water tender apparatus should also be developed as needed by the local department. The order in which these exercises are conducted may be modified. The exercises include the following **(Figures 6.17 a-e, p. 202-203)**:

- Alley dock
- Serpentine
- Three-point turn
- Diminishing clearance
- Slow-speed lane change
- Braking while turning
- Straight line braking
- U-turn
- Right-side road turn
- Left-side road turn

Emergency equipment has and will continue to improve with regards to safety equipment that enables the operator to better handle the vehicle in all situations. Antilock braking system (ABS) braking coupled with engine braking systems can wreak havoc on an unfamiliar operator who lacks the knowledge of the vehicle's specific braking system. The turning radius of various vehicles may differ, as can the turning ratio (steering-wheel revolutions to angle-of-wheel turn), which can affect the ability for gaining access and egress, especially in off-road operations, steep grades, or slippery conditions. Improper upshifting/downshifting can cause the loss of traction, loss of power, and possible driveline failure if the vehicle is not driven and operated in the correct manner.

Defensive-Driver Training

Sound defensive-driving skills are one of the most important aspects of safe driving. Every driver/operator should be familiar with the basic concepts of defensive driving. They include the following:

- anticipating other drivers' actions
- estimating visual lead time

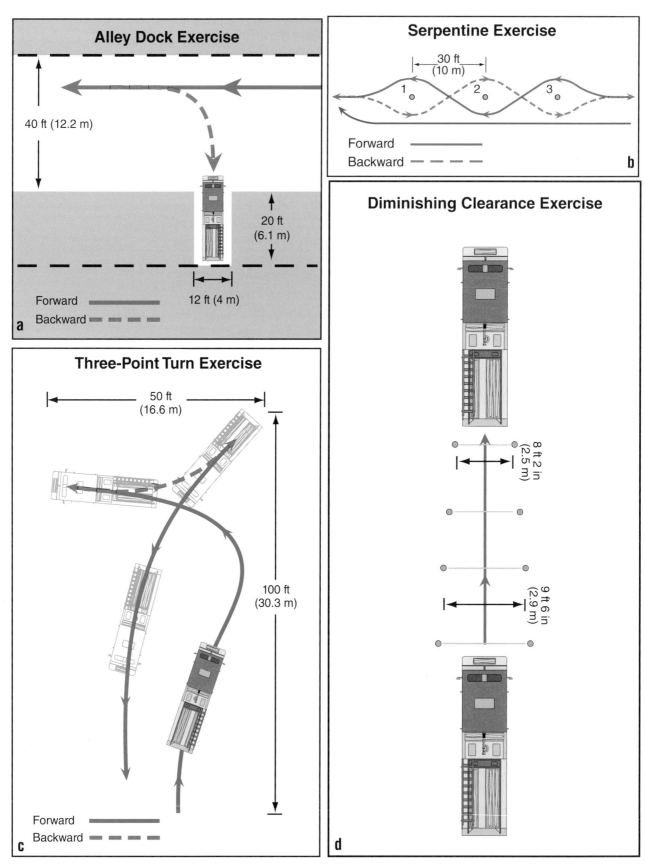

Figures 6.17 a-e NFPA® specifies five driving exercises that can be used to train and test driver/operator skills including a) alley dock; b) serpentine; c) three-point turn; d) diminishing clearance; and e) lane change, p. 203.

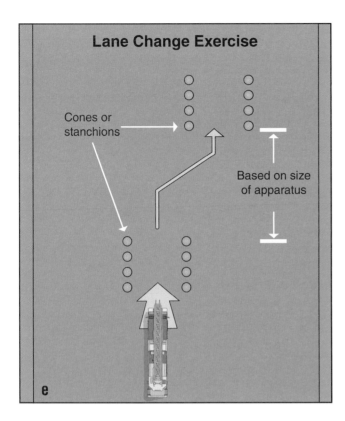

Lane Change Exercise

Cones or stanchions

Based on size of apparatus

Figure 6.17 e

- knowing evasive tactics
- knowing braking and reaction times
- combating skids
- having knowledge of weight transfer

The first three topics are covered in this section. The last three may be found in the section on apparatus response.

Driver/operators should know the rules that govern private vehicles when emergency vehicles are responding with warning lights and audible devices operating. Motor vehicle laws in most jurisdictions provide that private vehicles must pull to the right, stop, and remain at a standstill until the emergency vehicle has passed. However, this does not mean that driver/operators can ignore stopped vehicles — especially when two or more pieces of fire apparatus are responding in the same direction on the same street. After the first piece of apparatus passes, the citizen driver may pull back into the traffic lane without looking.

In addition, private vehicle drivers may panic at the sound of an approaching siren and simply accelerate to stay ahead of the emergency vehicle, pull to the left, or stop suddenly in the traffic lane. Other drivers may not hear sirens or air horns because of loud car radios, wearing headphones, closed windows, air-conditioning, or other noise-producing devices. Also, some drivers may be distracted by cell phones **(Figure 6.18, p. 204)**, manipulating radio controls, or tending to small children. Or they may simply choose to ignore emergency warning devices.

Traffic waiting to make a left-hand turn may pull to the right or left, depending upon the driver. In situations where all lanes of traffic in the same direction as the responding apparatus are blocked, a departmental SOP may allow the

Figure 6.18 Sending a text message is a distracting and dangerous activity that results in an increasing number of motor vehicle accidents.

apparatus to be driven into the opposing lane of traffic to proceed through the intersection at an extremely reduced speed. In some departments, an SOP requires that the street or road be blocked by law enforcement before fire apparatus is driven in an opposing traffic lane. Oncoming traffic must be able to see the approaching apparatus. Full use of warning devices is essential.

Driving in the oncoming lane is *not recommended* in situations where oncoming traffic is unable to see the apparatus for any reason (for example, traffic approaching from a side street or when approaching the crest of a hill). Be alert for traffic that may enter from side roads and driveways. Traffic coming over the crest of a hill, slow-moving traffic in both directions, and other responding emergency vehicles must be closely monitored.

Even though the use of sirens, lights, and other warning devices is essential, fire apparatus driver/operators should realize that these signals might be obscured or muffled by other warning devices and street noises. This is especially true when two or more emergency vehicles are approaching the same intersection from different directions. Serious collisions and fatalities have been caused by overreliance on warning devices.

Anticipating other drivers' actions. Never assume what another driver's actions will be. Instead, expect the unexpected. Anticipation is the key to safe driving. Always remember the following control factors:

- *Aim high in steering*: Find a safe path well ahead.

- *Get the big picture*: Stay back and see it all.

- *Keep your eyes moving*: Scan — do not stare.

- *Leave yourself a way out*: Do not expect other drivers to leave you an out (escape route). Be prepared by expecting the unexpected.

- *Make sure others can see and hear you*: Use lights, horn, and signals in combination.

Visual lead time. The concept of visual lead-time refers to driver/operators scanning far enough ahead of the apparatus, for the speed it is being driven, to assure that evasive action can be taken if it becomes necessary. For example, if a driver/operator is concentrating on vehicles that are 100 feet (30 m) in front of the apparatus, and based on the speed the apparatus is being driven it would take 200 feet (60 m) for it to stop or perform an evasive maneuver, a collision is likely to occur. The driver/operator needs to learn to match the speed of travel with the distance being surveyed ahead of the vehicle. Visual lead time interacts directly with reaction time and stopping distances. By *aiming high in steering* and *getting the big picture*, it is possible to become more keenly aware of conditions that may require slowing or stopping. Some departments train their driver/operators to practice looking 12 seconds ahead on city streets and 20 seconds ahead on highways and estimating where they will be in that time.

Visual lead time — The concept that driver/operators should scan far enough ahead of the apparatus, for the speed it is being driven, to assure that evasive action can be taken if it becomes necessary.

Driver/Operator Testing

After driver/operators have been selected and trained, their performance should be evaluated by some standard method. These evaluations should occur before the driver/operator is allowed to operate the apparatus under emergency conditions. NFPA® 1002 provides some specific directions on how driver/operator candidates should be tested. These directions need to be followed by agencies that certify their personnel to the standard. Other agencies should at least follow the standard to avoid possible civil law liabilities should the driver/operator be involved in a collision.

Figure 6.19 Driver/operators are tested on their skills in handling the apparatus.

Most agencies use a combination of written and practical testing for driver/operator candidates **(Figure 6.19)**. On volunteer fire departments, successful completion of this testing may simply result in making the person qualified to drive the fire apparatus. On career fire departments, this may be a promotional exam.

All fire apparatus training and testing should follow the requirements contained in NFPA® 1451, *Standard for a Fire Service Vehicle Operations Training Program*.

Driver-Training Programs

Besides the driver-training programs developed by the fire and emergency services organization, other programs are available through both nonprofit and for profit organizations. Among these are the following:

- **Smith System** — Commercially available training organization that provides custom-made driver/operator training programs as well as online training. They offer the concept of the *Five Keys of Space Cushion Driving*, which focuses on the core driving fundamentals of space, visibility, and time.

- **National Safety Council** — NSC offers a defensive driving course (DDC) in 4-, 6-, 7-, or 8-hour packages. These courses are also offered online. In addition the NSC offers the *Coaching the Emergency Vehicle Operator*™ (CEVO™) course.

- **Volunteer Firemen's Insurance Services® (VFIS®)** — The VFIS offers its members an emergency vehicle operator's training program.

- **DOT** — Federal and state departments of transportation in the U.S. offer driver-training courses and materials. The U.S. DOT has also developed the National Standard Curriculum for Ambulance Drivers.

- **National Highway Traffic Safety Administration (NHTSA)** — The NHTSA offers the Emergency Vehicle Operations Course (EVOC) for ambulance drivers. This program meets the requirement that all ambulance drivers undergo 40 hours of drivers' training that incorporates classroom training.

Medical

Operating vehicles during emergency conditions requires personnel to be physically fit and healthy. The ability to control large, heavy apparatus while contending with self-imposed stress can be difficult at best. Driver/operators must be physically and mentally fit to perform this function.

Medical fitness is facilitated by good nutrition, frequent physical examinations, and preventive precautions to illnesses and infections. You must remember, however, that some nonprescription and many prescription medications can impair your ability to operate a vehicle. If you are taking medication, it may be necessary to request a temporary assignment that does not involve the operation of a vehicle or power tools.

Physical fitness is essential to all firefighters. Fatalities due to heart attacks and injuries resulting in strains and sprains can all be reduced by good physical fitness. Model physical fitness programs are available based on job-related activities. These programs should include exercises that will improve cardiovascular functions, range of motion, upper body strength, and pulmonary capacity. Other programs for stress reduction are also important to good health.

Regardless, it is the responsibility of each member of the company to monitor their personal health and fitness testing and programs. If your organization does not promote health and welfare of its employees, take the lead and bring in a program that is equitable to the organization.

Vehicle Operation

Each and every one of us who respond to emergencies or incidents knows that it is one part of the many inherent risks we take as emergency services providers. We must also remember that as apparatus driver/operators we must require ourselves and the other vehicle occupants to take part in the safe and efficient operation when it comes to vehicle operation whether it is a crew of two, company of four, or a bus loaded with fire control laborers.

It is generally required that all occupants riding in vehicles equipped with seat belts are seated and using seat belts prior to the vehicle moving to the incident. Yet, that is not always the case, as indicated in the case history at the beginning of this chapter. Whether it is negligence, lack of communication, or lack of experience, there are reasons why every year, emergency service personnel are killed while a vehicle is in motion. In some organizations, standing on an apparatus while the vehicle is moving is still tolerated. This is highly hazardous behavior and has lead to injuries and fatalities. It should not be tolerated by agency administration.

In Motion

To ensure the safety of all personnel, certain actions must be taken when the vehicle is in motion. These actions must occur whether en route to or returning from an emergency, during training, or during any nonemergency function. Responsibility for these actions falls first on the individual, then on the driver/operator, and finally on the supervisor of the unit.

Riding in Apparatus

Since 1991, the use of seat belts or three-point restraints has been mandatory in all fire apparatus. Personnel must be secured prior to the movement of the vehicle **(Figure 6.20)**. Not only do seat belts prevent personnel from being ejected if the vehicle is involved in a collision, it greatly reduces the chance that they will be thrown from the vehicle on a curve or if the vehicle leaves the road surface. Being strapped into an SCBA while it is still in the mounting bracket is not considered to be wearing a seat belt or being restrained.

Figure 6.20 All personnel must be secured with a seat belt when the vehicle is in motion.

Standing on the tailboard of moving apparatus, on the running boards, or within the crew or passenger compartment of the apparatus is also prohibited by NFPA® 1500. Only two potential exceptions are allowed in NFPA® 1500: hose loading activities and tiller operator training. In both cases the emergency services organization is required to develop a policy that addresses the safety requirements for these exceptions. A possible alternative is the use of a safety harness that secures the firefighter to the vehicle while performing the task. NFPA® 1500 also requires that personnel in the patient compartment of ambulances either be secured with a seat belt or safety harness while providing patient care.

Company Officer

The company officer has ultimate authority over the safe operation of the emergency vehicle. While the driver/operator has control of the vehicle, the company officer must apply good judgment over the safe operation. If the officer believes that the driver/operator is exhibiting unsafe behaviors, it is the officer's responsibility to correct those behaviors immediately. Likewise, under the crew resource management (CRM) concept discussed in Chapter 3, it is also the responsibility of other crew members to alert the company officer and driver/operator of potential unsafe acts or conditions.

Backing Apparatus

Backing emergency apparatus is one of the most hazardous activities performed. It should be avoided whenever possible. If backing is unavoidable, then spotters should be used to safely direct the operation from outside the vehicle. When backing the apparatus, all members (excluding the driver/operator and tiller operator) should dismount and act as spotters for the backing operation **(Figure 6.21, p. 208)**. Spotters should not be permitted to ride the apparatus from an exposed position such as the tailboard.

The spotters should discuss a backing plan (consistent with the SOP) with the driver/operator and agree on the communication or warning process prior to backing. Communication could be verbal, by a remote electronic signaling device, hand signals, or radio. The vehicle should not be moved until the spotters are in position and have communicated their approval to start backing. Once the backing begins, the spotters should remain visible to the

Figure 6.21 Spotters are used to direct the driver/operator in the rearward movement of the fire apparatus.

driver. Spotters who are not wearing PPE should always wear high-visibility reflective vests. At least one spotter should be positioned at the left rear corner and operate as the primary spotter. Whenever the driver cannot see the primary spotter, the backing operation should immediately stop.

To reduce the risk associated with operating a vehicle in reverse, additional mechanical or electronic aids can be added. Fire apparatus could be equipped or retrofitted with additional mirrors, commercially available video cameras, or sensing devices. However, these devices cannot take the place of spotters who assist and direct the driver/operator while backing.

Water Tender Issues

Referred to variously as *mobile water supplies, tankers,* or *water tenders* (as specified under NIMS-ICS), these vehicles account for most of the firefighter response-related fatalities in fire apparatus. More firefighters are killed in water tender accidents than in accidents involving pumpers and ladder apparatus combined. In response to the alarming numbers of fatalities occurring in water tenders, the USFA published *Safe Operations of Fire Tankers* (FA-248) in 2003. This document, available on the FEMA Internet web page, provides an analysis of the causes of water tender accidents and suggests solutions to them.

Some of the primary contributing factors are the physical characteristics that are unique to water tender apparatus. Physical characteristics (**Figure 6.22**) include the following:

- *Weight* — Water tenders are heavier than the majority of other types of apparatus. Straight-frame water tenders may weigh as much as 25 tons (22 679.6 kg) while tractor-trailer water tenders can weigh much more. The additional weight makes the vehicle more difficult to control and stop.

- *High center of gravity* — Water tenders have a high center of gravity, making them more unstable than other vehicles. The lack of stability can result in the vehicles being difficult to control in curves and on uneven road surfaces.

- *Improper design or alteration* — Improper design of new water tenders or modification to existing vehicles can also create unsafe conditions. New apparatus chassis must be designed to carry the total weight of the tank filled with water. Consideration must also be given to the added weight of tools, equipment, and personnel plus a safety factor. Modifications to vehicles used for transporting other liquids can also cause water tender accidents. For instance, trucks used to carry liquid food products lack the interior tank baffles that prevent movement of the liquid that can cause rollovers. Cargo tank trucks used for carrying fuel are designed for a lighter load since gasoline weighs less than water. This weight difference must be taken into consideration. Finally, the braking systems of all modified vehicles must be capable of handling the added weight of the water in the tank.

- *Liquid surge* — The movement of water in an unbaffled tank has been shown to be the cause of many water tender accidents. Liquid surge can add to the forward momentum of the vehicle at the same time the driver/

operator is attempting to stop. Tanks must be designed or modified to conform to the requirements detailed in NFPA® 1901 for mobile water supplies.

NOTE: It is common practice that water tenders are driven with water tanks either full or empty to reduce the possibility of surges.

Besides the physical characteristics of water tenders, driver/operator behavior has been shown to contribute to water tender accidents. Lack of training, being unfamiliar with water tender driving, exceeding the safest speed for the water tender, and ignoring water tender operation SOPs and state/provincial motor vehicle laws are all behavioral factors resulting in accidents.

Figure 6.22 Water tenders weigh more than pumpers and have a higher center of gravity. They can also be more prone to turning over due to liquid surge in unbaffled tanks.

Reserve Apparatus

Reserve apparatus may be unsafe for a variety of reasons. Factors that may contribute to accidents involving reserve vehicles include the following:

- *Age* — Older apparatus exhibit the result of hard service and wear. Metal fatigue, reduced braking capacity, and electrical problems tend to affect the operation of reserve vehicles.

- *Maintenance* — Because reserve vehicles are not in continuous service, maintenance may not be as frequent or thorough.

- *Infrequent use* — Lack of use can cause a deterioration of parts and mechanical systems.

- *Design* — Reserve vehicles may not meet the most recent safety requirements. In addition, older vehicles may have manual transmissions and lack power steering or air or power brakes.

- *Lack of familiarity* — Driver/operators, especially personnel new to the position, may not be familiar with older reserve vehicles and their operation and design.

Privately Owned Vehicles

Firefighters who are killed in POVs during the course of their duties account for the largest percentage of vehicle-related deaths. These are typically volunteer firefighters who are responding to or returning from emergency calls.

Most volunteer departments do not require personnel to stand by in the fire station; members are allowed to respond directly to incidents from their homes or workplaces, often in their POVs in order to improve response time. Some organizations provide emergency lights for personnel to respond to incidents. State/provincial motor vehicle laws and local policies determine the type, location, and use of emergency lights on POVs.

A contributing factor (perhaps the most significant one) to accidents involving POVs, is the attitude of the driver. If the firefighter has the attitude that the emergency response is of greater importance than safety, then the chances increase that the driver will be involved in an accident. Speeding, violating traffic laws, and failing to stop at controlled intersections are the results of such an attitude. This is not only unsafe, it places both the individual and the department in a position of legal liability. Emergency warning lights may not alleviate an emergency responder traveling to an emergency in their POV from obeying all traffic laws and the increased liability involved. Some automobile insurance carriers may deny automobile damage claims if the responder's POV was not insured for that type of use. Check with your state motor vehicle agency and insurance carrier to determine emergency response requirements.

Just as training apparatus driver/operators is important, establishing training programs for all personnel who respond in POVs is equally as important. If you are a chief officer in your organization, you must take the responsibility to provide adequate training in defensive-driving skills to your personnel who operate their own private vehicles to an incident. Many programs teach defensive-driving skills to emergency service workers. For example, in the San Francisco Bay Area, the Alameda County Sheriffs Department instructs an Emergency Vehicle Operations Course (EVOC) The staff includes current and retired Sheriff and California Highway Patrol Officers who have many years of driving experience in emergencies. Instruction includes anticipation of other drivers, effective braking techniques, confidence courses, and skidding-recovery techniques.

Figure 6.23 Many career departments have large fleets of staff vehicles that are used for duties ranging from emergency response to conducting inspections and picking up supplies.

Nonemergency Operations

The majority of operations by fire and emergency services vehicles are nonemergency in nature. Staff vehicles are used to transport personnel to fire and life safety education programs, make building inspections, deliver documents, and attend meetings. Support services deliver fuel and supplies, provide facility and apparatus maintenance in the field, and move reserve apparatus around the jurisdiction. Ambulances are sometimes used to perform nonemergency patient transfers and pick up medical supplies from hospitals and supply facilities. Fire apparatus should be operated in a nonemergency manner when returning from an incident or when it is used for service training, preincident inspections and surveys, driver/operator familiarization, for repairs in the shop, and to pick up fuel and supplies.

All of these activities, though nonemergency, are essential to the operation of all fire and emergency services organizations **(Figure 6.23)**. Driver/operators must adhere to all state/provincial and local motor vehicle laws when performing these activities. The safety of other crew members and the general public depends on the ability of the driver/operator to control and maneuver the vehicle through traffic and constricted streets. In the case of oversized vehicles, such as tractor-trailer aerial devices, care is essential. Aerial devices equipped with a tiller require coordination and communication between the driver/operator and the tiller operator.

Emergency Response

During an emergency response, it is the driver/operator's responsibility to safely deliver the apparatus, its equipment, and crew to the scene of the emergency incident. Making this transit must be done within the limits of the department's response policies and the state/province's motor vehicle laws. You must remember that no emergency is worth the lives of your unit or members of the public that may be endangered by unsafe acts.

Operational concerns during emergency responses include the following:

- **Right-of-way** — Motor vehicle laws generally require drivers to give way to emergency vehicles. On two-lane roads or streets, vehicles in both lanes must pull to the curb or off the roadway to permit the apparatus to pass. On divided streets and highways, vehicles in the same lanes as the apparatus must pull to the shoulder of the road.

- **Speeds** — The maximum speed an emergency vehicle may attain is established by state/provincial motor vehicle laws and department policy. Traffic, roadway, and weather conditions will have direct effects on this speed. The U.S. DOT recommends operating at or below the posted speed limits.

- **Intersection safety** — All intersections pose a risk to emergency vehicles. With vehicles approaching from three (and sometimes more) directions, as well as overtaking, multiple turn lanes, and the possibilities of pedestrians, intersections require the full attention of all crew members. Some departmental policy may require that emergency vehicles come to a complete stop before entering any controlled intersection, even those controlled by preemptive devices. Audible warning devices must be used when approaching and entering the intersection. Pulling to the center lane prior to entering the intersection provides better visibility for the driver/operator and for other drivers to see the apparatus. Ladder apparatus, both straight-frame and tractor-trailer will require a wider turning radius and must have increased clearance. If necessary, the driver/operator can move into the opposite travel direction lane if the primary lane is blocked. Do not proceed into the intersection until you are certain that every other driver sees you and is allowing you to proceed. Simply slowing when approaching an intersection and then coasting through it is not an acceptable substitute for coming to a complete stop before proceeding when safe to do so. When proceeding through the intersection, attempt to make eye contact with each of the other drivers to ensure that they know you are there and about to proceed.

- **One-way streets** — One-way streets can pose a problem during emergency responses. They may require that an indirect route be taken to the scene or that the apparatus approach along the one-way street traveling against traffic. Departmental policies for operating against traffic must be established and followed by driver/operators. Care must taken when entering the one-way street to ensure that drivers see the apparatus and stop or pull to the side. In urban areas with heavy traffic, operating against traffic may be both more efficient and safer.

- **Highway access ramps** — Local policy should be established for converging responses on highways. Generally, apparatus should not attempt to enter a highway against the flow of traffic unless ordered to do so by the incident commander or law enforcement personnel who have stopped the flow of traffic on the highway.

- **Optical and audible warning devices** — The use of optical and audible warning devices is essential and mandated during emergency responses. Depending on local policy, their use may be prohibited for nonemergency-type responses. Most departments require that audible warning devices be turned off once the vehicle is at the incident scene. Optical warning devices may be reduced to just those that alert the public to the presence of the apparatus when parked. Driver/operators and those responsible for apparatus design must also be aware that the rear windows of some cars block the driver's view of warning lights mounted on the top of the apparatus.

- *Driver/operator multitasking* — Driver/operators should not attempt to perform more tasks than their primary duty: driving the apparatus. If the driver/operator is a member of a crew, then the company officer should be responsible for operation the radio, consulting the map book, and operating the air horn and siren. If the driver/operator is alone, or responding in a POV, then the driver should stop the vehicle when looking at the map book or involved in other distracting functions.

- *Tillering issues* — The operation of tractor-trailer aerial devices equipped with tiller positions requires training for both the vehicle driver/operator and the tiller operator **(Figure 6.24)**. Coordination and communication are essential. Only trained and certified tiller operators should be assigned tiller duties.

- *Skid recovery* — Avoiding conditions that lead to skidding is as important as knowing how to correct skids once they occur. The most common causes of skids involve the following driver errors:
  - Driving too fast for road conditions
  - Failing to properly appreciate weight shifts of heavy apparatus
  - Failing to anticipate obstacles (these range from other vehicles to animals)
  - Using auxiliary braking devices improperly
  - Failing to maintain proper tire pressure and adequate tread depth

When an apparatus that is *not* equipped with an antilock braking system goes into a skid, the driver/operator should release the brakes, allowing the wheels to rotate freely. Turn the apparatus steering wheel so that the front wheels face in the direction of the skid **(Figure 6.25)**. If using a standard transmission, do not disengage the clutch (by pushing in the clutch pedal) until the vehicle is under control and just before stopping the vehicle. Once the skid is controllable, gradually apply power to the wheels to further control the vehicle by giving traction.

Proficiency in skid control may be gained through practice at facilities having skid pads. These are smooth-surface driving areas that have water directed onto them to make skids easier to initiate. All training should be done at slow speeds to avoid damaging the vehicle or injuring participants. Some jurisdictions choose to use reserve apparatus or other older vehicles for this part of the training process.

- *Towing trailers* — Most fire and emergency services own or have access to some type of trailer. It may be used to transport a rescue boat, haul potable (drinking water) water, transport supplies, or provide lights or air to an emergency scene **(Figure 6.26)**. Towing a trailer behind a car, light truck, or apparatus requires skills similar to those required to operate a tractor-trailer. The driver/operator must take into account the different handling characteristics of driving and backing when towing a trailer. Additionally,

---

CAUTION

Using paved parking lots for skid training with heavy fire apparatus may damage the pavement surface and possibly result in rollover accidents. Safety dictates that such training be conducted only at proper facilities and under the supervision of qualified instructors.

Figure 6.24 Operating a tractor-trailer aerial device from the tiller position requires special training and certification. *Courtesy of Ron Jeffers.*

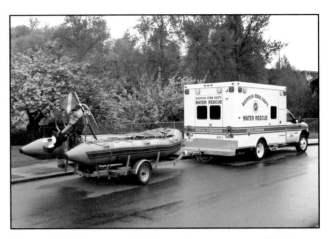

Figure 6.26 Equipment trailers, air compressors, lighting units, and boat trailers are often towed behind fire apparatus.

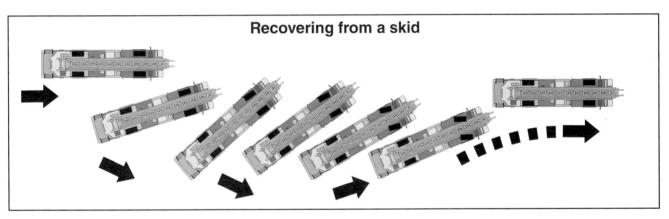

Figure 6.25 Recovering from a skid.

stopping distances are increased by the weight of the trailer and the potential for sliding on wet or icy roads or in windy conditions is greater. Training and practice are required for the safe operation of towing a trailer.

- *Vehicle recovery techniques* — Rollovers and loss of apparatus control have resulted from situations where the driver/operator attempted to pull back onto a road surface after running off. Driving off the roadway may be the result of attempting to avoid a crash or by accidentally driving off the road. At other times, a vehicle malfunction requires that the driver/operator pull off the road surface and then pull back onto it later.

  — *Initial reaction:* The most common reaction when leaving the road surface is to quickly try to get back onto the road. This reaction can cause the vehicle to rollover or cause a head-on collision with an oncoming vehicle because the driver/operator overcorrected and crossed the center line into the oncoming lane.

  — *Control:* To avoid over-correcting, immediately remove your foot from the accelerator and continue to drive the vehicle. Do not brake heavily or you may be pulled further from the road. If no obstacles are directly ahead, continue to slow and regain steering control in preparation for pulling back onto the roadway.

Total stopping distance — Sum of the driver/operator reaction distance and the vehicle braking distance.

Reaction distance — The distance a vehicle travels while a driver is transferring the foot from the accelerator to the brake pedal after perceiving the need for stopping.

Braking distance — The distance the vehicle travels from the time the brakes are applied until the apparatus comes to a complete stop.

— *Pulling back onto the road:* Check your side view mirrors and signal drivers behind you that you plan to return to the roadway. Smoothly turn the steering wheel and drive back onto the road. In the event an additional hazard (such as a pole or guard rail) is directly ahead on the shoulder of the road, you may have to decide whether it is safer to stop or drive immediately back onto the road. To do so, smoothly turn the steering wheel toward the road, causing the minimum angle turn necessary to avoid road shoulder obstructions, then immediately straighten the wheels as the second front tire makes contact with the edge of the roadway. The brief period between the steering input and the resulting action should allow the rear tires to also climb onto the road. This maneuver will allow the travel direction to be corrected before the apparatus is driven into oncoming traffic.

- *Vehicle retrieval* — Vehicle retrieval may be required when the apparatus leaves a hard surface road and becomes mired in mud or soft soil. Departments usually have a prohibition for taking certain types of apparatus, such as Type I pumpers, off roadways. Even so, it is always possible that a unit will require retrieval. Retrieval operations may rely on tow trucks or wreckers to attach lifting devices or cables to the vehicle frame, or tow hooks if provided, to pull it out. If the stuck vehicle is equipped with bumper-mounted winches, the winch cable may be attached to another department vehicle's frame or tow hook, acting as the anchor point, for self-extraction. Towing or winching apparatus can be extremely dangerous since the tow chain or cable can break striking any bystanders with lethal force.

- *Braking and reaction time* — Driver/operators should know the braking characteristics for the particular fire apparatus they are driving, including total stopping distance. The *total stopping distance* is the sum of the driver/operator's reaction distance and the vehicle's braking distance. Driver/operator *reaction distance* is the distance a vehicle travels while a driver is transferring the foot from the accelerator to the brake pedal after perceiving the need for stopping. Reaction distance can be affected by the driver/operator's age, aptitude, emotions, and fatigue as well as any intoxicating substances ingested. The *braking distance* is the distance the vehicle travels from the time the brakes are applied until the apparatus comes to a complete stop. **Tables 6.1** and **6.2** show driver reaction distances, vehicle braking distances, and total stopping distances for various sizes of vehicles. These tables indicate approximates for vehicles, and the statistics may vary for different fire apparatus. Each department should conduct braking distance tests with its own apparatus. Apparatus manufacturers may also be able to provide this information for specific apparatus.

There are a number of other factors that influence a driver/operator's ability to stop an apparatus such as the following:

- Condition and slope of the driving surface
- Speed being traveled
- Weight of the vehicle
- Type and condition of the vehicle's braking system
- Weather conditions
- Type and condition of tires
- Auxiliary traction devices

Table 6.1 (Customary)
Braking and Stopping Distances (dry, level pavement)

Speed (mph)	Average Driver Reaction Distance (feet)	Braking Distance (feet)				Total Stopping Distance (feet)			
		Vehicle A	Vehicle B	Vehicle C	Vehicle D	Vehicle A	Vehicle B	Vehicle C	Vehicle D
10	11		7	10	13		18	21	24
15	17		17	22	29		34	39	46
20	22	22	30	40	50	44	52	62	72
25	28	31	46	64	80	59	74	92	108
30	33	45	67	92	115	78	100	125	148
35	39	58	92	125	160	97	131	164	199
40	44	80	125	165	205	124	169	209	249
45	50	103	165	210	260	153	215	260	310
50	55	131	225	255	320	186	280	310	375
55	61	165	275	310	390	226	336	371	451
60	66	202	350	370	465	268	426	436	531

Typical Brake Performance
A–Average automobile   C–Heavy two-axle trucks
B–Light two-axle trucks   D–Three-axle trucks and trailers

Table 6.2 (Metric)
Braking and Stopping Distances (dry, level pavement)

Speed (km/h)	Average Driver Reaction Distance (meters)	Braking Distance (meters)				Total Stopping Distance (meters)			
		Vehicle A	Vehicle B	Vehicle C	Vehicle D	Vehicle A	Vehicle B	Vehicle C	Vehicle D
16	3.4		2.1	3.0	4.0		5.5	6.4	7.3
24	5.2		5.2	6.7	8.8		10.4	11.9	14.0
32	6.7	6.7	9.1	12.2	15.2	13.4	15.8	18.9	21.9
40	8.5	9.4	14.0	19.5	24.4	18.0	22.6	28.0	32.9
48	10.1	13.7	20.4	28.0	35.1	23.8	30.5	38.1	45.1
56	11.9	17.7	28.0	38.1	48.8	29.6	39.9	50.0	60.7
64	13.4	24.4	38.1	50.3	62.5	37.8	51.5	63.7	75.9
72	15.2	31.4	50.3	64.0	79.2	46.6	65.5	79.2	94.5
80	16.8	39.9	68.6	77.7	97.5	56.7	85.3	94.5	114.3
88	18.6	50.3	83.8	94.5	118.9	68.9	102.4	113.1	137.5
96	20.1	61.6	106.7	112.8	141.7	81.7	129.8	133.0	161.8

Typical Brake Performance
A–Average automobile   C–Heavy two-axle trucks
B–Light two-axle trucks   D–Three-axle trucks and trailers

A flat, dry, paved road provides the optimal stopping surface. The ability of a driver/operator to stop the apparatus is negatively affected by steep, wet, snowy, icy, or unpaved roads. Driver/operators must compensate for these conditions by reducing their speeds by an appropriate amount to match the conditions.

The correlation between vehicle weight and speed and stopping distance should be obvious. At the same speed, it will take a greater distance to stop a fully-loaded, three-axle water tender than a lighter vehicle. It will also take a greater distance to stop a vehicle that is going 50 mph (80 km/h) than the same vehicle when it is traveling at 30 mph (48 km/h).

The type and condition of the braking system also have a tremendous impact on the ability to stop the fire apparatus. Several serious fire apparatus accidents have been traced to poor maintenance of the braking system. Obviously, a vehicle that has a properly maintained braking system will stop faster than one that has a system in disrepair.

The condition of tires can also be a significant factor in stopping. Tires with worn tread will have less road grip that tires with full tread, especially on wet roads. Tires with treads designed for snow conditions will perform better in those conditions than general-use tires. However, some "snow" tires are at a disadvantage with dry road conditions.

Auxiliary traction devices, such as cable chains, link chains, and automatic traction devices, all require special driving and handling skills. If the road conditions require the use of these devices, slower speeds and increased stopping distances must be used.

Following distance. Whether responding to an emergency or simply driving your personal car on an errand, it is important to maintain a safe separation from the vehicle in front of you. To determine the proper following distance use the 2-4-12 rule **(Figure 6.27)**. Using this rule permits you to see conditions well ahead of your apparatus and to take action early. The 2-4-12 rule states the following:

- Maintain a 2-second interval between your vehicle and the vehicle ahead for speeds below 55 mph (88.5 km/h).

- Maintain a 4-second interval between your vehicle and the vehicle ahead for speeds above 55 mph (88.5 km/h). This allows for increased stopping distances required by higher speeds.

- Maintain a visual lead time of 12 seconds; that is, look ahead for possible hazards and potential alternate paths of travel in case an emergency occurs. The 12 second visual lead time permits you to make constant and smooth adjustments to speed and position.

The Volunteer Firemen's Insurance Services®, Inc. (VFIS®) recommends a slightly different means of calculating following distance. It suggests that driver/operators allow 1 second of following distance for every 10 feet of vehicle length for speeds under 40 mph and add 1 additional second for each 10 mph for speeds over 40 mph. That would translate into 4 seconds for speeds under 40 mph and 5 seconds for 50 mph.

Weight transfer. The effects of weight transfer must be considered in the safe operation of fire apparatus. Weight transfer follows the Law of Inertia which states that "objects in motion tend to remain in motion; objects at rest tend

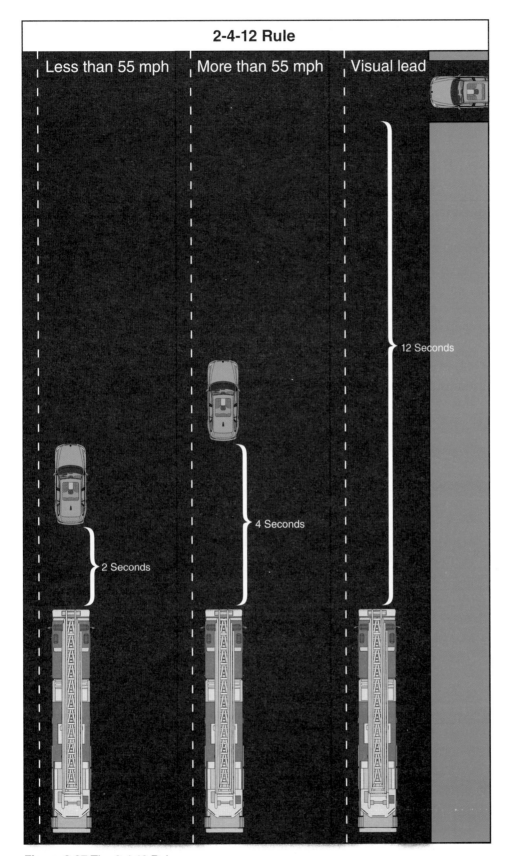

Figure 6.27 The 2-4-12 Rule.

to remain at rest unless acted upon by an outside force." Whenever a vehicle undergoes a change in speed or direction, weight transfer takes place relative to the rate and degree of change. Apparatus driver/operators must be aware that the weight carried on most fire apparatus can contribute to skidding or possible rollover due to lateral weight transfer. These hazardous conditions can result from too much speed in turns, harsh or abrupt steering action, or driving on slopes too steep for a particular apparatus. This is of particular concern with apparatus that have large water tanks that are improperly baffled and partially filled with liquid (water or foam concentrate).

Use only as much steering as needed to keep weight transfer to a minimum. Steering should be smooth and continuous. Also, maintain a speed that is slow enough to prevent severe weight transfer from occurring. This is particularly important on curves.

Passing other vehicles. In general, it is best to avoid passing vehicles that are not pulling over to yield the right-of-way to the fire apparatus. However, in some instances, the need to pass will occur, and driver/operators must be prepared to do it in the safest manner possible. The following guidelines should be used to ensure safe passing:

- Always travel in the innermost lane (closest to the centerline) on multilane roads. Wait for vehicles in front of you to move to the right before proceeding.

- Avoid passing vehicles on their right sides. It is normal for drivers to move to the right when an emergency vehicle is approaching. Thus, they could turn into your path if you are passing on the right. Some departments have strict SOPs prohibiting this practice.

  EXCEPTION: On multilane divided roads (such as interstate highways), drivers may move to the left if they believe their access to the right may be blocked by other vehicles. This may also be the case when vehicles are in the process of making a left turn from a multilane street. In these cases, and other more unusual situations, it will be necessary to pass on the right of those vehicles, but great care should be taken to make sure other drivers understand your intention.

- Make sure you can see that the opposing lanes of traffic are clear of oncoming traffic if you must cross the centerline to pass a vehicle. If this situation occurs at an intersection, care must be taken that vehicles approaching from the intersecting street(s) are aware of your presence.

- Avoid passing other emergency vehicles if at all possible. However, in some cases, it may be desirable for a smaller, faster vehicle (such as a chief's vehicle) to pass a larger, slower vehicle (such as an aerial apparatus). In these cases, the lead vehicle should slow down and move to the right to allow the other vehicle to pass. This maneuver should be coordinated by radio communication when possible

Adverse weather. Weather is another factor to consider in terms of safe driving. Rain, snow, ice, and mud make roads slippery. Driver/operators must recognize these dangers and adjust apparatus speed according to the crown of the road, the sharpness of curves, and the condition of road surfaces. They should decrease speed gradually, slow down while approaching curves, keep off low or soft shoulders, and avoid sudden turns. Driver/operators should recognize areas that first become slippery such as bridge surfaces, northern slopes of hills, shaded spots, and areas where snow is blowing across the roadway.

Because the stopping distance is greatly increased on slippery-road surfaces, it is sometimes prudent to apply the brakes carefully while in an area free of traffic to find out how slippery the road is. Speed and the use of auxiliary braking systems must be adjusted to road and weather conditions so that the apparatus can be stopped or maneuvered safely. Windshield wipers and defrosters should be kept in good repair in order to keep the windshield clean and clear.

Tire chains and other similar traction devices may increase the stopping distance and considerably increase traction for starting and hill climbing on snow or ice **(Figure 6.28)**. Apparatus may be equipped with the traditional, manually applied devices or the newer automatic variety. The manually applied systems generally include traditional tire chains and cable-type traction devices. Automatic tire chains consist of short lengths of chain that are on a rotating hub in front of each rear wheel. The hubs swing down into place when a switch on the dashboard is activated. The rotation of the hub throws the chains under the rolling tires. These chains tend to lose their effectiveness in snow that is deeper than 8 inches (200 mm) or when the vehicle is moving at very slow speed or in reverse.

During slippery-road conditions, the safe following distance between vehicles increases dramatically. Tire chains and traction devices provide traction for acceleration but do little to assist in deceleration. Remember that it takes 3 to 15 times more distance for a vehicle to come to a complete stop on snow and ice than it does on dry concrete.

Many of the manufacturers of tire chains and traction devices include a maximum speed allowed while using tire chains. Excessive speed or use on hard dry pavement may damage the chains and consequently damage the apparatus, including its tires, brakes, and body. Chains may also adversely affect stopping distances on dry pavement.

WARNING!

Auxiliary braking devices such as engine brakes, transmission retarders or, electromagnetic driveline brakes may cause the rear wheels to lock on slick roadways with very little pressure applied to the brake pedal. Based upon the organization's policy, performance of the vehicle, and the experience of the driver/ operator, the auxiliary brake may be turned off at the dash-mounted switch.

Figure 6.28 Automatic chains can be engaged when needed to provide additional traction on ice and snow.

Some emergency vehicles are equipped with an auxiliary traction control system called the inter-axle differential lock, power divider, or third differential. This is another type of switch that may be activated from the cab of an apparatus that has tandem rear axles. It allows for a difference in speed between the two rear axles, while providing pulling power from each axle. This is intended to provide greater traction for each axle.

Under normal operating conditions, the inter-axle differential switch should be in the unlocked position. Move the switch to the locked position when approaching or anticipating slippery-road conditions to provide improved traction. Always unlock the switch again when road conditions improve. You must lift your foot from the accelerator when activating the inter-axle differential lock. Do not activate this switch while one or more of the wheels are actually slipping or spinning because damage to the axle could result. Also, do not spin the wheels with the inter-axle differential locked because damage to the axle could result.

Some vehicles equipped with ABS, now have the ability to be provided with a feature frequently referred to as traction control. When manually switched on, the system uses the ABS system to apply the brake to the spinning wheel. Power is then directed to the wheel that is not spinning and forces it to help move the vehicle. Typically used only to assist the vehicle out of the mud or to accelerate a vehicle on slick roadways, the system should be turned off once the vehicle successfully moves or accelerates.

Multiple unit responses. There have been numerous collisions involving fire apparatus and other emergency vehicles. It is not always possible to hear the warning devices of other emergency vehicles when the audible warning devices on both vehicles are being sounded. When more than one emergency vehicle is responding along the same route, units should travel at least 300 to 500 feet (90 m to 150 m) apart. Some fire departments rely upon designated response routes. This practice can be hazardous if a company is delayed or detoured for some reason. SOPs may call for radio reports of location and status, particularly when you might be approaching the same intersection as another emergency vehicle. Regardless of the system or pattern used, always take precautions to ensure a safe, collision-free response. This includes coming to a complete stop to ensure it is safe before proceeding at any intersection that has a stop sign or a red signal light.

Apparatus Placement at the Incident Scene

Apparatus, crew, and driver/operator safety continue after the unit has arrived at the incident scene. Safety can be enhanced by apparatus placement, control of optical warning devices, and other activities.

Once the apparatus has arrived at the emergency scene, the driver/operator's exposure to injury changes. The driver/operator becomes vulnerable to the actions of other drivers. Standing at the pump panel, attaching hoselines, or removing equipment along the roadside makes the driver/operator and other personnel targets for inattentive drivers. It is essential that traffic cones and other control and warning devices be placed at the rear of the apparatus to direct traffic away from the work area.

Also apparatus placement to protect the driver and other personnel should always be taken into consideration. The pumper driver/operator must also be careful of changes in water pressure that may cause hoses to break or pump caps to be blown off. Aerial apparatus driver/operators must be aware of overhead power lines when positioning the apparatus. In addition, the slope of the grade and condition of the soil surface will affect the position of the aerial stabilizer jacks.

The driver/operator, under the supervision of the company officer, is responsible for placing or positioning the apparatus at the emergency incident scene. The placement of the apparatus is dependent on the type of incident and the location. The sections that follow describe suggested placement for the following scenes:

• Structural incidents

• Wildland incidents

- Hazardous materials incidents
- High-Rise incidents
- Technical rescue incidents
- Medical incidents
- Highway incidents

Structural Fire Scenes

When operating at a structural fire scene, vehicles directly involved in fire suppression (engines, aerial apparatus, etc.) should be positioned in locations that facilitate fire-suppression operations **(Figure 6.29)**. These vehicles provide the tools and equipment needed to gain access, affect search and rescue, apply extinguishing agents, and protect property through ventilation and loss-control activities. Additional apparatus that provide power for electric or hydraulic tools or incident scene lighting may also be found in this area.

Figure 6.29 Once a pumper is positioned and connected to a water supply, it is very difficult to relocate. *Courtesy of Ron Moore, McKinney (TX) Fire Department.*

Aerial devices should be strategically positioned so that they can be quickly placed into operation to provide access to upper stories or place elevate fire streams in operation. The length and type of aerial device as well as the existence of overhead obstructions such as power lines, streetlights, or traffic control devices must be taken into consideration when placing these devices. Of considerable importance is the need for the aerial device to be properly stabilized before the aerial device is operated. Narrow, crowded streets or uneven terrain can affect the stability of these vehicles. See the IFSTA **Aerial Apparatus Driver/Operator Handbook** for additional information.

Companies/units arriving subsequent to truck (ladder) companies should avoid parking to the immediate rear of these apparatus as this may interfere with the removal of ground ladders from their racks and increase the chance of injuries.

If not required for incident operations, support vehicles such as rescue apparatus and ambulances should be positioned in a staging area where they will not interfere with fire-suppression operations but their equipment will still be readily available. Tools and equipment that might be needed from support vehicles can be carried to the scene.

Wildland Fire Scenes

Placement of apparatus at wildland fires will be determined by the operational mode of attack established by the incident commander (IC) and off-road capability of the apparatus. An offensive operational mode will require that involved apparatus be placed close to the fireline (area where fire-suppression activities occur) Apparatus thus placed will be able to operate from the burned and upwind side of the fire **(Figure 6.30)**.

Figure 6.30 When attacking a wildland fire, operate from the burned side of the fireline.

This location can result in a direct attack on the flanks (sides) of the fire. Some apparatus may not be suitable for off-road operations, but in that case, they may be used for offensive operations from a hard-surface road on the burned side of the fire.

When the IC implements a defensive operational mode such as protecting structures, any type of apparatus may be used. The apparatus should be parked facing the escape route for quick withdrawal if necessary and in a position where it is protected from the fire front by a structure or topographical feature. The commanding officer (company officer, Strike Team Leader, Staging Officer, etc.) at the site may locate and announce a safety zone where apparatus can be parked so that they do not block or inhibit operations.

NOTE: You should remember that residential driveways may not be designed to support the weight of a fire apparatus.

Apparatus should not be parked in and around tall trees or heavy brush (potential fuel for the fire), on soft ground, near small creeks and streams, in open trenches, on residential lawns or yards (with possible septic tanks and lateral lines), and on roads adjacent to unburned vegetation that are halfway up slopes and susceptible to a fire moving uphill rapidly. Basically, apparatus should be positioned so that it is safe and convenient for responders.

Proper apparatus placement and positioning can be accomplished by using the following guidelines:

- Park apparatus off roadways or back into driveways to avoid blocking other fire apparatus or evacuating vehicles; notice soft shoulders or deep edges along roads.

- Use amber flashers when parking apparatus on a roadway or shoulder. Mark both the front and rear of the apparatus with traffic cones to warn motorists of the presence of apparatus and personnel.

- Park apparatus in a designated safety zone when utilized and do not leave it unattended at fires.

- Park apparatus to minimize its exposure to heat and blowing embers.

- Avoid blocking access ways with hoselines.

- Do not park apparatus next to or under hazards such as the following:
  — Power lines
  — Flammable trees or snags (standing dead trees)
  — Liquefied petroleum gas (LPG) tanks or other pressure vessels

- Use wheel chocks when apparatus is parked.

In addition to proper positioning, other guidelines that can protect the apparatus include the following:

- Keep apparatus compartment and cab doors closed and windows rolled up to keep out burning materials.

- Ensure that, when available, fire hose bed covers capable of protecting fire hose from blowing embers are utilized.

- Ensure that a fire-protection hoseline is connected and available for rapid deployment from the apparatus.

- Maintain communication/coordination with the rest of the fireground organization to ensure company operations are safe and effective.

- When required, lay supply hoselines only on road shoulders and only when a protracted operation is envisioned.

- When protecting structures, a "hit & run" tactic is typically used. Booster tank water is utilized to facilitate these types of operations.

- Do not lock apparatus doors or leave the apparatus unattended.

- Place the air-conditioning system of the apparatus in recirculation mode to avoid drawing in smoke from the outside.

- Leave the engine running and maintain a high engine idle to reduce apparatus stalling potential.

- Keep apparatus headlights turned on to maintain its visibility.

- Do not remove or deploy nonessential equipment from the apparatus.

- Remain on the burned side of the fireline when feasible.

You should remember that some structural fire-suppression apparatus are not designed for off-road use. Often the angle of approach and departure and ground clearance on the underside of the apparatus are inadequate to clear some obstacles, and the height of the apparatus can cause it to overturn on steep inclines due to a higher center of gravity. Even though all-wheel- or four-wheel-drive vehicles have superior climbing ability, they typically also have a high center of gravity that makes them susceptible to rollover. The weight of an apparatus can also result in it becoming stuck in soft, loose, or wet soil.

Hazardous Materials Incidents

Hazardous materials incidents can pose very complex problems that can change rapidly. The company officer must determine apparatus placement and positioning based on information that may be incomplete, incorrect, or altered. General apparatus placement and positioning guidelines for hazardous materials incidents include the following:

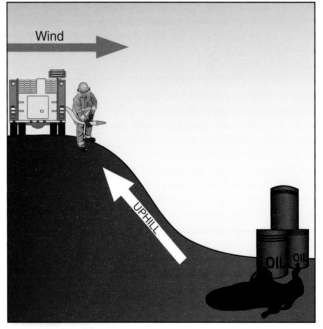

- Park apparatus upwind of the incident scene and uphill whenever possible **(Figure 6.31)**.

- Position apparatus well away from downed power lines, damaged transformers, or escaping flammable gas.

- Place apparatus so that it can be repositioned easily and quickly.

- Place or position apparatus so that its hoselines can protect the route of egress through the spill when rescue is required.

Figure 6.31 At hazardous materials spills or leaks, park the apparatus uphill and upwind of the spill.

High-Rise Incidents

Incidents in multistory or high-rise structures create special apparatus placement and positioning requirements. The structures may be located in a series or row of similar structures with only the front and rear of the structure accessible. They may also be independent of other structures and located on individual parcels of land surrounded by parking lots or connected to parking structures.

High-rise structures are generally considered to be 7 stories (70 feet [21 m]) or more in height from grade level. Some of the characteristics that influence apparatus placement and positioning at high-rise structures are as follows:

- Limited ground-level access with personnel-access doors located only in the front and rear walls (when the building is in a series of structures)
- Intermediate parking garage areas between grade level and upper occupied floors, which increase the distances that personnel must climb and equipment must be transported (parking levels will also be filled with vehicles that can be a source of ignition or fire)
- Fences, planting beds, decorative walls, or other barriers near the base of independently located structures
- Limited street frontage
- Building setbacks from the street, parking areas, or driveways, which may be obstructed by barriers in the setback area

Apparatus must be located away from any potential collapse zones. A collapse zone is described as 1½ times the height of the wall or structure.

Technical Rescue Incidents

Proper placement and positioning of emergency apparatus at technical rescue incidents depends upon a number of variables. Placement and positioning of apparatus and/or resources is generally different from that normally found at a fire scene, hazardous materials incident, or high-rise incident.

At a technical rescue scene (water, ice, confined-space, structural collapse, trench, or high-angle), rescue vehicles (or apparatus carrying rescue tools and equipment) should be positioned nearest the incident. Rescue and extrication equipment may be the most important items in an emergency situation and need to be readily available. Apparatus used for fire-suppression operations (engines, ladders/trucks) should be located in the staging area.

While it is important to place apparatus close enough to unload or support powered tools and equipment, it is also important that the weight of the apparatus not add to the problem. The weight of the apparatus and engine vibration can cause soil in a trench-collapse area to cave in. Vehicles that do not need to be close to the scene, including Command vehicles and POVs, should be parked in a staging area farther from the incident.

Some general guidelines for the placement and positioning of apparatus at technical rescue incidents **(Figure 6.32)** are as follows:

- Park apparatus between an incident and oncoming traffic to protect both personnel and victims when the incident is on or adjacent to a roadway or bridge.
- Park apparatus close enough to the incident to ensure its equipment is readily available with a minimum carrying distance.
- Do not position apparatus so close to the incident that it exposes victims or emergency responders to vehicle exhaust, vibration, or noise.

- Do not block an emergency scene. Allow access for ambulances and other emergency vehicles and normal flow of traffic when the incident is not in the roadway.

- Coordinate closely with on-scene law enforcement officials to address safety issues and concerns for emergency personnel, victims, bystanders, and other traffic.

Medical Incidents

The placement and positioning of apparatus and ambulances at medical incidents are critical factors to efficient and immediate patient care. Many medical emergencies are time-critical, requiring that a specific level of care be provided as quickly as possible. Cardiac arrest, asphyxiation, and massive blood loss are just three medical emergencies that demand immediate care.

Figure 6.32 Aerial devices may have to be positioned in such a way that the ladder can be used to assist in the technical rescue.

To reduce the amount of time required to reach patients, carry life-support equipment, and transport the patient to the ambulance, carefully consider the placement of the apparatus.

Some considerations about apparatus and ambulance placement and positioning during medical incidents include the following:

- Place apparatus and ambulances as close to the incident as possible.

- Units that are designed for patient transport should be parked in such a way as to provide clear access to the loading door.

- Do not place or position apparatus so that it blocks the entry to or exit from the emergency site, making movement of the patient gurney difficult or time-consuming.

- Do not position apparatus so close to victims or emergency responders that it expose them to vehicle exhaust, vibration, or noise.

- Initiate positive traffic control, preferably with a law enforcement official if apparatus must be positioned so that it blocks part of the roadway. Place traffic cones a sufficient distance from the apparatus to alert oncoming motorists of the hazard and continue use of red/amber flashers.

- Use amber flashers when parking apparatus on a roadway or shoulder. Mark both the front and rear of the apparatus with traffic cones to warn motorists of the presence of apparatus and personnel **(Figure 6.33, p. 226)**.

- Place apparatus on the side of the road and position to protect the victim and emergency responders while the patient is being stabilized and loaded for transport.

Highway Incidents

Highway incidents pose special problems for emergency vehicles. Traffic congestion, higher speeds, and inattention on the part of other drivers add to the risk posed to emergency services personnel. It is important that there be a safe response followed by protection of personnel, patients, and other drivers at an incident. This should be followed by an attempt to allow continued traffic flow with a minimal amount of disruption. In accordance with some

Figure 6.33 Traffic cones and flashing lights are essential to protect personnel working along a roadway.

states' requirements, law enforcement officers (highway patrol) may be the designated ICs during highway incidents. Operations should be coordinated with them in a spirit of cooperation to resolve the incident.

Fire and emergency responders are exposed to many hazards at highway incidents. The following is a summary of causal factors that have been identified in incident reports and through experience to be responsible for firefighters and other emergency responders coming in contact with other vehicles at a roadway incident scene.

- *Lack of training* — Responders are not trained on the hazards associated with roadway incidents and the proper ways to minimize their occurrences.

- *Lack of situational awareness* — Responders fail to recognize the dangers associated with a particular roadway situation that they are facing due to insufficient training or lack of experience.

- *Failure to establish a proper temporary traffic-control zone* — Many fire departments do not have sufficient training, equipment, or SOPs for the correct way to set up a properly marked work area when operating at a roadway incident scene. Cases have also been noted where the responders did have good training, equipment, and SOPs, but for whatever reason failed to use or follow them.

- *Improper positioning of apparatus* — Numerous cases have been cited where apparatus were not positioned to the fullest advantage of the incident. In some cases an apparatus was not positioned in a manner that protected the work area. In other cases apparatus were unnecessarily positioned in the roadway.

- *Inappropriate use or lack of scene lighting* — Inappropriate use of vehicle headlights, warning lights, and floodlights can confuse or blind approaching motorists. This causes them to strike an emergency vehicle, responder, or other vehicle in the incident area.

- *Failure to use safety equipment* — Responders working in the roadway must wear appropriate protective garments and use all available traffic-control devices in order to prevent being struck by oncoming traffic.

- *Careless, inattentive, or impaired drivers* — even when we try to do everything correctly, we must be cognizant of the fact that there are drivers who will not react correctly to the altered traffic pattern that occurs at a roadway incident. This may result in them driving into our work space.

- *Reduced vision driving conditions* — Although firefighters may be struck by vehicles in virtually any condition, the chances of an incident occurring are greater during obscured vision conditions, including darkness, fog, rain, snow, and blinding sunshine.

- *Altered traffic patterns* — Drivers may be confused by the traffic-control measures used at an incident scene or those being employed in a construction zone.

In response to this information, the USFA and the DOT/FHWA formed a partnership with the International Fire Service Training Association (IFSTA) to research and identify effective technical guidance and training programs for fire and emergency service providers in Traffic Incident Management Systems (TIMS).

The purpose of the project was to enhance responder safety and provide guidance to local-level fire departments on compliance with the DOT's *Manual of Uniform Traffic Control Devices* (MUTCD) and the National Incident Management System Consortium (NIMSC) (formerly the National Fire Service Incident Management System Consortium's [NFSIMSC]) *Model Procedures Guide for Highway Incidents*. The information contained in this document should help enhance firefighter operational effectiveness, reduce potential liability, and enhance responder safety at roadway emergency scenes.

Briefly, some of the highway incident recommendations include the following:

- Always establish an incident command organization at the scene. In more complex operations it may be necessary to establish a formal "unified command" structure to ensure close cooperation among the involved agencies

- Establish a *temporary traffic control* (TTC) zone or *traffic incident management area* (TIMA). This zone is divided into three areas: advanced warning, transition, and activity. Folding signs or lighting is used in the advanced warning area to get the attention of drivers, cones are used in the transition area to redirect traffic into an alternate lane, and the apparatus is positioned as a barrier in the activity area to protect workers and patients **(Figure 6.34)**.

- If necessary, assign a member of the unit to direct traffic into an alternate lane. This person must wear a reflective vest or PPE with reflective trim, be equipped with a hand light at night and a flag during the day, and have been trained as a flagger.

- Turn off white spotlights or direct them into the activity area rather than at oncoming traffic. Spotlights can blind drivers.

- Use directional arrows to indicate the required lane change. Generally, these arrows are amber in color.

- All personnel must wear reflective vests or PPE with reflective trim.

- Position the apparatus to provide an effective barrier to the work area.

- For incidents located in the inner lanes of a divided highway or in the center median, position apparatus in the inner lanes of both directions of travel if necessary to protect the incident scene. (In some instances, on highways with very wide dividing medians or dividing walls, placing apparatus on both sides of the median would be counterproductive.)

General Safety Concerns

There are other general operational safety concerns that may not fall into a specific category. These include the securing of tools and equipment, the addition of cabinets or lockers designed for medical equipment, and the refueling procedures for apparatus. These concerns overlap design requirements and crew activities.

Securing Equipment

In a motor vehicle accident involving a fire or emergency vehicle, tools and equipment can become deadly projectiles. NFPA® requires that all tools be securely mounted when carried in the crew compartment **(Figure 6.35)**. SCBAs must be firmly clamped into the passenger seat back or into a quick-release bracket. Spare air cylinders must also be secured in crash-resistant brackets.

In the body of the apparatus, tools and equipment should be secured to compartment walls, slide-out panels, or trays. In these areas, securing the tools is less of a safety issue and more of an issue of efficiency and damage control. Heavy equipment, such as generators or ventilation fans, should be secured to the cabinet floor or to roll-out brackets to improve access and prevent the equipment from shifting during travel.

Figure 6.34 Signs, lights, cones, and apparatus can be combined to provide a safe work area at a highway incident.

Figure 6.35 Tools and equipments should be secured in compartments to eliminate damage and possible injury.

In the patient compartment of ambulances, the gurney is required to be secured to the floor and the passenger and attendant seats have seat belts. All cabinet doors must have positive latches to prevent them from opening in an accident.

Refueling

Apparatus can be refueled at fixed fueling sites, from fuel trucks, or at temporary dispensing sites during emergency operations. In any case, care must be taken in transferring the fuel into the apparatus. Refueling should never be assigned to anyone who has not been trained to perform the task. A Class B extinguisher should always be immediately available.

Fixed Fueling Site

The person filling the apparatus remains at the pump nozzle during the filling operation. The possibility of static electricity exists. Therefore, the person filling the tank should not reenter the apparatus until the operation is complete. Smoking must be prohibited during fueling operations. Spilled fuel should be cleaned up immediately.

Fuel Truck

Transferring fuel from the fuel tank truck to an apparatus may take place at the station, emergency scene, or any other location. When refueling at the station, the operation should not take place inside the apparatus bay to prevent vapors building up in the structure. The fuel truck and apparatus should be connected with a grounding wire to prevent the creation of static charge. Spills should be cleaned up as soon as possible if they occur.

Temporary Dispensing Sites

During long-duration emergency operations, it may be necessary to establish temporary dispensing sites. Fuel may be transferred from fuel trucks, 55-gallon (208.20 L) drums with hand pumps, or from 5-gallon (18.93 L) fuel cans. Safety precautions include prohibition of smoking during the operation, spill or leak prevention, grounding or bonding to eliminate static electricity, and cleaning up all spills immediately.

Summary

When called to assist a citizen, we need to be able to arrive in one piece. Apparatus safety is paramount when responding to an emergency incident. Annually, firefighters are injured or killed, because of unsafe acts due to a lack of knowledge, skills, abilities, and attitudes. Apparatus placement on scene can be helpful or a hindrance if apparatus is improperly placed. Cooperation with law enforcement agencies is imperative when placing a unit in the flow of traffic to protect the scene. Regular maintenance and inspections make for a long vehicle life and safe operations. Reserve apparatus must be inspected and checked for safety equipment or the lack thereof. Defensive driving is a large section of this chapter, because of its importance. Anticipating other drivers' actions and driving safely in adverse weather conditions contribute to safe practices when asking for the right-of-way from other vehicles during a response. Good health is not only required to legally operate a fire apparatus, it is ethical towards your company members.

Unsafe behavior is a contributing factor in most vehicle accidents whether involving emergency or nonemergency vehicles. To aid in preventing vehicle accidents, you should take into consideration the following keys to safe driving:

- *Attitude* — Your personal attitude can affect how you drive. If you have the attitude that your apparatus or POV has the right-of-way because you are involved in an emergency response, then you will be a hazard on the road, increasing your chances of having an accident. If you allow yourself to be emotionally or mentally distracted, you will not be able to focus on the safe operation of the vehicle.

- *Situational awareness* — As mentioned in Chapter 3, you must be aware of your surroundings at all times. The weather and road conditions, other traffic, apparatus speed and condition, and the severity of the emergency incident are all part of those surroundings.

- *Physical condition* — You must be aware of your own physical condition. Lack of sleep, stress, fatigue, or the consumption of alcohol, drugs, or medication will impair your ability to operate a vehicle. If you are not physically able to drive, do not do it. You will jeopardize your own life, the lives of your crew, and the general public.

- *Training/certification* — Never operate any vehicle or piece of equipment that you have not been trained to operate and/or are certified to operate. This action will place you and your organization under a legal liability and potentially result in injury or death if an accident occurs.

Review Questions

1. What are some basic safety components found on a fire apparatus?

2. How does the concept of visual lead time effect stopping distance, reaction distance, and braking distance?

3. What are five common causes of fire apparatus crashes?

4. What specific legal requirements affect the design of a fire apparatus?

5. How is corrective maintenance different from preventative maintenance?

6. What are the differences between the operation of a fire apparatus for a company officer versus that of a driver/operator?

7. How can one determine the difference between an acceptance test and a service test?

8. What aspects of defensive driving can help prevent fire fighters from coming into contact with other vehicles at roadway incidents?

Tool and Equipment Safety

Chapter Contents

chapter 7

Key Terms

Job Performance Requirements

This chapter provides information that addresses the following job performance requirements of NFPA® 1500, *Standard on Fire Department Occupational Safety and Health Program:*

NFPA® References

6.5.1	6.5.5	6.5.9	6.5.12
6.5.2	6.5.6	6.5.10	6.5.13
6.5.3	6.5.7	6.5.11	6.5.14
6.5.4	6.5.8		

FESHE Learning Objectives

Occupational Safety and Health for Emergency Services
Describe the relevance for safety in the work place including the importance of PPE.

Tool and Equipment Safety

Learning Objectives

After reading this chapter, students will be able to:

1. Describe the two leading causes of equipment accidents.

2. Summarize the general inspection points for tools and equipment.

3. Summarize the specific inspection points for ground ladders, ropes, fire hose, portable fire extinguishers and electronic equipment.

4. Describe procedures that reduce potential hazards in a maintenance area.

5. Summarize the general safety considerations for battery charging and changing.

6. Describe the general procedures for cleaning and maintaining unpowered tools.

7. Describe the general procedures for cleaning and maintaining power-operated equipment.

8. Describe the general procedures for cleaning and maintaining portable fire extinguishers.

9. Describe the general procedures for disinfecting tools and equipment.

10. Summarize the procedures for storage and use of hazardous materials.

11. Describe the general procedures for safe operation of tools and equipment.

Chapter 7
Tool and Equipment Safety

Case History

On July 10, 2003, a 50-year-old male volunteer fire fighter/fire service products salesman (the victim) died after being struck by a rescue airbag (high-pressure lifting device). The victim was demonstrating the use of rescue airbags when the incident occurred. A set of three airbags had been placed under the rear bumper of a front-end loader, coupled together and inflated. The airbags were being deflated when they separated and burst out from under the loader. The middle airbag struck the victim on the head and chest. The victim was immediately treated at the scene by volunteer emergency medical technicians and transported by helicopter to a local hospital where he was pronounced dead later that day.

NIOSH investigators concluded that, to minimize the risk of similar incidents, fire departments and fire service products distributors should:

- *provide training to all fire fighters and employees expected to use or demonstrate rescue airbags*
- *follow safety rules for airbag lifting operations (e.g., use cribbing)*
- *provide and enforce the use of personal protective equipment during airbag demonstrations and use*

Additionally, fire service product manufacturers should:

- *ensure that information and photographs contained in catalogs, operator's manuals and internet web sites correspond with one another and provide consistent safety information*

Source: NIOSH Firefighter Fatality Investigation Report, F2003-34

Tools and equipment are vital to efficient, effective, and safe operations at both emergency and nonemergency incidents. Power tools have improved both efficiency and effectiveness of emergency and nonemergency operations, and when used properly, can minimize hazards to the firefighter. But tools can, if not used and maintained properly, increase the risk of injury during operations. See **Table 7.1, p. 236,** for a list of hand tools and **Table 7.2, p. 236,** for a list of power tools commonly used in the fire service.

Table 7.1
Hand Tools

Cutting	Striking	Prying	Pushing/Pulling	Wildland Tools	Misc. Tools
Pick-head axe	Sledgehammer	Crowbar	Standard pike pole	Pick-head axe	Shovel, scoop
Flat-head axe	Maul	Halligan bar	Clemens hook	Flat-head axe	Broom, push
Bold cutters	Battering ram	Pry (pinch) bar	Plaster hook	McLeods	Mop
Carpenter's saw	Pick	Hux bar	Drywall hook	Pulaskis	Squeegee
Keyhole saw	Flat-head axe	Claw tool	San Francisco hook	Brush hook	Mop Bucket
Drywall saw	Mallet	Kelly tool	Multipurpose hook	Pick	Water key
Hack saw	Hammer	Pry axe	Roofman's hook	Fire rakes	
Wire cutters	Punch	Flat bar (nail puller)	Rubbish hook	Hoes	
Burning bars	Chisel	Rambar		Wire Brooms	
Cutting flares				Flails	
				Combination tools	
				Shovel, pointed tip	

Table 7.2
Power-Operated Tools

Saws	Cutting	Prying	Ventilation	Wildland Tools	Power Plants
Circular saw	Rebar cutter	Hydraulic door opener	Fans, electric, gas, hydraulic	Chain saw	Gasoline generator
Reciprocating saw	Plasma cutter	Extrication or rescue tools		Power (leaf) blower	Wire reels
Chain saw	Oxy-gasoline cutting torch			Portable pumps	
Ventilation saw	Oxyacetylene cutting torch				
Rotary saw					

This chapter describes recommended tool and equipment design criteria, operations, inspections, and maintenance practices. The chapter will also address the consequences of unsafe practices and the resulting injuries and deaths caused by improper use, lack of maintenance and the lack of knowledge, skill, and attitude in regard to tools and equipment.

Equipment Accidents

Unsafe behaviors while using tools and equipment can result in injuries, fatalities, and damage to or loss of tools and equipment and delayed completion of assigned tasks. At the same time, tool or equipment failure caused by a lack of periodic inspections, preventive maintenance, or repairs can also be the cause of equipment accidents or failures. Recognizing the results and causes of tool and equipment accidents is the first step in reducing or eliminating such accidents.

Results

Due to the methods of data collection that are currently used to document fire and emergency responders' injuries, it is almost impossible to statistically link injuries to the use of tools and equipment. We do know that numerous injures such as cuts, abrasions, contusions, and broken bones occur in the fire and emergency services. It is reasonable to assume that many of these injuries are due to the unsafe use of tools and equipment or to the tools and equipment themselves.

Some injuries, such as hearing loss, eye injuries, or cuts, are due as much to the lack of proper personal protective equipment (PPE) as unsafe behavior or tool failure. While no objective (factual) sources exist, one subjective source can provide an indication of potential for injuries. Firefighter Near-Miss.com and EMS Near-Miss.com contain anecdotal stories about situations that could have resulted in an injury or death. These stories range from the failure of a coupling caused by a design flaw to the result of a water hammer in a hoseline caused by unsafe behavior.

Fatalities are easier to associate with tool or equipment usage since incident-related fatalities are typically thoroughly investigated and reports are made public. However, the relationship between a tool or piece of equipment and a fatality may be indirect and may not be reported. For instance, the failure of a forcible entry tool may prevent a firefighter from a successful self-escape, but the cause of the fatality may be listed as asphyxiation.

Appropriate data collection regarding incidents of tool and equipment accidents and/or failure can provide the basis for tool and equipment safety training and preventative maintenance.

Causes

Unsafe behavior is the documented cause of most accidents. Tool or equipment failure is a secondary cause of accidents. You could also make the argument that tool or equipment failure also resulted from unsafe behavior in the form of poor inspection or maintenance skills or administrative decisions that did not provide proper preventive maintenance or appropriate specification development.

Unsafe Behavior

Unsafe behavior when using tools and equipment can place you and others around you in danger. Factors that contribute to unsafe behaviors include the following:

- Improper use of tools or equipment
- Use of incorrect tool or equipment for specific task (**Figures 7.1 a and b, p. 238**)
- Lack of knowledge or training regarding proper operation of tools or equipment
- Lack of safety measures taken by the operator to ensure the safest operation possible

During crises, firefighters and emergency responders tend to improvise if the correct tool or equipment is not available. For instance, utility rope may be used in place of rescue rope or an axe blade maybe used as a prying tool.

Figures 7.1a and b There is an unsafe way to use a tool and a safe way to use it: a) using a cheater bar can result in a damaged hydrant valve, broken hydrant wrench, or operator injury; b) using the hydrant wrench correctly will prevent damage or injury.

Both of these examples are unsafe behaviors that will and have resulted in injuries and fatalities. Changes in the current culture are necessary to cause fire and emergency services personnel to take the time to select and employ the correct tool or equipment for the correct task.

Equipment Failure

Tools and equipment can fail for a number of reasons. Some of the reasons may be directly or indirectly linked to unsafe behaviors on the part of the tool operator, the maintenance personnel, the administration, or the manufacturer. Tool and equipment failures may be caused by the following:

- Inappropriate specifications
- Design flaws
- Unreported/unrepaired damage
- Used beyond design, life, or safety limits
- Material fatigue **(Figure 7.2)**
- Lack of preventive maintenance
- Lack of periodic or scheduled inspections

When an accident analysis indicates that the cause was equipment failure due to manufacturing error or inappropriate design, the manufacturer should be notified. If the manufacturer agrees with the report, then changes in the tool or equipment will be made or a product safety warning will be issued to all organizations that own similar tools or equipment. Some accidents will result in a product recall. It is imperative that the fire and emergency services administration monitor product recalls and information published in fire and emergency services trade journals and Internet sites. In one case listed on Firefighter Near-Miss.com, a 10-year old coupling failed under water pressure. The fire department had not received a manufacturer's recall that would have repaired the unit before it failed.

Tool and Equipment Design

The majority of tools and equipment used by fire and emergency services organizations are designed specifically for emergency operations. Many are required to meet NFPA® design standards, while others must meet standards established by federal agencies. If you are assigned the task of selecting or purchasing tools and equipment for your emergency organization, you must be familiar with the standards and regulations that apply.

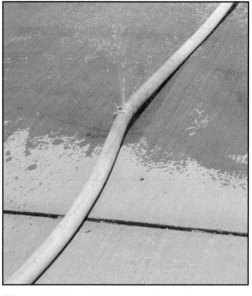

Figure 7.2 Material fatigue to hose jackets or couplings can result in loss of water pressure at the nozzle.

Because tool and equipment manufacturers must meet design standards and regulations, as well as industry standards for materials, they have the responsibility to recommend (and require, in some cases) how their products are used, stored, cleaned, and maintained. It is your obligation to follow the manufacturer's recommendations when using their tools and equipment. The emergency services organization has the responsibility for training employees in the proper use, care, and maintenance of tools and equipment and for providing the proper fuel, parts, and support for them.

Tools and equipment that may be used in an atmosphere that contains flammable vapors must not be able to create a spark or ignition source. Hand tools such as scoop shovels must be constructed of materials that will not cause a spark. Communications equipment such as radios and ventilation fans must be intrinsically safe (**Figure 7.3**).

Intrinsic safety (IS) is a protection technique for safe operation of electronic equipment in explosive atmospheres. The theory behind IS is to ensure that the available electrical and thermal energy in the system is always low enough that ignition of the hazardous atmosphere cannot occur. This is achieved by ensuring that only

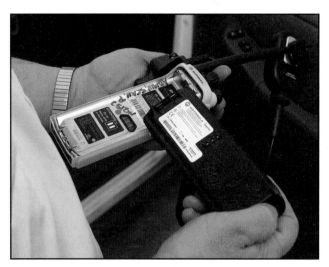

Figure 7.3 Intrinsically safe radios are designed for use where flammable vapors may be present.

low voltages enter the hazardous area and that all electric supply and signal wires are protected by zener safety barriers. Sometimes an alternative type of barrier known as a *galvanic isolation barrier* may be used. In addition, no single field device or wiring is intrinsically safe by itself (except for battery-operated, self-contained devices), but is intrinsically safe only when employed in a properly designed IS system.

Intrinsic safety (IS) — A protection technique for safe operation of electronic equipment in explosive atmospheres.

The majority of power-operated tools are designed with safety guards and features on them. Safety guards, depending on the tool, are intended to protect hands and fingers from contact with running blades or using chains on saws while the operator still maintains control of the saw. Automatic power shutoff switches, known as *dead man switches*, are required on all power tools to prevent them from operating if the operator releases the switch. Eye guards are also provided on tools that produce sparks or metal shavings, such as grinders. These safety guards and switches are required and must not be removed or tampered with.

Apparatus and portable power supply units may have ground fault circuit interrupter (GFCI) receptacles installed on them. These receptacles are designed to shut off the flow of electricity through the system if a short circuit occurs. GFCIs on this equipment are similar to those found in buildings in areas that contain water or moisture such as kitchens and bathrooms.

Tool and Equipment Inspection

When new tools and equipment are purchased, they must be inspected and in some cases tested to determine if they meet the purchasing specifications. Hand tools are generally inspected when they are received and placed into service. Power-operated equipment, fire hose, respiratory protection equipment, and generators must be acceptance tested when received. Requirements for acceptance testing are usually included in the purchase contract or agreement.

Following the acceptance tests, periodic inspections and tests are performed to ensure that tools and equipment remain in good condition. Inspection schedules are established by the local jurisdiction based on federal regulations, NFPA® standards, and manufacturer's recommendations. These mandated inspections, which may include testing by third-party organizations, apply to equipment such as respiratory protection and breathing air equipment, power rescue tools, ground ladders, fire hose, and generators.

In addition, the individual fire and emergency services organizations will require frequent inspections performed by fire and emergency responders. These inspections occur at the beginning of each work shift, weekly, monthly, or after each use of the tool of equipment. The following sections present general inspection points for all tools and equipment and inspection points for specific tools and equipment.

General Inspection Points

At the beginning of each work shift, weekly, monthly, and after each use of the tools and equipment assigned to you or your unit, you must inspect their condition. A thorough inspection requires that you look for the following items:

- Damage
  - Broken parts
  - Chipped blades
  - Missing parts
  - Dents, cracks, or splits
- Wear
  - Worn surface
  - Chipped paint
  - Metal fatigue; cracks or splits
  - Worn belts
  - Corrosion

- Lack of cleanliness
  - Dirt, grime, or debris
  - Contamination; medical or hazardous material
- Proper operation of power-operated equipment
  - Unit starts and runs per normal operation
  - Unit opens or closes per normal operation
- Effects of heat or direct-flame contact
  - Blackened surface
  - Singed surface, bubbled paint or material
  - Charring
- Cylinders hydrostatic testing date
  - Includes all self contained breathing apparatus (SCBA) cylinders
  - Includes fuel and fire extinguisher cylinders

Specific Inspection Points

By nature of their design and use, some tools and equipment require inspections that exceed the general points mentioned previously. You should be familiar with the manufacturer's recommended inspection practices as well as the NFPA® standards for these tools and equipment. The following sections address tools and equipment generally used in the fire service.

Ground Ladders

All fire service ground ladders are required to meet the inspection requirements of NFPA® 1932, *Standard on Use, Maintenance, and Service Testing of In-Service Fire Department Ground Ladders.* This standard states that ground ladders must be service tested annually, inspected monthly and after each use, and inspected according to the manufacturer's recommendations.

When you perform a visual inspection after each use or monthly, you should ensure that the following items are checked:

- Heat-sensor labels on metal and fiberglass ladders, and on wood ladders, if provided, have not changed color, indicating heat exposure **(Figure 7.4)**

- All rungs are snug and tight where they connect to the beams

- All bolts and rivets are tight; bolts on wood ladders are snug and tight without crushing the wood

- Welds are free of any cracks or apparent defects

- Beams and rungs are free of cracks, splintering, breaks, gouges, checks, wavy conditions, or deformation

- Butt spurs are free of excessive wear, are firmly attached to the ladder, or are free of other defects

- Halyards are free of fraying or kinking

Figure 7.4 Metal and fiberglass ground ladders as well as some wood ladders have heat sensors attached to the tip end of beams. These sensors will change color if the ladder has been exposed to high temperatures.

- Roof hooks are sharp and operate properly without binding
- Rungs are free of punctures, wavy conditions, worn serrations, or deformation
- Surfaces are free of corrosion

Ropes

Fire service rope falls into two classifications: *life safety rope* and *utility rope*. Life safety rope is used to support rescuers and/or victims during emergency incidents or training exercises. Utility rope is used to hoist tools and equipment, secure items in place, or as a barrier to cordon off an area (**Figures 7.5 a and b**).

All life safety (rescue) rope must conform to NFPA® 1983, *Standard on Life Safety Rope and Equipment for Emergency Services*. This standard specifies that only rope of block creel construction (without knots or splices in fibers) using continuous filament virgin fiber for load-bearing elements is suitable for life safety applications. Rope made of any other material or construction must not be used to support firefighters or victims.

Utility rope is made from natural fibers such as hemp (manila or sisal) or cotton or from synthetic material such as nylon. There is not a fire service standard for utility rope selection, care, and maintenance. However, you should be aware that natural fiber ropes can lose up to 50 percent of their strength when they become wet and then dry. This is an important issue when selecting a rope for use in hoisting tools and equipment.

In addition to life safety rope, there is a category known as *escape rope*. Escape rope is a one-use/onetime rope and is destroyed after a single use. Escape rope is not considered a life safety rope.

All types of ropes can be damaged by high stresses, impact loading, abrasions, kinking, and exposure to heat and chemicals. Life safety ropes should be inspected according to manufacturer recommendations by qualified personnel only. A record-keeping program must be kept for all life safety ropes. Information regarding inspection, after-incident review, and in-service date are kept in the record.

Inspect all types of rope after each use. Unused ropes should be inspected at least annually. Inspect rope visually and by touch. When making inspections, use the following methods described for the various ropes and note any observations. Rope should be inspected for imbedded shards of glass, metal shavings, wood splinters, or other foreign objects that could cause damage. If any of these are found, the rope should be taken out of service. Document all inspections in a rope log.

Kernmantle rope. Inspecting kernmantle rope for damage is somewhat difficult because the damage may not be obvious. The inspection can be performed by putting a slight tension on the rope while feeling for any lumps, depressions, or soft spots. A temporary soft spot resulting from hard knots or sharp bends in the rope may be felt; however, the fibers within the core may realign themselves over time if the rope is undamaged. The only way to determine whether such a soft spot is damage or just temporarily misaligned core fibers is by carefully inspecting the outer sheath. Any damage to the outer sheath

---

Life safety rope — Rope that meets the requirements of NFPA 1983®, *Standard on Fire Service Life Safety Rope, Harness, and Hardware*, and is dedicated solely for the purpose of constructing lines to be used for the raising, lowering, or supporting people during rescue fire fighting or other emergency operations, or during training. Also called lifeline.

Utility rope — Rope to be used in any situation that requires a rope — except life safety applications. Utility ropes can be used for hoisting equipment, securing unstable objects, and cordoning off an area.

indicates probable damage to the core. The core of a kernmantle rope can be damaged without visible evidence on the outer sheath. If there is any doubt about the rope's integrity, it should be downgraded to utility status **(Figure 7.6)**.

In addition to inspecting rope for damage to the core and sheath, inspect the rope for irregularities in shape or weave, foul smells, discoloration from chemical contamination, roughness, abrasions, or fuzziness. A certain amount of fuzziness is normal and is not necessarily a cause for concern. If there is a great amount of fuzziness in one spot or if the overall amount is excessive based upon the inspector's judgment and experience, the rope should be removed from life safety use.

Laid rope. Inspect synthetic laid ropes for soft, crusty, stiff, or brittle spots; for areas of excessive stretching; for cuts, nicks, or abrasions; for chemical damage; for dirt or grease; and for other obvious flaws. Laid rope should be untwisted and checked internally for these flaws. In synthetic laid rope the presence of mildew does not necessarily indicate a problem; nonetheless, the rope should be cleaned and reinspected. In natural fiber rope, a foul smell might indicate rotting or mildew.

Figures 7.5 a and b There are two classifications of fire service ropes: a) Life-Safety; and b) Utility.

Figure 7.6 Kernmantle rope consists of an inner core of fibers covered by an outer sheath.

Braided rope. Visually inspect braided rope for exterior damage such as heat sears (caused by friction or fire), nicks, and cuts. Also, visually inspect for excess or unusual fuzziness. Tactilely inspect for permanent mushy spots or other deformities.

Braid-on-braid rope. Inspect braid-on-braid rope for heat sears, nicks, and cuts. Also inspect for the sheath sliding on the core. If sliding is found, cut the end of the rope and pull off the excess material; then seal the end. Inspect for lumps that indicate core damage. A reduction in the rope's diameter may indicate a break in the core. Carefully examine any type of damage or questionable wear to the sheath.

Fire Hose

Annual service tests are required for all fire hose. The tests are intended to ensure that the hose is able to maintain pressure without leaking or breaking at a coupling. Hose is inspected monthly, usually when the apparatus hose load is removed and repacked, or after use as it is reloaded onto the apparatus. Fire hose should be inspected for the following conditions **(Figure 7.7)**:

- Dented or damage couplings
- Damaged coupling threads
- Missing or damaged coupling gaskets
- Abraded or damaged exterior hose jacket
- Discoloring of exterior hose jacket due to heat exposure
- Presence of mildew or mold on hose jacket or lining
- Lining detached from jacket
- Lining detached from coupling
- Unclean or contaminated jacket

Portable Fire Extinguishers

Annual inspections and tests of extinguishers are listed in NFPA® 10, *Standard for Portable Fire Extinguishers.* The fire and emergency service organization will also require that extinguishers be inspected weekly or monthly and after every use. Items to look for when inspecting extinguishers include the following:

- Damage to the extinguisher shell or cylinder
- Damage to discharge nozzle, horn, or hose

Figure 7.7 Fire hose should be inspected monthly, when the hose is removed from the hose bed and repacked, or when it is repacked following its use.

- Obstructions in nozzle, horn, or hose
- Locking pin or tamper seal are in place
- Correct pressure on gauge of stored-pressure extinguisher
- Hydrostatic test date on cylinder to ensure that it is current
- Evidence that the extinguisher has been discharged and not refilled

Electronic Equipment

Every year, new electronic equipment becomes available for the fire and emergency service. Higher levels of technology and sophistication mean that equipment may be more sensitive to abuse, damage, or the environment. Therefore, electronic equipment must be inspected and, if necessary, recalibrated at the start of each work shift and after each use. Although some equipment will require a trained and certified technician to calibrate it, the majority will be simple enough for users to calibrate. It is essential the personnel be trained in this activity prior to attempting to calibrate any type of electronic equipment. Examples of electronic equipment include thermal imaging cameras (TICs), dosimeters, and gas meters.

Tool and Equipment Maintenance

Tool and equipment maintenance is the responsibility of fire and emergency responders as well as maintenance technicians employed by, or under contract to, the jurisdiction. When an inspection reveals the need for maintenance, it should occur as soon as possible. Local policy and procedures will dictate how tools and equipment are removed from service and repaired or discarded.

Use of Maintenance Tools

While some tool and equipment maintenance, such as respiratory protection equipment and radio maintenance, must be performed by trained and certified technicians, much of the work can be done by fire and emergency responders. However, you must be trained to use any power-operated maintenance equipment safely and to repair tools and equipment according to the manufacturer's recommendations. All required PPE, including hand, eye, respiratory, and hearing protection, must be worn while doing the maintenance.

The potential for injuries in maintenance areas can be reduced through the design of the space and by keeping it clean and orderly. The preferred method for ensuring safety is through risk reduction during the design of the shop or maintenance area. Other safety hazards can be controlled by adding safeguards to the individual equipment. Some of these design features include the following:

- Eye shields and mechanical safeguards installed on all power tools
- Large tools such as grinders and drill presses secured to the floor, workbench, or wall
- Noise-reducing materials installed on walls, floors, and ceilings where possible
- General room ventilation provided to filter and remove particulates, vapors, and odors from the area

- Spray-paint booths provided in any area where spray painting takes place
- Designated rooms for the maintenance of SCBA regulators and oxygen regulators **(Figure 7.8)**
- Adequate lighting; fixtures equipped with screens to prevent or contain lightbulb breakage
- Electrical receptacles equipped with ground fault circuit interrupters (GFCIs) to reduce shock hazards in the work area

Potential hazards can also be controlled in the maintenance area through the following policies and procedures that promote cleanliness in the area:

- Keep maintenance areas clean and neat.
- Store tools in a designated area when not in use.
- Provide waste and trash containers and mark them according to their contents to ensure that solid waste is not contaminated with hazardous waste.
- Remove all waste from the maintenance area daily and from the facility at least weekly.
- Remove hazardous waste, oils, batteries, grease, etc., and dispose of them properly in accordance with OSHA and EPA requirements.
- Keep walkways clear to prevent tripping hazards.
- Mark floors to indicate walkways and segregate equipment areas.
- Clean up oil, grease, and liquid spills when they occur to prevent slipping and contaminating of other areas.
- Make periodic inspections of hoses, couplings, fittings, and joints on all power-operated equipment to detect possible damage.
- Follow the manufacturer's lubrication and maintenance schedules for all power-operated equipment.

Battery Charging and Changing

Battery-operated tools and equipment have become common in the fire service. They allow increased flexibility in areas where electrical power may not be available. The following items may be powered by a battery:

- Radios
- Thermal imaging cameras (TICs)
- Medical monitors/Automatic External Defibrillators (AEDs)
- Flashlights
- Gas/radiation detectors
- SCBA Personal Distress Locators/PASS devices
- Drills
- Saws
- Computers/mobile data terminals (MDTs)
- Global Positioning Systems (GPSs)

Batteries come in various shapes, sizes, and capacities. They may be designed for single use and then discarded or they may be rechargeable. Rechargeable batteries have portable chargers that are kept in a facility or mounted on the

Figure 7.8 Designated maintenance areas for self-contained breathing apparatus (SCBA) contain spare parts, repair tools, testing equipment, and cleaning areas. *Courtesy Cherry Hill (NJ) Fire Department.*

Figure 7.9 Radio batteries are recharged as needed or when the battery must be recycled by draining the charge completely.

apparatus. Recharging generally requires a certain amount of time for the battery to gain a full charge **(Figure 7.9)**. Some rechargeable batteries must be completely depleted before recharging. If the battery is not completely discharged, it will establish a memory based on the recharged amount, thereby reducing its ability to hold its complete charge.

Batteries contain lead acid, nickel-cadmium, alkaline, Lithium-ion, or nickel zinc. All batteries can become dangerous when overheated or exposed to the elements.

Battery inspection and replacement periods vary. Batteries used in portable radios, SCBA personal locator devices, and TICs should be inspected at the beginning of each work shift. Batteries should be changed or charged if they appear to be weak or the unit fails to operate properly.

Charging batteries can be hazardous. Older batteries can leak contents, because of damage to the casing, repeated use, and exposure to heat. You should ensure that the battery matches the charging system. Case configurations may look the same, but contacts may not match the battery bank's contacts. Read the manufacturer's recommendations for charging and changing batteries. A department policy in your organization may dictate when to change and how to charge a battery safely. Facilities that house electric forklifts and golf carts will contain charging stations.

Battery charging should take place in areas devoted exclusively to this type of activity and isolated from other employees who are not involved in battery charging. Wood-slat floorboards or rubber mats should be utilized in the area to minimize slips and falls and the danger of electric shock. Area ventilation is required to remove potential toxic fumes from the area. Water for fire protection and for flushing spills of electrolyte on the floor must be available in the charging area. Flammable liquids must not be located in the charging area. Other safety precautions include:

- No Smoking is allowed in charging areas
- Use proper lifting techniques or mechanical hoists to lift batteries
- Only trained personnel should charge or change batteries
- Do not place metal tools on top of batteries
- Prevent metal tools from contacting both battery terminals and causing a short circuit
- Check the fluid levels of wet-cell batteries regularly

Batteries release oxygen and hydrogen gases when they are charging and can create a potentially explosive atmosphere. Locations used for charging must be well ventilated to prevent concentrations of hydrogen and oxygen from reaching volatile levels. General or local ventilation can be provided by a fume hood or an exhaust fan.

Sulfuric acid is a common component in batteries. In the event of a battery acid spill, neutralizing agents should be spread on the spill. After the spill is neutralized, it can be safely cleaned up and disposed of by trained personnel in accordance with local ordinances.

When jump-starting a vehicle, forklift, or cart with a dead battery, follow these steps (**Figure 7.10**):

Step 1: Ensure the vehicle is correctly positioned and the brakes are set.

Step 2: Connect the red clamp of the jumper cables to the positive terminal of the good battery first.

Step 3: Connect the other red clamp to the positive terminal of the dead battery.

Figure 7.10 Occasionally, it may be necessary to jump-start a vehicle. A safe procedure should be followed when performing this task.

Step 4:   Connect the black clamp to the negative terminal of the good battery.

Step 5:   Connect the other black clamp to the frame or ground of the dead vehicle away from the battery.

Step 6:   Start the vehicle with the good battery. Run the engine at a fast idle between 1,200 to 1,500 rpm. This action helps charge the dead battery, making starting it easier.

Step 7:   Now start the vehicle with the dead battery. If the cable connections to the batteries are good, the engine should start if there are no other electrical problems such as a bad starter or ignition circuit problem.

Step 8:   When the dead engine starts, disconnect both jumper cables. Do not allow the jumper cables to touch each other or the red cables to touch anything metal on either vehicle.

Step 9:   Allow the engine to run 20 to 30 minutes to recharge the battery.

Unpowered, Powered, and Extinguisher Maintenance

As a member of an emergency response unit, you will be responsible for the care and upkeep of tools and equipment assigned to you and to the unit. You may also be assigned the task of caring for tools that are kept in the station or maintenance area. Most of these tools will consist of nonpowered hand tools. However, power-operated tools and equipment will also require frequent care and maintenance that you may be able to do. In addition, you may be responsible for maintaining portable fire extinguishers carried on the apparatus and located in the fire station. This section includes general information on maintaining most types of unpowered and powered tools, equipment, and portable extinguishers. Remember that you must always follow the manufacturer's instructions for cleaning and maintaining any tool or equipment.

Unpowered

Unpowered, or hand, tools are the most common types of tools used in the fire and emergency services. Most are designed to be carried and used by one person although some require two people to operate. Examples of unpowered tools include but are not limited to

- Pike poles
- Sledge hammers
- Axes, flat-head and pick-head
- Claw hammers and roofing hammers
- Picks
- Shovels
- Bolt and wire cutters
- Pry bars
- Halligans
- Battering rams

Unpowered tools may be constructed from a single material such as metal or a combination of materials such as metal and wood. The working end is usually made from a very hard grade of steel while the handles are generally wood or fiberglass. Metal parts may be natural (unpainted) or painted.

Wooden Handles. Most wooden handles are found on hammers, sledgehammers, axes, pike poles, and wildland tools. Care and maintenance of wooden handles include the following:

- Inspect the handle for cracks, blisters, or splinters.
- Sand the handle if necessary to eliminate splinters **(Figure 7.11)**.
- Wash the handle with mild detergent and rinse; wipe dry. Do not soak the handle in water because it will cause the wood to swell.
- Apply a coat of boiled linseed oil to the handle to preserve it and prevent roughness and warping. Do not paint or varnish the handle.
- Check the tightness of the tool head.
- Limit the amount of surface area covered with paint for tool marking. Some departments use self-adhesive bar codes on the handle for identification.

Fiberglass Handles. Fiberglass handles have become a popular replacement for wood handles because they are sturdy and do not require the amount of maintenance that wood handles do. When caring for fiberglass handles, you should use the following procedure:

- Wash the handle with mild detergent, rinse, and wipe dry.
- Check for damage or cracks.
- Check the tightness of the tool head.
- Do not paint or varnish fiberglass handles.

Cutting Edges. Cutting edges are found on hand tools such as axes, bolt and wire cutters, some tools used for opening walls, and wildland tools. Inspect the tool's cutting edge for chips, cracks, or burrs.

Figure 7.11 Tools with wood handles, such as axes, must be sanded and refinished to remove splinters and prevent permanent damage to the wood surface.

If the edge is damaged, replace it. While hand sharpening is the preferred method, some cutting edges may require the use of a grinding wheel or stone. Care must be taken when using a grinding wheel because it will generate enough heat to weaken the metal of the cutting tool.

The choice of sharpening tools is a matter of preference and may be regulated by local policy. Sharpening tools include the following:

- **Whetstones** — Most common sharpening tool, available in a variety of surfaces and sizes

- **Diamond-coated flat file** — Durable, long lasting; this type of file only requires water for lubrication and remains flat while sharpening

- **Sharpening steel** — Useful for finishing or for repairing minor dullness or damage

You should follow the manufacturer's instructions or departmental procedures for sharpening the cutting edge of a tool. Remember that a blade that is extremely sharp and ground too thin may break when cutting gravel roofs or striking nails in roof decking or flooring. General guidelines for sharpening a cutting edge are as follows:

Figure 7.12 A file or a rough whetstone can be used to sharpen the cutting edge of an axe blade.

Step 1: Secure the tool in a vice or clamp to support it.

Step 2: Using a file or rough whetstone, push the file inwards from the cutting edge to the back edge **(Figure 7.12)**.

Step 3: Using a smooth whetstone, finish the edge with a circular motion.

Step 4: For bolt or wire cutters, sharpen only the inclined edge.

Step 5: For axes or hatchets (two-sided cutting edge), turn the tool over and sharpen the other side of the cutting edge.

Step 6: Wipe off the cutting edge to remove metal shavings and coat with a thin layer of oil or lubricant

NOTE: DO NOT PAINT AXE HEADS! Painting hides faults in the metal. Paint also may cause the cutting surface to stick and bind.

Plated Surfaces. Plated surfaces are those that are protected by chromium or other metal applied by an electroplating process. General maintenance includes inspecting for damage, wiping plated surfaces clean, or washing with mild detergent and water. Conducting repairs to a plated surface may release cadmium fumes that require the use of respiratory protection equipment. Plating materials that are flaking can produce a sharp edge that can result in cuts or abrasions.

Unprotected Metal Surfaces. Unprotected metal surfaces are the blades and other tool components that are not electroplated to protect them from rust or corrosion. General maintenance includes the following:

- Keep the surface clean, dry, and free of rust.

- Oil the metal surface lightly. Light machine oil works best. Avoid using any metal protectant that contains 1-1-1-trichloroethane because this chemical may damage and weaken the handle.

- Do not paint metal surfaces — paint hides defects.
- Inspect the metal for chips, cracks, or sharp edges, and file them off when found.

Power-Operated Equipment

Care and maintenance of power-operated equipment must follow the manufacturer's recommendations. Read the instructions carefully. Maintenance for power-operated equipment includes the following **(Figure 7.13)**:

- Check all fluid levels.
- Use only the recommended types of lubricants, hydraulic fluids, and grades of fuels in the equipment.
- Be sure that battery packs are fully charged.
- Inspect power tools at the beginning of the work shift and ensure they will start manually.
- Check blades for damage or wear.
- Replace blades that are damaged or worn.
- Check all electrical components (cords, portable receptacles, etc.) for cuts or other damage.
- Ensure that all guards are functional and in place.
- Ensure that fuel is fresh. A fuel mixture may separate or degrade over time.
- Inspect hydraulic and fuel supply hoses for damage.
- Inspect hydraulic hose couplings (quick-disconnect fittings) to ensure that they are clean and functional.
- Ensure that all parts and support items are available for immediate access.

Figure 7.13 Power tools such as generators and fans are inspected regularly.

Portable Fire Extinguishers

General fire extinguisher maintenance may be performed by shop personnel. Fire extinguishers share a few common characteristics regarding safe operation such as the following:

- Extinguishers expel contents under pressure. Care must be taken to avoid directing streams to the faces of firefighters or occupants.

- Extinguishers do not all operate in the same manner. Training in the use of extinguishers must be thorough.

- Extinguisher contents must match the type of hazard they are intended to control.

- The operator must be aware of the environment during operation: for example, expelling carbon dioxide in an enclosed room or expelling dry chemical powder in windy conditions or in crowds should be avoided.

- Avoid splashing or spraying flammable liquid when discharging the extinguisher on Class B spills or fires.

Pump-Type Water Extinguishers. Pump-type water extinguishers are intended for extinguishment of small Class A fires only. There are several kinds of pump-type water extinguishers, but all operate in a similar manner. They are equipped with either a single- or double-acting pump. A stream of water is emitted from a nozzle with each action of the pump.

Stored-Pressure Water Extinguishers. Stored-pressure water extinguishers, also called *air-pressurized water (APW) extinguishers,* or *pressurized water extinguishers* are useful for all types of small Class A fires and are often used for extinguishing confined hot spots during overhaul operations.

Water is stored in a tank along with either compressed air or nitrogen. A gauge located on the side of the valve assembly shows when the extinguisher is properly pressurized. When the operating valve is activated, the stored pressure forces water up the siphon tube and out through the hose.

Water-Mist Stored-Pressure Extinguishers. Although very similar in appearance to standard stored-pressure water extinguishers, water-mist extinguishers use deionized water as the agent and nozzles that produce a fine spray instead of a solid stream. Because it is the impurities in water that make it electrically conductive, the deionized water makes these Class A extinguishers also safe to use on energized electrical equipment (Class C). The fine spray also enhances the cooling and soaking characteristics of the water and reduces scattering of the burning materials.

Wet Chemical Stored-Pressure Extinguishers. Also similar in appearance to standard stored-pressure water extinguishers, wet chemical (Class K) units are specifically designed to control and extinguish fires in deep fryers. These extinguishers contain a special potassium-based agent formulated to cool and suppress fires in unsaturated cooking oils.

Aqueous Film Forming Foam (AFFF) Extinguishers. Aqueous film forming foam (AFFF) extinguishers are suitable for use on Class A and Class B fires. They are particularly useful in combating fires in or suppressing vapors from small liquid fuel spills.

AFFF extinguishers are different from stored-pressure water extinguishers in two ways. The AFFF extinguisher tank contains a specified amount of AFFF concentrate mixed with the water, and it has an air-aspirating nozzle that aerates the foam solution, producing a better-quality foam than a standard extinguisher nozzle provides.

The water/AFFF solution is expelled by compressed air or nitrogen stored in the tank with the solution. To prevent the disturbance of the foam blanket when applying the foam, it should not be applied directly onto the fuel but should be applied in front of the spill and allowed to roll onto the surface of the fuel.

NOTE: AFFF is corrosive and can remove paint from tools and apparatus.

Clean Agent Extinguishers. Designed specifically as a replacement for Halon 1211, these newer extinguishers use hydrochlorofluorocarbon (HCFC), hydrofluorocarbon (HFC), perfluorocarbon (PFC) or fluoroidiocarbon (FIC), so-called *clean agent*s discharged as a rapidly evaporating liquid that leaves no residue. These agents effectively cool and smother fires in Class A and Class B fuels, and are nonconductive so that they can be used on energized electrical equipment (Class C) fires. Pressurized with argon gas, Halotron extinguishers are approved by the U.S. Environmental Protection Agency (EPA).

Carbon Dioxide (CO_2) Extinguishers. Portable carbon dioxide (CO_2) fire extinguishers are found as both handheld units and wheeled units. CO_2 extinguishers are most effective in extinguishing Class B and Class C fires. Because their discharge is in the form of a gas, they have a limited reach and the gas can be dispersed by wind. They do not require freeze protection.

CO_2 is stored under its own pressure as a liquefied gas ready for release at anytime. The agent is discharged through a plastic or rubber horn on the end of either a short hose or tube. The gaseous discharge is usually accompanied by dry ice crystals or CO_2 *snow*. This snow sublimes — changes into a gaseous form without becoming a liquid — shortly after discharge. When released, CO_2 gas displaces available oxygen and smothers the fire. Even though CO_2 discharges at subzero temperature, it has little if any cooling effect on fires. CO_2 produces no vapor-suppressing film on the surface of the fuel so reignition is always a danger.

CO_2 wheeled units are similar to the handheld units except that they are considerably larger. Wheeled units are most commonly used in airports and industrial facilities. After being wheeled to the fire, the hose (usually less than 15 feet [5 m] long) must be deployed or unwound from the unit before use. The principle of operation is the same as in the smaller handheld units.

Dry Chemical Extinguishers. The terms *dry chemical* and *dry powder* are often incorrectly used interchangeably. Dry chemical agents are for use on Classes A-B-C fires and/or Classes B-C fires; dry powder agents are used on Class D fires only. Dry chemical extinguishers are among the most common portable fire extinguishers in use today. There are two basic types of dry chemical extinguishers: (1) regular B:C-rated and (2) multipurpose and A:B:C-rated. Unless specifically noted in this section, the characteristics and operation of both types are the same. The following are commonly used dry chemicals:

- Sodium bicarbonate
- Potassium bicarbonate
- Urea-potassium bicarbonate
- Potassium chloride
- Monoammonium phosphate

During manufacture, these agents are mixed with small amounts of additives that make the agents moisture-resistant and prevent them from caking. This process keeps the agents ready for use even after being stored for long periods, and it makes them free-flowing.

Some dry chemical agents are not compatible with foam. Monoammonium phosphate and some sodium bicarbonate agents will break down the foam blanket when applied in conjunction with or after foam.

The dry chemical agents themselves are nontoxic and generally considered quite safe to use. However, the cloud of chemicals may reduce visibility and create respiratory problems like any airborne particulate. Some dry chemicals are compatible with foam, but others will degrade the foam blanket. On Class A fires, the discharge should be directed at whatever is burning in order to cover it with chemical. When the flames have been knocked down, the agent should be applied intermittently as needed on any smoldering hot spots. Many dry chemical agents are corrosive to metals, so it may be better to use another agent such as CO_2 on them.

Portable dry chemical extinguishers are produced in two different styles: handheld and wheeled. Handheld units weigh from 1 to 30 pounds (0.5 to 14 kg) and are designed to be carried and used by one person. Wheeled units, mounted on carts, weigh 50 pounds (23 kg) or more and can be moved by one or more persons.

There are two basic designs for handheld dry chemical extinguishers: *stored-pressure* and *cartridge-operated*. The stored-pressure type is similar in design to the air-pressurized water extinguisher, and a constant pressure of about 200 psi (1 400 kPa) is maintained in the agent storage tank. Cartridge-operated extinguishers employ a pressure cartridge connected to the agent tank. The agent tank is not pressurized until a plunger is pushed to release the gas from the cartridge. Both types of extinguishers use either nitrogen or CO_2 as the pressurizing gas. Cartridge-operated extinguishers use a CO_2 cartridge unless the extinguisher is going to be subjected to freezing temperatures; in such cases, a dry nitrogen cartridge is used **(Figure 7.14, p. 256)**.

Dry chemical wheeled units are similar to the handheld units but are on a larger scale. They are rated for Class A, Class B, and Class C fires based on the dry chemical in the unit.

Operating the wheeled dry chemical extinguisher is similar to operating the handheld, cartridge-type dry chemical extinguisher. The extinguishing agent is kept in one tank, and the pressurizing gas is stored in a separate cylinder. When the extinguisher is in position at a fire, the hose should be stretched out completely first. Once the agent storage tank and hose are charged, it can make removing the hose more difficult, and the powder can sometimes pack in any sharp bends in the hose. The pressurizing gas should be introduced into the agent tank and allowed a few seconds to fully pressurize the tank before the nozzle is opened. The agent is applied in the same way as that described for the handheld, cartridge-type dry chemical extinguishers.

Extinguishers and Agents for Metal Fires. Special extinguishing agents and application techniques have been developed to control and extinguish metal fires. No single agent will control or extinguish fires in all combustible metals. Some agents are effective against fires in several metals; others are effective on fires in only one type of metal. Some powdered agents can be applied with portable extinguishers, but others must be applied by either a shovel or a scoop. The appropriate application technique for any given dry powder is

WARNING!
Never mix or contaminate dry chemicals with any other type of agent because they may chemically react and cause a dangerous rise in pressure inside the extinguisher.

CAUTION
When pressurizing a cartridge-type extinguisher, do not have your head or any other part of your body above the top of the extinguisher. If the fill cap was not properly screwed on, the cap and/or a cloud of agent could forcibly discharged.

CAUTION
As with handheld units, the top of the extinguisher should be pointed away from the operator and any other nearby personnel when pressurizing the unit. Because of the size of the nozzle, the operator should be prepared for a significant nozzle reaction when it is opened.

Figure 7.14 Fire extinguisher maintenance: refilling, charging, and repair. Some testing will require certified technicians to perform the work.

described in the manufacturer's technical sales literature. Firefighters should be thoroughly familiar with the information that applies to any agent carried on their emergency response vehicles.

Portable extinguishers for Class D fires come in both handheld and wheeled models. Whether a particular dry powder is applied with an extinguisher or with a scoop, it must be applied in sufficient depth to completely cover the area that is burning to create a smothering blanket. The agent should be applied gently to avoid breaking any crust that may form over the burning metal. If the crust is broken, the fire may flare up and expose more uninvolved material to combustion. Care should be taken to avoid scattering the burning metal. Additional applications may be necessary to cover any hot spots that develop.

Cleaning and Disinfecting Tools and Equipment

Both Title 29 CFR 1910.1030 (Subpart Z_Toxic and Hazardous Substances, Bloodborne Pathogens) and NFPA® 1500 and 1581 require new and existing fire stations to have two separate areas designated for cleaning and disinfecting. One area is designated for the cleaning of equipment and protective clothing and the other is for disinfecting contaminated emergency medical equipment. It is the responsibility of the health and safety officer (HSO) or the infection control officer to designate these areas.

Local health department codes as well as federal and state/provincial environmental protection regulations will determine the appropriate filtration, storage, or disposal of contaminated wastewater from the cleaning and disinfecting areas. Cleaning materials used in these two areas, such as scrub brushes, disinfectant solutions, soaps, and sponges, should be stored within easy reach of the sink. To prevent further exposure or contamination, chemical goggles, aprons, and vinyl medical gloves must be provided by the fire and emergency services organization. A written protocol for disinfecting and cleaning of protective clothing, medical equipment, and fire-fighting tools and apparatus must be implemented. This protocol should be posted in the disinfecting area and included in the health and safety training for all personnel.

The HSO or the infection control officer must ensure that all personnel wear proper clothing when involved in cleaning or disinfecting activities. Equipment that has been cleaned or disinfected should be allowed to drain into the sink and not onto the floor. This prevents the possibility of tracking contaminated water into the living quarters and prevents a slipping hazard. Well-ventilated drying areas for clothing and equipment should be provided.

The disinfecting, cleaning, and drying of SCBA facepieces are extremely important. If they are not properly cleaned, SCBA facepieces can spread infection through the eyes, nose, and mouth of the wearer. Disinfecting and cleaning facepieces must follow the manufacturer's recommendation and the guidelines found in the current edition of NFPA® 1981, or other applicable respiratory protection standards. The written protocol for cleaning and disinfecting should designate the location and method for maintaining the SCBA facepieces. However, the cleaning and disinfecting must never be done in the kitchen or bathroom of the fire station.

Designated Cleaning Area

The designated cleaning area, which should be located in the apparatus bay, must be equipped with proper ventilation, lighting, and drainage facilities (**Figure 7.15**). Drainage must be connected to the sanitary sewer system or septic tank and located away from the kitchen, sleeping and living quarters, bathroom, and designated disinfecting area. Personal protective clothing, fire-fighting equipment, and portable equipment are cleaned in this area. The immediate cleaning of contaminated clothing and equipment prevents the potential contamination of other personnel or other portions of the facility.

Facility cleaning equipment may also be cleaned in this area. You should check with your local health department, water/sewer department, and state/local EPA to ensure there are no problems putting contaminated water into the public sewer.

Figure 7.15 Designated equipment cleaning areas are located in the apparatus bay and equipped with sinks for washing equipment, as well as laundry equipment.

Figure 7.16 Areas used to disinfect medical equipment are maintained in separate areas from the living quarters.

Designated Disinfecting Area

The designated disinfecting area is primarily intended for the disinfecting of medical equipment exposed to bloodborne pathogens **(Figure 7.16)**. It should be designated for disinfecting only and not used for the cleaning of mops, other materials used for general cleaning in the station, or equipment and clothing that have not been contaminated by blood or other body fluids. The area should be equipped with a stainless steel, two-bay utility sink with both hot and cold water service, nonporous rack shelving, adequate drainage, proper lighting, and proper ventilation. The disinfecting area should be located in or adjacent to the apparatus bay to encourage use immediately upon return to quarters. Immediate disinfecting of contaminated medical equipment reduces the possibility of exposure to bloodborne pathogens by other personnel. You should check with your local health department, water/ sewer department, and state/local EPA to ensure there are no problems putting contaminated water into the public sewer.

Biological waste should be disposed of in properly marked containers. Used needles and other disposable sharp objects should be placed in *sharps containers* and disposed of according to local protocol.

Hazardous Materials Storage and Use

Generally, limited quantities of hazardous materials in the form of flammable/ combustible liquids, corrosives, and compressed/liquefied gases are stored and used in fire and emergency services facilities. The following sections describe the types of storage containers and safe practices for using the flammable/ combustible liquids, corrosives, and compressed/liquefied gases.

Flammable/Combustible Liquids

Flammable and combustible liquids may be found in a variety of containers:

- Glass containers
- Metal containers
- Polyethylene containers
- Safety cans

Because flammable and combustible liquids have ignition temperatures of varying degrees, they are divided into classes. The classes are used to determine the type and capacity of storage containers **(Tables 7.3 and 7.4)**. In addition to storing them in the proper containers, flammable/combustible and corrosive liquids should also be kept in approved storage cabinets or rooms. These cabinets and rooms are designed to contain any spill or fire involving the materials or that they may be exposed to.

Glass containers. Glass containers or bottles may be found in sizes up to 1 gallon (4 L). Because glass will not react with acids and bases, it is suitable for storing corrosives. Generally, the types of contents may include cleaning or housekeeping solutions, antiseptic or medical solutions, and pesticides.

Table 7.3
ICC® and NFPA® Flammable Liquids Classes

Class	Flash Point	Boiling Point	Examples
Class IA	Below 73°F (23°C)	Below 100°F (38°C)	Acetaldehyde Cyclohexane Ethel ether
Class IB	Below 73°F (23°C)	At or Above 100°F (38°C)	Acetone Benzene Gasoline
Class IC	At or Above 73°F (23°C)	Below 100°F (38°C)	Hydrazine Styrene Turpentine

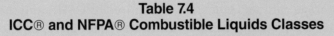

Table 7.4
ICC® and NFPA® Combustible Liquids Classes

Class	Flash Point Minimum	Flash Point Maximum	Examples
Class II	At or Above 100°F (38°C)	Below 140°F (60°C)	Acetic acid Diesel fuel Naphtha Stoddard solvent
Class IIIA	At or Above 140°F (60°C)	Below 200°F (93°C)	Cyclohexanol Formic Acid Nitrobenzene
Class IIIB	At or Above 200°F (93°C)	N/A	Formalin Motor oil Picric acid

Due to the potential for breakage, the quantity of flammable and combustible liquids is restricted by NFPA® 30. The maximum quantities for glass containers are as follows:

- *Class IA* — 1.05 pints (0.5 L)
- *Class IB* — 1.05 quarts (1 L)
- *Class IC* — 1.3 gallons (5 L)
- *Class II* — 1.3 gallons (5 L)
- *Classes IIIA* — 5.3 gallons (20 L)
- *Class IIIB* — 5.3 gallons (20 L)

Metal containers. Metal containers include cans up to 10 gallons (40 L) and drums up to 55 gallons (275 L). Metal containers may be made from aluminum, steel, stainless steel, or tin. Corrosion, in the form of rust, can form on the exterior causing the containers to leak especially along weld or seam lines. Some metal containers are coated on the inside to prevent corrosion and reactions between the contents and the container.

Safety cans. An inspector most often encounters flammable or combustible liquids stored in containers that are a maximum of 5 gallons (20 L) or less in size. Although there are other acceptable methods of storing these small amounts, the safest containers are approved safety cans. Safety cans are constructed to reduce the chance of leakage or container failure. They are also designed to virtually eliminate vapor release from the container under normal conditions. Safety cans use self-closing lids with vapor seals and contain a flame arrester in the dispenser opening. The self-closing lid also acts as a pressure-relief device when the can is heated.

Polyethylene containers. Polyethylene or plastic containers come in all sizes ranging from small containers used for household products to bulk packaging used for transport and storage. All types of hazardous materials may be stored in plastic when the contents are compatible with the container materials. Plastic can break, split, or crack if dropped or struck. Leaks along seam lines or around caps may also occur.

Storage cabinets. Designed primarily for flammable and combustible liquids, the use of storage cabinets or lockers is required by OSHA and the model fire codes **(Figure 7.17)**. Storage cabinets are intended to safely contain hazardous materials. Storage cabinet safety characteristics are as follows:

- Provides a fire-resistant steel enclosure that permits materials to be stored near the point of use, and reduces the risk of fire

- Provides spill containment by a leak-proof sill under the cabinet doors and welded seams at all joints

- Uses vents with flame arrestors to prevent fire from extending into the cabinet and provide ventilation

- Improves visibility and recognition of the cabinet under reduced light conditions with high-visibility reflective warning labels and yellow painted exteriors

- Prevents the accumulation of static electric charges with built-in grounding connectors

Storage rooms. Facilities storing flammable or combustible liquids containers in excess of the capacity of storage cabinets must be stored in storage rooms (called *control areas* by the model fire codes) that are specially constructed of fire-resistive materials in order to protect the

Figure 7.17 Flammable and combustible liquids must be properly stored in approved storage cabinets.

containers from any nearby fire exposures. These storage rooms must have appropriate secondary containment such as closed bottles or safety cans as required by the locally adopted code requirements. Storage rooms must have some form of ventilation, either natural or mechanical.

Corrosives

Corrosives include acids and bases that can cause harmful injures to human flesh. They must be stored and used carefully. Generally, storage requirements for corrosives are the same as that for flammable/combustible liquids. Corrosives will be found in glass and plastic containers and should be kept in cabinets or storage rooms that will retain the material if it is spilled. Proper PPE must be worn when dispensing or using corrosives, including eye and hand protection.

Compressed/Liquefied Gases

Compressed/liquefied gases are stored in U.S. Department of Transportation (DOT) approved cylinders. These cylinders must be secured in racks or chained to the structure to prevent them from falling or being struck. Cylinders must be hydrostatically tested with the test date stamped in the surface of the cylinder. Remember that SCBA cylinders are also compressed gas cylinders and must meet the same safety requirements. To ensure the safety of compressed/liquefied gases, you must perform the following procedures:

- Ensure that the hydrostatic test date is current.
- Inspect cylinders carefully for any physical damage.
- Store and use according to the manufacturer's recommendations.

General Safe Usage Practices

A number of important, preventative safety measures can be taken to prevent a fire or explosion where flammable/combustible liquids and compressed/liquefied gases are stored and used. You should take the following measures:

- Eliminate all sources of ignition such as the following:
    — Electrical, mechanical, or frictional sparks
    — Static electricity
    — Open flames
    — Hot surfaces
    — Incompatible materials
- Store flammable/combustible liquids and compressed/liquefied gases in approved closed containers.
- Ensure that all storage areas are ventilated to prevent the accumulation of vapors within flammable ranges or above flammable limits.
- Ensure that all storage containers or areas are properly marked for the type of material stored in them.
- Wear the correct personal protective equipment (PPE) as indicated by the manufacturer in the material safety sheet/safety data sheet (MSDS/SDS) or by department policy when using or handling hazardous materials

- Provide proper ventilation when using these materials, including exhaust hoods, fans, and openings in the structure.
- Separate products containing chlorine from ammonia or acid-based cleaners; mixing these incompatible chemicals can produce a dangerous toxic gas.
- Keep hazardous materials in their original containers or in approved and marked secondary containers; all containers must be tightly sealed and stored in an upright position.
- Ensure that different products are *never* combined or mixed together.
- Keep storage and work areas clean and free of trash.
- Ensure that numbers or marks stamped on compressed/liquefied gas cylinders remain in place and unchanged with the exception of those made by hydrostatic testing organizations.
- Protect compressed/liquefied gas cylinders from cuts or other physical damage.
- Do *not* tamper with safety devices on the valves of cylinders.
- Mark empty compressed/liquefied gas cylinders with the word *Empty* or the letters *MT* (phonetic equivalent of *Empty*), and secure their valve caps in place.
- Cover compressed/liquefied gas cylinder valves with safety caps when cylinders are not in use.

- Anchor stored compressed/liquefied gas cylinders to a wall or other secure object with chains around the tanks so that they cannot be knocked over accidentally **(Figure 7.18)**.
- Always disconnect a cutting torch head or instrument from the supply hose before storing it in a toolbox or cabinet.
- Ensure that oxygen cylinders, hoses, and connections are free of oil, grease, or other hydrocarbon (petroleum-based) lubricants.
- Store incompatible gases in separate areas.

Figure 7.18 Compressed gas cylinders are secured in place to prevent them from being knocked over accidentally.

- Store compressed/liquefied gas cylinders in an area that is clean and free of accumulated combustible materials.

- Ensure that cylinders that have been abused, dented, arc-struck, damaged, or are severely corroded are removed from service and replaced.

- Report compressed/liquefied gas cylinders that have expired hydrostatic test dates and remove them from service.

- Ensure that the proper type, size, and number of portable fire extinguishers are located in the storage and work areas and that they are visible and mounted to the structure.

Testing Following All Major Repairs

Prior to returning a tool or equipment to service, it must be inspected and, in some cases, tested. Simple hand tools and ladders may only need to be inspected to ensure that the repairs have been properly made. Fire hose should be inspected, pressure-tested, and checked for leaks, especially if couplings have been replaced.

Power-operated equipment should be tested using the initial acceptance or annual test criteria. Those requirements may include a minimum operating time, pressure, or other criteria.

Records should be kept on tool and equipment repairs and replacement. Records serve a multitude of uses; among these are the following:

- Help keep track of department property

- Assist in developing the department's budget based on actual cost of repair and replacement

- Are essential for providing an accurate history of the tool or equipment

- May be required by law or ordinance

Disposal of Damaged or Obsolete Tools and Equipment

Tools and equipment that are obsolete, have reached the end of their life cycle, are damaged beyond repair, or have accumulated a high operating cost must be properly disposed of. Any disposal of damaged or obsolete tools should be in accordance with local, state/provincial, and federal laws. Most jurisdictions have policies in place that regulate how equipment is disposed of. In some cases, the equipment must be destroyed due to the liability potential in the event of an accident by a future owner.

Tool and Equipment Operation

When you are at the site of an emergency incident, you must use the tools and equipment assigned to you or your unit safely. As mentioned at the beginning of this chapter, injuries can result from the unsafe use of tools and equipment. This section addresses selecting the proper tool for the task, the proper way to carry tools and equipment, and the proper way to use them.

Selecting the Proper Tool

The majority of tools are designed to perform specific functions. Very few universal tools can perform multiple tasks. Always select the tool based on the task that you need to accomplish. Training prepares you for making the appropriate choice at an emergency incident. Practice helps you to remember what you learned in training; experience helps you to make the selection quickly.

Remember, pry tools are designed and constructed to allow force to be applied to one end in order to move something on the other end. Cutting tools are designed to cut or sever an object. However, specific types of materials, such as steel and concrete will require specialized cutting blades on power-operated equipment. Utility rope is designed to lift light loads and not for lifting humans or making rescues.

It is important for you to follow your organization's policies, procedures, and the manufacturer's recommendations for the use of all tools and equipment. Safety must be the first consideration when selecting and using tools and equipment.

 ## Carrying Tools

Firefighters must carry tools and equipment in the safest manner possible. Precautions should be taken to protect the carrier, other firefighters, and bystanders. Some recommended safety practices for carrying specific tools are as follows:

- *Axes* — If not in a scabbard, carry the axe with the blade away from the body. With pick-head axes, grasp the pick with a hand to cover it. Never carry an axe on the shoulder. A best practice is to carry an axe in a scabbard when ascending a ladder. This practice frees both hands to hold onto the ladder.

- *Prying tools* — For safety, carry these tools with any pointed or sharp edges away from the body. With multiple surfaces, this will be somewhat difficult.

- *Combinations of tools* — Strap tool combinations together. Halligan bars and flat-head axes can be nested together and strapped. Short sections of old hose can be slipped over the handles of some tools and smaller prying tools inserted into the hose **(Figure 7.19)**.

- *Pike poles and hooks* — For safety, carry these tools with the tool head down, close to the ground, and ahead of the body when outside a structure. When entering a building, carefully reposition the tool and carry it with the head upright close to the body to facilitate prompt use. These tools are especially dangerous because they are somewhat unwieldy and can severely injure anyone accidentally jabbed with the working end of the tool.

- *Striking tools* — Keep the heads of these tools close to the ground. Maintain a firm grip. Mauls and sledgehammers are heavy and may slip.

- *Power-operated tools* — The weight of the tool will dictate the number of firefighters that will be required to carry it. Regardless of the number, always lift with the legs by crouching to grasp and raise the tool. Never bend over to lift an object: this leads to back strains and injuries. Heavy objects that are located in apparatus compartments that are above shoulder height can cause back or shoulder injuries when lifting them off the vehicle. These items should be carried in lower compartments or equipped with some form of mechanical assistance to lower to the ground **(Figure 7.20)**.

Figure 7.20 Heavy power tools, such as a cutting saw kit, require two persons to lift and carry. Heavy tools should always be stored in lower cabinets of apparatus.

Figure 7.19 A *Set of Irons* generally consists of a prying tool and an axe that can be easily carried together.

— Refer to department policy in regard to carrying a running power-operated tool. In most cases, the safest method is to transport the tool to the area where the work will be performed and then safely start it.

— Prior to starting any power-operated equipment, determine if the atmosphere is free of flammable vapors that the tool might ignite.

- *Ground ladders* — Ground ladders may be carried and raised by one or more firefighters. To prevent injuries, proper lifting and carrying techniques must be employed. PPE, including gloves, must be worn when handling ladders. Special care to avoid electrical wires must be exercised with metal ladders. Information on ladder lifts and carries is contained in IFSTA **Essentials of Fire Fighting 5th** edition.

- *Portable Fire Extinguishers* — Carry the extinguisher by its handle. Do not pull the operating pin until you are ready use the extinguisher. Protect the operating mechanism from impact or damage.

Safe Tool Usage

You should use all tools and equipment with safety in mind. Improperly used tools can result in injuries, unnecessary damage, and loss of the tool. Improper use can also result in inefficient use of your energy and time that may endanger the people you are trying to help.

Hand Tools

A key to hand tool safety is situational awareness. You must be aware of your immediate surroundings, the location of other emergency responders, and how the object you are using the tool on will react to your efforts. Each tool and situation will be different, and you will need to take time to evaluate it before using the tool. For instance, cutting a roof made of wet plywood with an axe can result in the axe bouncing off the surface rather than penetrating it. Extremely dry wood may splinter causing pieces of wood fly out in many directions. In both cases, the axe or the splinters could strike you or another firefighter.

As with all tools, using prying tools incorrectly creates a safety hazard. For example, it is not acceptable to use a *cheater bar* or to strike the handle of a pry bar with other tools. A *cheater bar* is a piece of pipe slipped over the handle of a prying tool to lengthen the handle, thus providing additional advantage. Use of a cheater bar can put forces on the tool that are greater than the tool was designed to withstand. This can cause serious injury to the operator if the tool slips, breaks, or shatters. Furthermore, such action can damage the tool. If a job cannot be done with a particular tool, do not strike the handle of the tool; use a larger tool. Also, do not use a prying tool as a striking tool unless it has been designed for that purpose.

Power-Operated Tools

When operating power equipment, you must be even more aware of your own situation, the potential of the equipment, and your environment. Power-operated equipment can cause personal harm to you and those around you if you lose control of it. Wearing the appropriate PPE is mandatory when operating any power-operated equipment. You must always maintain your balance when using any of the power-operated saws, chisels, cutters, or spreaders.

Rotary saws must be used with extreme care to prevent injury from the rapidly rotating blade that continues to spin after the throttle has been released. Another hazard associated with the rotary saw is the twisting (*gyroscopic* or *torsion effect*) caused by the spinning blade. When using a rotary saw, it is important to start all cuts at full rpm to prevent the blades from binding into the material (**Figure 7.21**).

Remember that blades from different manufacturers may look alike, but they may not be interchangeable. When replacing blades, ensure that they are mounted with the correct direction of rotation. It is important to store blades in a clean, dry environment. Do not store composite blades in any compartment where gasoline fumes accumulate (such as where spare saw fuel is stored) because the hydrocarbons will attack the bonding material in the blades and make them subject to sudden and violent disintegration during use.

Using the following simple safety rules when operating any type of power saw will prevent most typical accidents:

- Match the saw to the task and the material; never force a saw beyond its design limitations.

- Always wear proper protective equipment, including gloves, hearing protection, and eye protection.

- Fully inspect the saw before and after use.

- Do not use any power saw when working in a flammable atmosphere or near flammable liquids.

- Maintain situational awareness.

- Keep unprotected and nonessential people out of the work area.

- Follow the manufacturer's guidelines for proper saw operation.

- Keep blades and chains well sharpened. A dull saw is more likely to cause an accident than a sharp one.

- Be aware of hidden hazards such as electrical wires, gas lines, and water lines.

- Start all cuts with the saw running at full rpm.

When using a power-operated extrication tool, such as a spreader, cutter, or ram, use the following safety precautions:

- Check the atmosphere for the presence of flammable vapor.

- When using a power-operated extrication tool, a charged hoseline must be in place.

- Wear full PPE.

- Determine if the vehicle has any passive restraints that might be activated by the cutter blade.

- Do not put hands or fingers between the blades or spreader tips.

- When using chains with a spreader to pull a vehicle apart, anchor the chain securely and work slowly.

- Keep all nonessential personnel away from the work area.

Figure 7.21 Complete personal protective equipment (PPE) must be worn when operating a power tool such as a rotary saw.

Providing lighting at an emergency scene improves the visibility and safety of all personnel working in areas that are dark whether during the day or night. Depending on the organization's size and resources, the auxiliary light units available may vary, including handheld battery-powered portable lights, apparatus-mounted lights, or portable light units powered by generators. Lights can become very hot and can cause severe burns. Gloves must be worn when handling or touching light units. Care must also be taken when handling energized lighting equipment. Extension or power cords must be of sufficient size and capacity to carry the required wattage and should have twist-locking devices to prevent disconnection. Cords must be kept out of the water to prevent electrocution and should be replaced if frayed or damaged. Grounding of all electrical equipment is important.

Nonemergency Tools and Equipment

Nonemergency tools and equipment must also be used safely. Sometimes, we overlook the need for appropriate PPE when using lawn or repair equipment. However, wearing the appropriate PPE when operating a lawnmower or drill press is every bit as important as wearing respiratory protection when working at a structure fire. For example:

- Wear work boots with safety toe caps when operating lawnmowers and edgers.
- Wear hearing and eye protection when operating power equipment.
- Wear respiratory protection (dust masks or respirators) when spray painting or sanding.
- Wear eye, face, and hand protection when using a cutting or welding torch.
- Never remove or alter guards on equipment.

Summary

Tools and equipment are essential for performing the duties assigned to fire and emergency services personnel. You must be able to recognize the various types of tools and equipment, know how to select and use them, and how to maintain them. In all situations, tools and equipment must be used with safety in mind. Always use the correct tool to fulfill the task for which it was designed.

Review Questions

1. What are two leading causes of equipment accidents?
2. What are some general inspection points for tools and equipment?
3. How are the inspections points for ground ladders, ropes, fire hose, portable fire extinguishers, and electronic equipment similar?
4. What procedures can be used to reduce hazards in the maintenance area?
5. What are the safety considerations for changing and charging batteries?
6. What are the general procedures used when cleaning unpowered tools?
7. What general procedures are used when maintaining power-operated equipment?
8. What general procedures should be used when maintaining portable fire extinguishers?
9. What procedures should be followed when disinfecting tools and equipment?
10. What are the procedures that should ensure safer storage and use of hazardous materials?

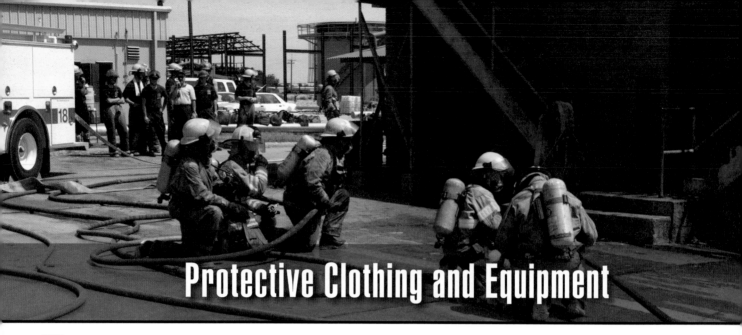

Protective Clothing and Equipment

Chapter Contents

Key Terms

Job Performance Requirements

This chapter provides information that addresses the following job performance requirements of NFPA® 1500, *Standard on Fire Department Occupational Safety and Health Program:*

NFPA® References

7.1.1	7.3.3	7.5.2.5.1	7.9.1	7.11.8	7.14.1
7.1.2	7.3.4	7.5.2.6	7.9.2	7.11.8.1	7.14.2
7.1.3	7.3.4.1	7.5.2.7	7.9.3	7.11.9	7.14.2.1
7.1.4	7.4.1	7.5.2.8	7.9.4	7.11.10	7.14.3
7.1.5	7.4.2	7.5.2.9	7.9.5	7.11.11	7.14.3.1
7.1.6	7.4.2.1	7.6.1	7.10.1.1	7.11.12	7.14.3.2
7.1.7	7.4.3	7.6.2	7.10.1.2	7.11.13	7.14.4
7.1.7.1	7.4.4	7.6.2.1	7.10.1.3	7.11.14	7.14.5
7.1.7.2	7.4.5	7.7.1	7.10.2	7.11.15	7.14.6
7.2.1	7.5.1.1	7.7.2	7.10.2.1	7.12.1	7.15.1
7.2.2	7.5.1.2	7.7.3	7.10.2.2	7.12.2	7.15.1.1
7.2.2.1	7.5.1.3	7.7.4	7.10.2.3	7.12.3	7.15.2
7.2.3	7.5.1.4	7.7.5	7.10.3.1	7.12.4	7.15.2.1
7.2.4	7.5.1.4.1	7.7.6	7.10.3.2	7.12.5	7.15.3
7.2.5	7.5.1.5	7.8.1	7.10.3.3	7.12.6	7.16.1
7.2.5.1	7.5.1.6	7.8.2	7.11.1	7.12.7	7.16.2
7.2.5.2	7.5.2.1	7.8.3	7.11.2	7.12.8	7.16.3
7.2.6	7.5.2.2	7.8.4	7.11.3	7.12.9	7.17.1
7.3.1	7.5.2.3	7.8.5	7.11.4	7.12.10	7.17.2
7.3.1.1	7.5.2.4	7.8.6	7.11.5	7.13.1	7.17.3
7.3.2	7.5.2.4.1	7.8.7	7.11.6	7.13.2	
7.3.2.1	7.5.2.5	7.8.8	7.11.7	7.13.3	

FESHE Learning Objectives

Occupational Safety and Health for Emergency Services
Describe the relevance for safety in the work place including the importance of PPE.

Learning Objectives

After reading this chapter, students will be able to:

1. Describe types of personal protective equipment (PPE) for various types of emergency responses.
2. Describe the characteristics of Levels A, B, C, and D protective clothing used at hazardous materials incidents.
3. Describe the differences between approach, proximity, and entry PPE.
4. Summarize the accessories that are used to supplement PPE.
5. Summarize the criteria used to select PPE.
6. Describe the general procedures used to inspect PPE.
7. Describe the general procedures for care and maintenance of PPE.
8. Summarize the types of information that must be recorded about PPE.

Chapter 8
Protective Clothing and Equipment

Case History

During the final evolution of evening vehicle rescue training, the assistant chief conducting the training was standing approximately 15 feet away from firefighters using spreaders to pry open a door on an overturned vehicle. The door popped open and a large bolt flew into the face of the assistant chief hitting him in the safety glasses. Happily, he had proper eye protection, which probably saved his eyes. During the conclusion of the class, he pulled out the bolt and explained what had happened to those who had not been watching the evolution. He stressed the importance of wearing the correct eye protection and personal protective equipment (PPE) at all times.

It is essential to wear the correct PPE when you are in or near a hazardous activity. This includes emergency scene operations, training, or maintenance. At the same time, you must always be aware of your surroundings and stay alert. Be diligent, be aware, and wear your PPE all the time. Safety is your number one priority.

Source: *National Fire Fighter Near-Miss Reporting System*

There are three means for protecting you from hazards: Eliminating the hazard through engineering, providing administrative controls that prevent exposure to the hazard, and providing personal protective equipment (PPE) that shields you from the hazard. Along with training and good judgment, your personal protective clothing and equipment is your protection against the hazards that you face during emergency operations. Collectively referred to as *personal protective equipment (PPE)*, it is designed to allow you to more safely fight most types of fires, enter collapsed structures, enter contaminated areas to perform essential tasks, and treat patients.

However, PPE does not provide complete protection to each hazard nor does one type of PPE protect you from all types of hazards. PPE is designed to protect you from specific types of hazards or in specific types of situations. This chapter presents the types of hazards and describes the PPE that should be selected for use in each.

As the types and variety of hazards has increased over the past 50 years, the protective clothing and equipment industry has attempted to develop equipment with increased levels of protection. At the same time, NFPA®

Figure 8.1 In the past 50 years, turnout coats and pants have evolved through many types of materials resulting in higher levels of thermal protection.

committees responsible for PPE design criteria continues to improve the level of protection as defined in the standards. Technological advances in heat-resistant and fire-retardant fabrics developed by the aerospace industry were adopted and applied to fire-fighting PPE. Of particular interest, were the lightweight fabrics that could reduce the weight of clothing while increasing the level of thermal protection **(Figure 8.1)**. PPE weight is a critical factor because medical research and field experience have shown that physiological stress due to equipment weight can have an effect on cardiovascular incidents among fire service personnel.

This chapter also provides you with information on the components of personal protective clothing and equipment, its selection, use, and inspection. Additional information is provided on maintenance and care, disposal of damaged, contaminated, or aged equipment, and the record keeping required by the NFPA® standards.

PPE Related Injuries and Fatalities

PPE, when used for the hazard it is intended to protect you from, should not result in any injuries or fatalities. Unfortunately, injuries and fatalities still result when PPE is not used appropriately. Selecting the correct type and level of protection is critical. At the same time, PPE can cause you to place too much confidence in it and expose yourself to dangers that the garment was not designed to withstand. For instance, structural fire fighting PPE can provide so much thermal protection that it prevents the wearer from noticing rapid temperature changes inside a structure that typically precede a flashover. This condition exposes the wearer to flashover conditions that the clothing cannot withstand.

Not providing the correct type of PPE or continuing to use out-of-date, contaminated, or damaged PPE can contribute to firefighter injuries or fatalities. Emergency service administrators continually evaluate the PPE currently provided to ensure that it is adequate to the challenges for the hazardous conditions and situations people in service will face.

Unfortunately, complete statistics on firefighter injuries and fatalities resulting from the lack or improper use of PPE are not available. However, fatalities that result from asphyxiation in contaminated atmospheres can generally be related to either the failure of a breathing air system such as self-contained breathing apparatus (SCBA) or supplied-air respirator (SAR) or the improper use of the system. Also, the lack or improper use of the personal alert safety system (PASS) device can be considered as a contributing factor in the deaths of some firefighters.

Types of PPE Based on Hazard

Personal protective clothing is designed to meet NFPA® standards and then is tested and certified by third-party testing organizations to ensure that it will protect firefighters and emergency responders from specific types of hazards. Certification may be provided for a single garment, such as a pair of fire-fighting gloves, or for an entire ensemble, such as an aircraft fire-fighting entry suit. Therefore, the person responsible for purchasing PPE must be aware of the specific use that a garment or ensemble is certified for when purchasing clothing. Personal protective clothing should be specified and purchased based on the individual department's intended use of the clothing.

While NFPA® provides the standards for the design, construction, and performance criteria for PPE, the use of the equipment is mandated by the federal, state/provincial, or local level of government. In the U.S., the Occupational Safety and Health Administration (OSHA), or the state equivalent, mandates the use of PPE meeting NFPA® standards. These agencies also require that the employer, whether public or private, provide employees with the correct type of PPE.

Remember, NFPA® standards establish the minimum protection requirements to meet each of the types of hazards. Some garments, however, may have dual certification for more than one function; that is, the garment or ensemble has been designed and tested to protect against multiple types of hazards. An example of a dual-certified garment would be one that protects the wearer from the hazards of a structural fire incident and also from biological hazards encountered in an emergency medical services (EMS) incident. When dual-certified garments are specified, purchased, and issued, they must meet the requirements of each NFPA® standard that applies to them **(Figure 8.2)**.

Another consideration is the effect that climate can have on the PPE design. Personal protective clothing certified for structural fire fighting, while protecting the wearer from high temperatures encountered in a fire, may also contribute to heat stress in the wearer. Garments that are acceptable for northern climates may not be safe for hotter southern climates. Heat stress factors must also be considered in selecting personal protective clothing.

The types of responses that PPE are designed to meet, specific hazards, and the applicable NFPA® standard(s) are found in **Table 8.1, p. 276**. Each response type and the PPE that is designed to address it are discussed in the following sections.

Figure 8.2 Some modern fire helmets are designed and constructed to meet NFPA® standards for both structural fire fighting and confined space rescue.

Medical Hazards

With the increase in the variety and quantity of infectious diseases encountered by fire and emergency responders, they should be very concerned with personal protection during emergency medical responses. Traditionally, personnel who perform both EMS and structural fire fighting duties have depended on their structural clothing for protection, believing that the multiple layers of fabric will protect against biological contamination. However, unless the ensemble is dual certified, it will not provide the required level of protection. This may create a difficult situation at an incident that requires both fire fighting and patient care or extrication.

EMS-certified ensembles that provide protection against liquid-borne pathogens and other bodily fluids consist of garments, gloves, and face protection that meet the requirements of NFPA® 1999. This standard is compatible with the other protective clothing standards in its requirements for testing, labeling, and third-party certification during manufacture **(Figure 8.3, p. 278)**.

Table 8.1
NFPA® Standards To Meet Specific Hazards and Responses

Response Type	Specific Hazard	NFPA® Standard
Medical	Contaminated or low-oxygen atmospheres	NFPA® 1981, *Standard on Open-Circuit Self-Contained Breathing Apparatus (SCBA) for Emergency Services*, 2007
	Chemical, Biological, Radiological, or Nuclear	NFPA® 1994, *Standard on Protective Ensembles for First Responders to CBRN Terrorism Incidents*, 2007
	Biological contamination	NFPA® ® 1999, *Standard on Protective Clothing for Emergency Medical Operations*, 2008
Structural	Heat, flame, sharp objects	NFPA® 1971, *Standard on Protective Ensembles for Structural Fire Fighting and Proximity Fire Fighting*, 2007
	Contaminated or low-oxygen atmospheres	NFPA® 1981, *Standard on Open-Circuit Self-Contained Breathing Apparatus (SCBA) for Emergency Services*, 2007
	Disorientation and unconsciousness	NFPA® 1982, *Standard on Personal Alert Safety Systems (PASS)*, 2007
	Contaminated or low-oxygen atmospheres	NFPA® 1989, *Standard on Breathing Air Quality for Emergency Services Respiratory Protection*, 2008
	Chemical, Biological, Radiological, or Nuclear	NFPA® 1994, *Standard on Protective Ensembles for First Responders to CBRN Terrorism Incidents*, 2007
Wildland	Heat, flame, burning embers, and sharp objects	NFPA® 1977, *Standard on Protective Clothing and Equipment for Wildland Fire Fighting*, 2005
Hazardous Materials	Contaminated or low-oxygen atmospheres	NFPA® 1981, *Standard on Open-Circuit Self-Contained Breathing Apparatus (SCBA) for Emergency Services*, 2007
	Contaminated or low-oxygen atmospheres	NFPA® 1989, *Standard on Breathing Air Quality for Emergency Services Respiratory Protection*, 2008
	Contact with dangerous or hazardous vapors or gases	NFPA® 1991, *Standard on Vapor-Protective Ensembles for Hazardous Materials Emergencies*, 2005

Continued

Table 8.1 *(Concluded)*

Response Type	Specific Hazard	NFPA® Standard
Hazardous Materials *(Concluded)*	Contact with dangerous or hazardous liquids or powders	NFPA® 1992, *Standard on Liquid Splash-Protective Ensembles and Clothing for Hazardous Materials Emergencies*, 2005
	Chemical, Biological, Radiological, or Nuclear	NFPA® 1994, *Standard on Protective Ensembles for First Responders to CBRN Terrorism Incidents*, 2007
Water Rescue	Falls, under currents, swift water, and extrication	NFPA® 1983, *Standard on Life Safety Rope and Equipment for Emergency Services*, 2006
Industrial	Heat, flame, sharp objects	NFPA® 1971, *Standard on Protective Ensembles for Structural Fire Fighting and Proximity Fire Fighting*, 2007
Proximity	Heat, flame, and sharp objects	NFPA® 1971, *Standard on Protective Ensembles for Structural Fire Fighting and Proximity Fire Fighting*, 2007
	Contaminated or low-oxygen atmospheres	NFPA® 1981, *Standard on Open-Circuit Self-Contained Breathing Apparatus (SCBA) for Emergency Services*, 2007
	Contaminated or low-oxygen atmospheres	NFPA® 1989, *Standard on Breathing Air Quality for Emergency Services Respiratory Protection*, 2008
Urban Search and Rescue(USAR)	Sharp objects	NFPA® 1951, *Standard on Protective Ensembles for Technical Rescue Incidents*, 2007
	Contaminated or low-oxygen atmospheres	NFPA® 1981, *Standard on Open-Circuit Self-Contained Breathing Apparatus (SCBA) for Emergency Services*, 2007
	Fall prevention and extrication	NFPA® 1983, *Standard on Life Safety Rope and Equipment for Emergency Services*, 2006
	Contaminated or low-oxygen atmospheres	NFPA® 1989, *Standard on Breathing Air Quality for Emergency Services Respiratory Protection*, 2008
Special Duty	None	NFPA® 1951, *Standard on Protective Ensembles for Technical Rescue Incidents*, 2007

Budget constraints may tempt financially challenged emergency response agencies to depend solely on structural fire fighting clothing for all types of responses, including EMS. Not only does this practice place firefighters at risk, but it places patients at risk as well. Fire fighting clothing exposes patients to contamination and risk of infection from hydrocarbons, soot, and chemicals on the clothing. The agency's health and safety officer should provide agency administrators with supporting documentation, including NFPA® and OSHA requirements, which justify the funding to provide the correct PPE for medical responses.

The increase in violence and personal attacks on fire and emergency services personnel has created the need for an additional type of PPE. Body armor similar to the types worn by law enforcement and military personnel is now becoming available for emergency responders. Body armor is designed to protect the torso from knives, sharp objects, and some pistol or rifle ammunition **(Figure 8.4)**. It is usually donned when there is an indication that the responders may be in danger of personal harm.

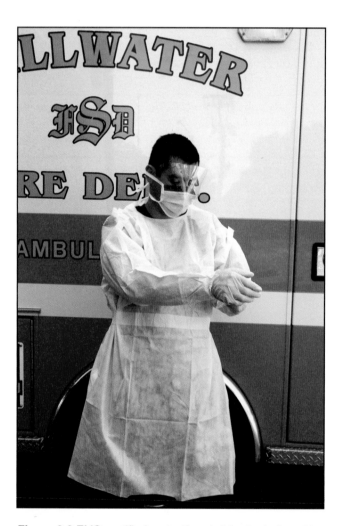

Figure 8.3 EMS-certified protective clothing is designed to provide a barrier against liquid-borne pathogens and other bodily fluids.

Figure 8.4 An increase in violence against EMS personnel has created a need for body armor.

Structural

A fully equipped structural firefighter should be equipped as follows **(Figure 8.5)**:

- *Helmet with face shield and ear cover* — The helmet is designed to protect the head from falling debris and water. The face shield provides eye protection, although it may not be necessary when an SCBA is in use. The ear cover, also referred to as neck protection, is intended to protect the area between the helmet and coat.

- *Protective hood* — Worn over the head, the hood protects the exposed portions of the face and neck and interfaces with the SCBA facepiece and coat collar. It can take the place of the ear cover.

- *Eye protection* — Provided by the SCBA facepiece, helmet face shield, or goggles.

- *Turnout or bunker coat* — Multilayered garment consisting of an outer shell of fire-resistant fabric such as polybenzimidazole (PBI), Kevlar®, Nomex®, or cotton and an inner liner or thermal barrier. The liner may or may not be removable from the shell. The coat provides protection from heat, debris, and momentary flame impingement.

- *Turnout or bunker pants* — Pants are the same as the turnout coat in construction and design.

- *Boots* — Waterproof boots are made of rubber or leather with reinforced shanks to protect the feet from puncture wounds and safety toe caps to protect the toes from crushing. Boots must be a minimum of 8 inches (203.2 mm) high and fit beneath the trouser cuffs.

- **Gloves** — Gloves protect the hands from steam burns and injury due to contact with sharp objects.

Although not required, hearing protection in the form of earplugs can be worn under the protective hood or ear cover. When operating power tools, ear-muffs may be used as part of the ensemble.

The health and safety officer and anyone responsible for purchasing protective clothing should be familiar with the requirements of NFPA® 1971, 2007 edition. The most recent changes to the standard have incorporated the requirements for aircraft rescue and fire fighting, including proximity clothing.

Wildland

Wildland fire fighting presents unique conditions that require specially designed PPE. The design and testing requirements are found in NFPA® 1977. The design of personal protective clothing for wildland fire fighting must balance the need for adequate protection from the hazards created by the fire with the need to wear the clothing for long periods in a very hot and hostile environment.

A fully equipped wildland firefighter is protected by the following equipment **(Figure 8.6, p. 280)**:

- *Protective helmet with shroud* — Protects the wearer from injury due to low-hanging vegetation or falling debris. Most wildland protective helmets resemble a typical construction-type hard hat although some lightweight structural fire-fighting helmets are available, one neck shroud is attached to the helmet and protects the ears and neck of the wearer.

Figure 8.5 Modern structural personal protective equipment (PPE) permits firefighters to work inside hostile environments for longer periods of time than older equipment.

WARNING!

Structural fire-fighting clothing is tested as a complete unit. Removing the liner from the coat or trouser shell and wearing the shell alone will void any manufacturer's warranty as well as place a legal liability on the wearer and organization. It can also result in injuries to the wearer due to steam and heat burns.

Figure 8.6 Wildland-certified personal protective equipment (PPE) must provide protection while still being very lightweight to prevent physical exhaustion and heat stress from wearing the equipment.

Chemical Protective Clothing (CPC) — Clothing designed to shield or isolate individuals from the chemical, physical, and biological hazards that may be encountered during operations involving hazardous materials.

- *Eye protection* — Protects eyes from smoke, soot, and cinders. Goggles can also help the wearer to see in smoky conditions. Helmet-mounted face shields provide some protection but do not keep smoke away from the eyes.

- *Protective clothing* — Jackets provide upper-torso protection from embers and cinders, while trousers protect lower torso and legs. Wildland PPE must have fire resistant qualities, be waterproof, and yet lightweight. A second layer of clothing, including long-sleeved shirts, is worn under the wildland PPE. Structural fire-fighting ensembles are not designed for wildland activities due to their weight and bulk.

- *Gloves* — Protect the hands from cuts and abrasions and permit the wearer to pick up or move warm objects in the burn area.

- *Protective footwear* — Protects and supports the feet and ankles. Boots designed for wildland fire fighting are similar to hiking boots with ankle supports and steel soles and toe caps. Boots should be waterproof and designed for comfort during long periods of wearing.

- *Respiratory protection* — Protects the wearer from particles in the air. Respiratory protection consists of dust masks, respirators, and bandanas. Structural SCBA are not used in wildland situations due to the weight and limited air capacity. Some new designs of respiratory protection provide increased face and lung protection. These devices are made of multiple layers of fire-resistant material and support an activated carbon filter inside the mask to capture smoke and ash particulates. Some of these masks ONLY filter particulates. They do not protect against carbon monoxide.

- *Fire shelter* — Provides some protection for the user from heat and direct flame contact when caught in a fire. This type of equipment is a last resort and is used as an alternative when timely evacuation is not possible.

- *Hearing protection* — Required when operating any noise-producing power operated equipment. Styles may include earmuffs or earplugs.

Hazardous Materials

You must wear the correct PPE whenever you respond to hazardous materials incidents. Many hazardous materials can cause a variety of injuries, including chemical burns, allergic reactions and rashes, diseases, and absorption of toxic materials into the body with skin contact. PPE is designed to prevent these. However, in the majority of incidents the first type of PPE that may be available is structural fire-fighting clothing. With proper monitoring, this equipment may provide sufficient protection to enable a rescue or an initial evaluation of the incident before other types of equipment are available or needed.

Similar to the hazardous materials ensembles, are PPE that is designed and certified for chemical, biological, radiological, nuclear, or explosive (CBRN[E]) incidents. These incidents are generally the result of acts of terrorism.

The purpose of chemical-protective clothing (CPC) and equipment is to shield or isolate individuals from the chemical, physical, and biological hazards that may be encountered at hazardous materials incidents. Design and testing standards generally recognize two types of CPC: liquid-splash protective clothing and vapor-protective clothing.

CPC is made from a variety of different materials, none of which protects against all types of chemicals. Each material provides protection against certain chemicals or products, but only limited or no protection against others. The manufacturer of a particular ensemble must provide a list of chemicals for which the suit is designed. Selection of appropriate CPC depends on the specific chemical and on the specific tasks to be performed by the wearer.

Liquid-Splash Protective Clothing. Liquid-splash protective clothing is primarily designed to protect users from chemical liquid splashes, but not against chemical vapors or gases **(Figure 8.7)**. Liquid-splash protective clothing can be encapsulating or nonencapsulating.

One limitation common to both encapsulating and nonencapsulating liquid-splash protective clothing is that such clothing is not resistant to heat or flame exposure. It does not protect against projectiles or shrapnel. The material of liquid-splash protective clothing is made from the same types of material used for vapor-protective suits.

Vapor-Protective Clothing. Vapor-protective clothing is designed to protect the wearer against chemical vapors or gases and offers a greater level of protection than liquid-splash protective clothing. Vapor-protective ensembles must be worn with positive-pressure SCBA or combination SCBA/SAR. These suits are also primarily used as part of an EPA Level A protective ensemble, providing the greatest degree of protection against respiratory, eye, or skin damage from hazardous vapors, gases, particulates, sudden splashes, immersion, or contact with hazardous materials **(Figure 8.8)**. Several limitations to vapor-protective suits are as follows:

- Do not protect the user against all chemical hazards
- Impair mobility, vision, and communication
- Do not allow body heat to escape; can contribute to heat stress, which may require the use of a cooling vest

The approach in selecting PPE must encompass an ensemble of clothing and equipment items that are easily integrated to provide both an appropriate level of protection and still allow one to perform activities involving hazardous materials. For example, simple protective clothing such as gloves and a work uniform in combination with a face shield (or safety goggles) may be sufficient to prevent exposure to certain etiological agents (such as bloodborne pathogens). At the other end of the spectrum, the use of vapor-protective, totally-encapsulating suits combined with positive-pressure SCBA is considered the minimum level of protection necessary when dealing with vapors, gases, or particulates of material that are harmful to skin or capable of being absorbed through the skin.

Figure 8.7 Liquid-splash protective clothing is designed to protect the wearer from physical contact with chemicals that will cause damage to the skin.

Figure 8.8 Vapor-protective clothing protects against hazardous gases and vapors as well as liquid splashes.

The U.S. EPA has established the following levels of protective equipment to be used at incidents involving hazardous materials: Level A, Level B, Level C, and Level D **(Figures 8.9 a-d)**. They can be used as the starting point for ensemble creation; however, each ensemble must be tailored to the specific situation in order to provide the most appropriate level of protection.

Level A

The Level A ensemble provides the highest level of protection against vapors, gases, mists, and particles for the respiratory tract, eyes, and skin. For flash or thermal protection, an additional layer of protection is required. The elements of Level A ensembles are as follows:

- *Components* — Ensemble requirements are as follows:
  - Positive-pressure, full-facepiece SCBA or positive-pressure airline respirator with escape SCBA approved by National Institute for Occupational Safety and Health (NIOSH)
  - Vapor-protective suits: Totally encapsulated suits constructed of protective-clothing materials that meet the following criteria:
    - o Cover the wearer's torso, head, arms, and legs
    - o Include boots and gloves that may either be an integral part of the suit or separate and tightly attached
    - o Enclose the wearer completely or in combination with the wearer's respiratory equipment, gloves, and boots
    - o Provide equivalent chemical-resistance protection for all components of a totally encapsulated suit (such as relief valves, seams, and closure assemblies)
- NFPA® 1991 requirements.
  - Coveralls (optional)
  - Long underwear (optional)
  - Chemical-resistant outer gloves
  - Chemical-resistant inner gloves
  - Chemical-resistant boots with steel toe and shank
  - Hard hat (under suit) (optional)
  - Disposable protective suit, gloves, and boots (can be worn over totally encapsulating suit, depending on suit construction)
  - Two-way radios (worn inside encapsulating suit)
- *Protection Provided* — Ensembles provide the highest available level of respiratory, skin, and eye protection from solid, liquid, and gaseous chemicals
- *Use* — Ensembles are used in the following situations:
  - Chemical hazards are unknown or unidentified
  - Chemical(s) has/have been identified and has/have high levels of hazards to respiratory system, skin, and eyes
  - Site operations and work functions involve a high potential for splash, immersion, or exposure to unexpected vapors, gases, or particulates of material that are harmful to skin or capable of being absorbed through the intact skin

— Substances are present with known or suspected skin toxicity or carcinogenicity

— Operations that are conducted in confined or poorly ventilated areas

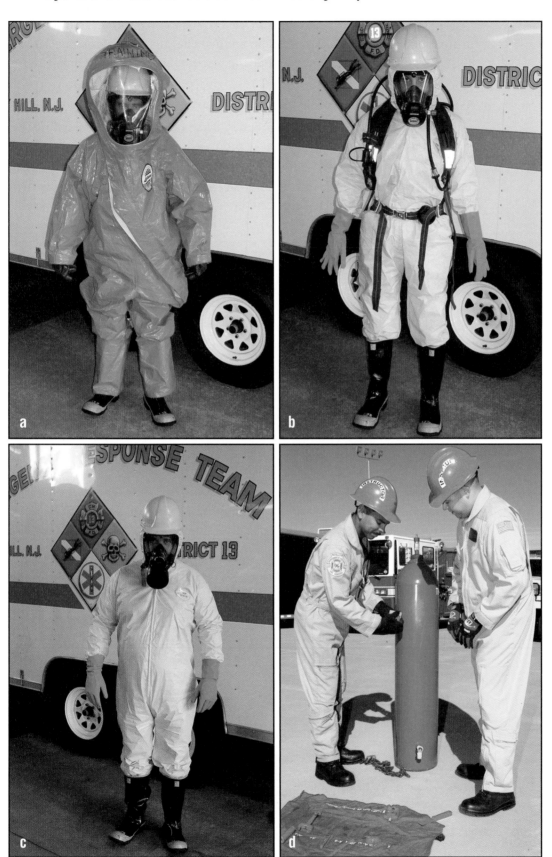

Figures 8.9 a-d The U.S. Environmental Protection Agency (EPA) has established four levels of hazardous materials protection: a) Level A; b) Level B; c) Level C; and d) Level D. *Photos a-c courtesy of Cherry Hill (NJ) Fire Department.*

Level B

Level B protection requires a garment that includes an SCBA or a SAR and provides protection against splashes from a hazardous chemical. This ensemble is worn when the highest level of respiratory protection is necessary but a lesser level of skin protection is needed. A Level B ensemble provides liquid-splash protection, but little or no protection against chemical vapors or gases to the skin. Level B CPC may be encapsulating or nonencapsulating. The elements of Level B ensembles are as follows:

- *Components* — Ensemble requirements are as follows:
  - Positive-pressure, full-facepiece SCBA or positive-pressure airline respirator with escape SCBA approved by NIOSH
  - Hooded chemical-resistant clothing that meets the requirements of NFPA® 1992, (overalls and long-sleeved jacket, coveralls, one- or two-piece chemical-splash suit, and disposable chemical-resistant overalls)
  - Coveralls (optional)
  - Chemical-resistant outer gloves
  - Chemical-resistant inner gloves
  - Chemical-resistant boots with steel toe and shank
  - Disposable chemical-resistant outer boot covers (optional)
  - Hard hat (outside or on top of nonencapsulating suits or under encapsulating suits)
  - Two-way radios (worn inside encapsulating suit or outside nonencapsulating suit)
  - Face shield (optional)

- *Protection Provided* — Ensembles provide the same level of respiratory protection as Level A but have less skin protection. Ensembles provide liquid-splash protection, but no protection against chemical vapors or gases.

- *Use* — Ensembles are used in the following situations:
  - Type and atmospheric concentration of substances have been identified and require a high level of respiratory protection but less skin protection
  - Atmosphere contains less than 19.5 percent oxygen or more than 23.5 percent oxygen
  - Presence of incompletely identified vapors or gases is indicated by a direct-reading organic vapor detection instrument, but the vapors and gases are known not to contain high levels of chemicals harmful to skin or capable of being absorbed through intact skin
  - Presence of liquids or particulates is indicated, but they are known not to contain high levels of chemicals harmful to skin or capable of being absorbed through intact skin

Level C

Level C protection differs from Level B in the area of equipment needed for respiratory protection. Level C is composed of a splash-protecting garment and an air-purifying respirator (APR) or powered air-purifying device (PAPR).

Level C protection includes any of the various types of APRs. The elements of Level C ensembles are as follows:

- **Components** — Ensemble requirements are as follows:
  - Full-face or half-mask APRs, NIOSH approved
  - Hooded chemical-resistant clothing (overalls, two-piece chemical-splash suit, and disposable chemical-resistant overalls)
  - Coveralls (optional)
  - Chemical-resistant outer gloves
  - Chemical-resistant inner gloves
  - Chemical-resistant boots with steel toe and shank
  - Disposable, chemical-resistant outer boot covers (optional)
  - Hard hat
  - Escape mask (optional)
  - Two-way radios (worn under outside protective clothing)
  - Face shield (optional)

- **Protection Provided** — Ensembles provide the same level of skin protection as Level B but have a lower level of respiratory protection. Ensembles provide liquid-splash protection but no protection from chemical vapors or gases on the skin.

- **Use** — Ensembles are used in the following situations:
  - Atmospheric contaminants, liquid splashes, or other direct contact will not adversely affect exposed skin or be absorbed through any exposed skin
  - Types of air contaminants have been identified, concentrations have been measured, and an APR is available that can remove the contaminants
  - All criteria for the use of APRs are met
  - Atmospheric concentration of chemicals does not exceed immediately dangerous o life or health (IDLH) levels. The atmosphere must contain between 19.5 and 23.5 percent oxygen

Level D

Level D ensembles consist of typical work uniforms, street clothing, or coveralls. This PPE level is used for nuisance contamination only. Level D protection can be worn only when no atmospheric hazards exist. The elements of Level D ensembles are as follows:

- **Components** — Ensemble requirements are as follows:
  - Coveralls
  - Gloves (optional)
  - Chemical-resistant boots/shoes with steel toe and shank
  - Disposable, chemical-resistant outer boot covers (optional)
  - Safety glasses or chemical-splash goggles
  - Hard hat
  - Escape device in case of accidental release and the need to immediately escape the area (optional)
  - Face shield (optional)

Level C Protection — Personal protective equipment that affords a lesser level of respiratory and skin protection than Levels A or B. Consists of full-face or half-mask APR, hooded chemical-resistant suit, inner and outer gloves, and chemical-resistant boots.

Level D Protection — Personal protective equipment that affords the lowest level of respiratory and skin protection. Consists of coveralls, gloves, and chemical-resistant boots or shoes.

- **Protection Provided** — Ensembles provide no respiratory protection and minimal skin protection.

- **Use** — Ensembles may not be worn in the hot zone and are not acceptable for hazardous materials emergency response above the Awareness Level. Level D ensembles are used when both of the following conditions exist:

  — The atmosphere contains no hazard

  — Work functions preclude splashes, immersion, or the potential for unexpected inhalation of or contact with hazardous levels of any chemicals

CBRNE

Since the beginning of this century, an emphasis has been placed on developing and using PPE that will protect you from various weapons of mass destruction (WMDs), commonly referred to as chemical, biological, radiological, nuclear, or explosive (CRBN[E]) devices. Both NFPA® 1971 and NFPA® 1994 provide requirements for the design and use of this type of equipment. Currently, PPE manufacturers are developing structural, medical, and hazardous materials PPE that will meet the new criteria.

Water Rescue

PPE used in water rescue should be designed for the intended purpose and for the qualifications of the rescuer. The PPE must match and fit the user/rescuer. Structural fire fighting clothing and SCBA should never be used for water rescues because they will retain water and even with a personal flotation device (PFD) the rescuer may drown. Generally, water rescue PPE must be worn by all personnel working within 10 feet (4 m) of the water's edge **(Figure 8.10)**. The rescuer MUST train with the PPE regularly to ensure a safe outcome in the event of a water rescue deployment. Minimum water rescue PPE should include the following:

- **Personal Flotation Device (PFD)** — PFDs are typed by the U.S. Coast Guard. Type 3 and Type 5 PFDs are the standard for rescue. A Type 3 (Flotation Aid) PFD is considered a general use PFD. The Type 5 (Wearable Special Use Device) PFD is designed for water rescue operations, and the wearer must be trained in its use. Type 5 PFDs generally allow the trained rescuer to make contact rescues with a releasable tether system. There are specialized PFDs for Helicopter Rescue Crewman for example. A buoyancy compensator commonly used by self-contained underwater breathing apparatus (SCUBA) rescuers is not an acceptable replacement for a typed PFD.

- **Water rescue helmets** — Water rescue helmets are made and designed for the purpose. General construction considerations are an impact-resistant shell, an impact-absorbing material or suspension system, and a sturdy chin strap. Holes in the helmet shell allow water to pass through the helmet without retaining the water. Some helmets are equipped with lanterns or lights.

- **Footwear** — Either specifically designed shoes/booties or tennis/basketball high-top shoes can be worn. The purpose of the footwear is to provide traction and protect against unseen objects underwater while not adding to the user's weight. The footwear must allow for the use of swim fins if the rescuer is going to deploy them. Structural fire-fighting boots should not be

Figure 8.10 Personal protective clothing designed for water rescue is intended to keep the wearer afloat as well as warm. *Courtesy of Michael Porowski.*

worn as an alternative because they will hold up to 8 gallons of water and act like an anchor.

- *Thermal protection* — It is generally considered that water will rob the body of heat 25 times faster than the ambient air temperature and under conditions such as wind chill. Thermal protection is designed to prevent hypothermia by retaining body temperature in the rescuer in the event of an intended or unintended exposure to the water. The amount and type of thermal protection must be matched to the water hazard. The most common types of thermal protection ensembles are wet suits and dry suits with a liner. Each has its place and must be matched to the hazard to which the rescuer will be exposed.

  - **Wet suits:** Come in many thicknesses to provide more or less thermal retention for the rescuer. Wet suits do allow the rescuer to be in contact with the water. In water that is contaminated by floods or sewage, this exposes the rescuer to contamination.

  - **Dry suits:** Can provide more thermal retention based on the thickness of the liner that the rescuer wears under the dry suit. Dry suits can be worn for many different water hazards and water temperatures. By varying the liner material or layers, the rescuer can increase or decrease the amount of thermal retention. Dry suits encapsulate most of the rescuer, except for hands and head and are preferred for cold water and flood-waters. Dry suits provide some measure of protection over wet suits in contaminated water.

- *Gloves* — Gloves should be worn to protect the hands because injury to the hands is a significant hazard to the rescuer. The issues with gloves are the tradeoff with thermal protection vs. dexterity. The rescuer will have to match the gloves to the intended use. Gloves that permit the handling of debris, ropes, and equipment during the search process may not provide enough thermal protection, thus rendering the rescuer ineffective. Both diving-type gloves and work gloves can be used interchangeably as long as the rescuer is aware of the greatest hazard to his hands. Without hands, fingers, and thumbs, the rescuer is ineffective.

- *Eye protection* — Safety glasses or goggles may be necessary depending on the given hazard. Operating in extreme weather or around aircraft may necessitate the use of eye protection. Water craft operations will also dictate the use of eye protection. There many different types of safety glasses and goggles that meet the American National Standards Institute Standard ANSI. Z87. The rescuer should evaluate his or her needs and obtain the eye protection that is comfortable and they will wear.

- *Clothing* — Generally, shorts, T-shirts, or swimming trunks/suits are worn with the water rescue equipment.

Personal protective clothing used for rescues in ice-covered rivers, ponds, lakes, and coastal areas are especially designed for use in extremely cold temperatures. Ice rescue PPE is similar to a dry suit and is designed to protect the wearer from hypothermia during rescue operations. Many states/provinces and local jurisdictions require that personnel performing ice rescue be certified Ice Rescue Technicians.

Industrial

There are a variety of levels of training and response capabilities established for industrial fire fighting (**Figure 8.11**). The type of PPE required for industrial fire fighting is based on the level of training. The levels and appropriate PPE include the following:

Industrial Training Level	Appropriate PPE
Incipient	Normal work clothing
Advanced Exterior	Structural PPE with respiratory protection
Interior Structural	Structural PPE with respiratory protection
Advance Exterior and Interior Structural	Structural PPE with respiratory protection

Proximity PPE may be used by Advanced Exterior personnel trained in its use for advancing on an exterior fire. It may not be used for interior structural fire fighting.

Specialized Fire Fighting PPE

Some fires produce high levels of radiant, convective, and conductive heat that can limit the ability of firefighters to gain access to the seat of the fire, to make rescues, or to control the fuel source of the fire. These types of fires include, but are not limited to, flammable liquids storage facilities, aircraft crash sites, and transportation accidents involving flammable/combustible liquids or compressed/liquefied gases.

There are three general classifications used for PPE that protect firefighters from high radiant, convective, and conductive heat: Approach, Proximity, and Entry (**Figure 8.12**). The design standard for proximity and entry level PPE is currently included in NFPA® 1971.

Figure 8.11 Personal protective clothing worn by industrial fire brigades for use in interior fire attack is similar to the clothing used by municipal departments.

Figure 8.12 Approach ensembles provide protection in areas of high radiant heat.

Approach

This category is described in some sources as PPE that will permit firefighters to approach or work in an area of high radiant heat. Although there are no design standards for this category of PPE, structural fire fighting PPE should provide the required protection. Control over approach operations, used to assess conditions or affect a rescue, should be established by the organization's administration, and provided in the standard operating procedure/ standard operating guideline (SOP/SOG).

Proximity

Proximity fire fighting involves operations at incidents such as aircraft fires, bulk flammable liquids fires, and bulk flammable gas fires that involve high levels of radiant, convective, or conductive heat. Proximity ensembles include coat, hood, footwear, gloves, and trousers that meet the requirements of NFPA® 1971. SCBA should be worn internally rather than outside the proximity clothing. This placement protects the SCBA and its fittings from extreme radiant heat.

Because of the specialized design and material used in proximity clothing to reflect high levels of heat, follow the manufacturer's recommendations for care, inspection, and cleaning. Like structural clothing, proximity clothing is required to be labeled and third-party certified.

You must remember that proximity clothing is not entry clothing and is not intended for protection against prolonged direct flame contact. The limitation of proximity clothing should be emphasized to all personnel during training.

Entry

Entry suits allow firefighters to work in total flame environments for short time periods. They provide short-duration and close-proximity protection at radiant heat temperatures as high as 2,000°F (1 093°C). Each suit has a specific use and is not interchangeable. Fire-entry suits are not designed to protect the wearer against chemical hazards nor are they intended for interior structural fire fighting.

There are several limitations to high-temperature proximity and entry PPE including the following:

- The suits prohibit the body from releasing excess heat thereby contributing to heat stress for the wearer
- The suits are bulky
- The suits limit wearer's vision
- The suits limit wearer's mobility
- The suits limit communication
- The suits require frequent and extensive training for efficient and safe use
- The suits are expensive to purchase

Search and Rescue

PPE used for search and rescue, including trench, rope rescue, confined space, and structural collapse, are defined in NFPA® 1951. While structural fire fighting PPE can be used to a limited extent, it is bulky and can prevent access to some areas. Wildland Fire PPE is commonly used in lieu of specialized Search and Rescue PPE, although this PPE lacks some of the desired performance and special needs for Search and Rescue. Search and rescue PPE includes the following (**Figure 8.13**):

- *Head protection* — Protects the head from falling or protruding objects. Although structural fire fighting helmets can be worn, specially designed rescue helmets that meet NFPA® 1951 standards are lighter and smaller and are intended for working in confined spaces. Brims on structural fire fighting helmets can be a hazard in that they catch on objects in confine spaces. In the event of fall or head impact, the brim strikes first thus causing neck trauma.

- *Eye protection* — Protects the eyes from debris, liquids, airborne particles, and bright light. Safety goggles and glasses should meet ANSI Z87.1-2003 design criteria for use in confined-space operations. When using cutting torches, approved eye protection must be worn.

Figure 8.13 Confined-space search and rescue operations prohibit the wearing of bulky structural PPE. Therefore, specialized search and rescue PPE has been designed for use in these tight spaces.

- *Hearing* — Earplugs or earmuffs should be worn to protect the hearing from power-operated equipment or noise that is amplified in confined spaces.

- *Footwear* — Boots with steel insoles and toe caps should be worn to protect the feet from possible puncture wounds.

- *Hand wear* — You should wear comfortable, close-fitting leather or synthetic gloves that allow dexterity but are sturdy enough to protect the hands. When performing emergency medical aspects of rescue, latex or other impermeable exam gloves should always be worn. In some situations, they should be worn inside protective leather or synthetic outer gloves.

- *Knee and elbow pads* — Because rescuers may be required to crawl into confined, debris-strewn spaces, knee and elbow protection is essential. Strap-on pads are commercially available, either specifically designed for these situations or for use in the construction trades.

- *Body protection* — Structural fire fighting PPE may be used in search and rescue operations, although the clothing is heavy and bulky. Full coveralls or jumpsuits may be more appropriate, providing complete coverage and protection while allowing flexibility for the wearer. Coveralls made from fire-resistant materials, such as Nomex®, are available and meet NFPA® 1971 requirements.

- *Respiratory protection* — Following a structural collapse, fine particles of concrete and other materials are released into the atmosphere. In addition, confined spaces may have toxic gases or low oxygen levels in them. Respiratory protection and atmospheric monitoring is mandatory and in some cases legally required in these situations. The type of respiratory protection will depend on the hazard assessment.

Special Duty

Fire fighting, search and rescue, response to hazardous materials incidents, and medical activities are not the only tasks that you perform. The majority of your time will be spent in station wear or work uniforms. Your department may also require the use of physical fitness uniforms and clothing specific to inspections or investigations. While these uniforms are not considered PPE, they may become part of your emergency response ensemble if they are worn beneath your wildland or structural PPE.

Station Wear and Work Uniforms

Station wear and work uniforms vary from uniform shirts and trousers made of fire-resistant fabrics to nonuniform T-shirts and jeans. The design and construction of approved station wear is found in NFPA® 1975, *Standard on Station/Work Uniforms for Emergency Services*, 2009. Clothing must be made of fire-resistant fabrics or 100 percent cotton and have third-party certification **(Figure 8.14)**. Clothing made of synthetic materials should not be worn due to the potential for melting under high temperatures. This includes underclothing that should be all cotton.

Figure 8.14 Station wear is intended to provide a minimum level of protection against the hazards encountered during nonemergency activities.

Figure 8.15 Career fire departments that have mandatory fitness programs generally provide clothing designed for these types of activities.

While the structural PPE will protect you in emergency situations, station wear is your only protection when working around the station or in the repair shop. Therefore, coveralls or overalls should be worn to protect you from sharp objects, chemical spills, or flash fires.

Physical Exercise Uniforms

Physical fitness clothing usually consists of shorts, T-shirts, or sweat suits, socks, and jogging or tennis shoes. If these items are intended for wear under PPE and are worn as night wear while sleeping, they must also be made of cotton **(Figure 8.15)**.

Inspections and Investigations Clothing

Inspectors and investigators generally wear dress or station wear uniforms or civilian clothing when performing their duties. However, some conditions may warrant the use of PPE. Inspectors should be provided with fire-resistant coveralls, gloves, hard hats, hearing, and eye protection when working at construction or industrial sites. Investigators must have structural PPE along with respiratory protection when investigating fire scenes or chemical spills.

PPE Accessories

Besides the PPE ensembles specific to the tasks firefighters perform, there are other items essential to your health and safety. They are addressed in this section and include the following:

- General duty gloves
- Eye protection device
- Personal alert safety system (PASS) devices
- Respiratory protection device
- Traffic control vests and PPE reflective striping
- Life safety rope, life belts, harnesses, and systems
- Infection Control PPE
- Personal lighting unit
- Heads up display unit
- Thermal imaging camera
- Lighted rope
- Rescue loop on PPE coat
- Truckman belt for tools and equipment
- Extrication equipment
- Rapid intervention team (RIT) equipment

General Duty Work Gloves

Gloves designed for fire fighting and medical responses have already been described in previous sections. Station and shop activities and apparatus operations also require various types of hand protection. The type of glove used should meet the type of activity being performed in order to afford the correct level of protection.

Work gloves may be cotton, leather, or vinyl depending on the task. Stripping paint, using welding or cutting equipment, or working with hazardous materials all require specific types of protection.

Driving apparatus requires that the driver/operator be in full control of the vehicle at all times. Driving gloves provide the ability to grip the steering wheel without slipping. Moisture, such as perspiration, is absorbed into the liner of the glove preventing it from making the steering wheel or gearshift knob slippery.

Eye Protection

During emergency operations, the helmet face shield, safety goggles, and the SCBA facepiece provide adequate eye protection. Safety goggles or glasses should also be worn when operating any tools or equipment that might create sparks or cause small pieces of material to fly off.

PASS Devices

The use of PASS devices is mandated by OSHA or equivalent state agency when operating in IDLH areas. Designed to alert other emergency responders when someone is incapacitated or lost, these devices only work when they are activated. Inspections of the units at the beginning of each work shift and prior to entering the IDLH area, along with constant training and supervision, are essential to the PASS device success. Some PASS devices are integrated into SCBA units and activate automatically when the air supply valve is activated **(Figure 8.16)**.

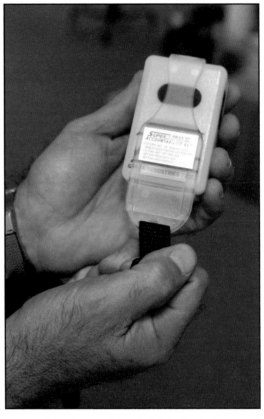

Figure 8.16 Personal Alert Safety System (PASS) devices come in a variety of styles and may be individual units or integrated into the respiratory protection system.

Respiratory Protection

Many emergency responses have the potential to become IDLH atmospheres. Respiratory protection should also be used in non-IDLH atmospheres (for example, wildland fires). To protect their personnel, fire and emergency services organizations should provide a wide variety of respiratory protection for both emergency and nonemergency use. Respiratory protection ranges from simple particle or dust masks to SCBA units. Selection of the appropriate type of protection is based on the existing hazard or potential hazard.

Personnel must be facepiece fit-tested based on specific criteria and trained in the use of all types of respiratory protection equipment **(Figure 8.17)**. Training includes the selection of the proper type of mask for the type of hazard and the use, care, cleaning, and disposal of masks.

Figure 8.17 NFPA® standards require that each person who must wear an SCBA must be fit-tested for the facepiece and issued their own facepiece.

Often overlooked as a form of respiratory protection, the particle or dust mask should be used when exposure to airborne particulates is poss ible. Particle masks are generally considered a one-time-use item and should be disposed of properly. You are responsible for selecting the appropriate type of particle mask for the task you are performing. Uses of the particle masks include the following:

- *Medical responses* — Protects against airborne germs and some body fluids.

- *Lawn mowing and yard work* — Protects against allergens such as pollen.

- *Painting* — Protects against paint fumes and spray paints.

- *Tool and equipment repair* — Protects against dust caused by sanding and grinding.

- *Building maintenance* — Protects against dust from demolition, sanding, or cleaning.

- *Salvage and overhaul* — Protects against small particles that are stirred up during post suppression activities; should not be used until the safety officer determines that no smoke or gases remain in the work area.

- *Postfire Investigations* — Protects against small particles that are stirred up during postsuppression activities; should not be used until the safety officer determines that no smoke or gases remain in the work area.

APRs are the next level of protection. APRs may be full or half face masks that have replaceable filters. They are more effective than particle masks and can be used for the same types of activities. They are also effective for wildland fire fighting.

SCBA and SARs are the most effective types of respiratory protection. Both are designed for IDLH atmospheres. However, SAR units are not intended for situations that involve fire due to the potential for damage to the air supply hose.

Traffic Control Vests

Accidents that involve fire and emergency responders being struck by vehicles while working along roadways in low visibility are all too frequent. The U.S. Department of Transportation (DOT) has issued procedures for highway operations that include the use of high-visibility clothing. Besides a structural fire fighting ensemble that is designed with reflective trim, safety vests are also available. These are lightweight vests that have reflective trim and colors that are intended to alert drivers to the presence of responders. Some incident command vests may also be used as safety vests. These vests must meet the requirements of ANSI/ISEA 207-2006, *High Visibility Public Safety Vests*, standards and federal and state department of transportation regulations **(Figure 8.18)**.

Figure 8.18 In an attempt to further protect fire and emergency service personnel, reflective traffic vests are now being issued for personnel who respond to roadway incidents.

Life Safety Rope, Life Belts, Harnesses, and Systems

Generally, life safety rope, life belts, harnesses, and similar equipment is only issued to and used by personnel who are trained and certified as Rescue Technicians. The equipment is designed, tested, inspected, and third-party certified for rescue operations.

Personal Lighting

Relatively new to the safety scene, personal lighting is becoming available as a means of identifying the location of personnel and making them visible in the dark. These units are battery-operated flashing lights that can be color-coded for rapid identification. They are worn on the coat, belt, or SCBA harness and are visible up to 100 feet (40 m). Personal lighting may also be used with flotation devices if approved by the U.S. Coast Guard.

New Technology

Improvements in electronic technology have provided many innovations to the fire service. Among these are a SCBA heads-up display, an electronic accountability system, lighted rope, built-in back support systems, and rescue harness loops built into the PPE.

Heads-up Display

The heads-up display is one of those innovations. Designed into the SCBA face piece, the heads-up display provides the wearer with a visual indication of the amount of air remaining in the SCBA cylinder. Four lights of different colors indicate when the cylinder is at 75 percent, 50 percent, 25 percent, and 10 percent capacity.

Electronic Accountability System

A new system designed to meet both accountability requirements and provide notification in the event a firefighter is trapped or injured is the electronic accountability system. Based on radio transmissions, the master unit located at the incident command post tracks the units worn by firefighters. If a wearer stops moving for an extended period or activates the remote unit, an alarm is transmitted to the control panel.

Lighted Rope

Designed to provide emergency responders with a marked egress route from a darkened area, lighted rope can be used in a number of situations. It can be used in search and rescue operations, including confined spaces, to mark a path for escape during fire suppression operations, or to define or delineate an area.

Built-in Back Support Systems

Originally developed in the 1990s, back support braces have been designed into protective trousers, coats, and as separate belts. They provide additional support when you are required to lift a heavy object or victim. The supports should not be activated until they are needed however. They have been shown to reduce the incidents of back injuries **(Figure 8.19, p. 296)**.

Rescue Loop

Mandated by the current edition of NFPA® 1971, rescue harnesses are required in all new protective coats. The harness has an attachment loop located at the back of the coat near the collar. If a firefighter is incapacitated, a rescue rope can be attached to this loop permitting the wearer to be pulled or lifted to safety.

Cooling Vests

To offset the stress-inducing heat and humidity contained within structural PPE, cooling vests have been developed. A variety of vest styles have been developed relying on fabrics that absorb and remove perspiration, use chemical packs that can be frozen prior to use, or have water cells that can be cooled prior to use. Some vests are designed with retroreflective strips for use as outer wear at night.

Selection

Because no single type of PPE will work for multiple types of emergencies, it is important that the equipment be carefully selected. Selecting the correct PPE actually occurs twice and by two different entities: your organization selects the PPE to purchase and issue and you select the correct one to use during a response. How you select the correct PPE is covered in the next section. This section will provide a general overview of the purchasing process including a model for selecting the proper PPE and the criteria that must be met **(Figure 8.20)**.

PPE must meet certain design standards. Among them are the following:

● NFPA® standards

● OSHA federal regulations

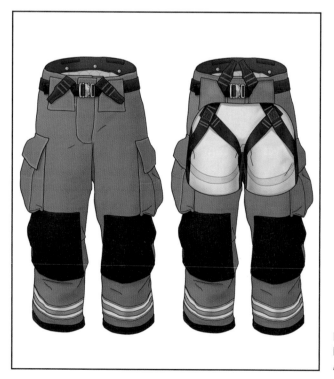

Figure 8.20 Personal protective equipment (PPE) is selected based on the intended use.

Figure 8.19 Current NFPA® standards require that rescue harnesses be built into turnout coats. *Courtesy of Globe Manufacturing Company.*

- State OSHA regulations such as CalOSHA (California)

- National Institute for Occupational Safety and Health/ Mine Safety and Health Administration (NIOSH/ MSHA)

Other sources for design criteria include both certification organizations **(Figure 8.21)** and professional organizations, including the following:

- American National Standards Institute (ANSI) certification for the intended use

- Unerwriters Laboratories Inc. (UL) and Underwriters Laboratories of Canada (ULC)

- FM Global

- Southern Area Fire Equipment Research (SAFER)

- Fire Industry Equipment Research Organization (FIERO)

- Safety Equipment Institute (SEI)

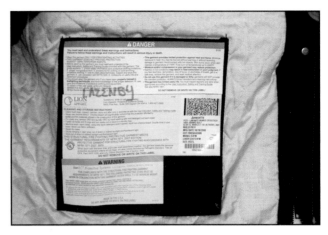

Figure 8.21 Turnout coats and pants that are intended for use in structural fire fighting must contain certification labels from the manufacturer.

Use

PPE only protects you when you use it properly. Even the best equipment cannot protect you if it is not worn or if it is not worn properly. You are responsible for determining when to wear the equipment. You must also understand that your PPE has limitations and that using it beyond those limitations increases your risk of being injured. Some safety issues that you must consider when using your PPE are as follows:

- Determining when to wear PPE

- Donning and doffing techniques for PPE

- Equipment distribution and load carrying

- Heat, hydration, and stress issues caused by the weight of the PPE

- Heat and moisture retention caused by moisture barriers

Determining when to wear your PPE is based on a number of factors. The first is the hazard you are going to face. You should be equipped with PPE for each type of emergency response you are trained to respond to. Experience will also help in making the correct selection. You will be able to base your decision on similar situations and training scenarios that you have participated in. Another factor will be the orders given by your supervisor or the officer in charge. That person will make the decision on what actions need to be taken and those actions will dictate the proper PPE.

Once you have determined the proper type of PPE to wear, you must put on or *don* the equipment. Your department's donning and doffing procedures, as well as training and practice, will provide you with the correct sequence for putting on your equipment. Some donning procedures for structural and wildland PPE that you should consider include the following:

- Cover all areas of exposed skin

- Secure all zippers, Velcro® fasteners, clasps, snaps, or buttons

- Use neck flap even when wearing protective hood **(Figure 8.22, p. 298)**

- Secure helmet chin strap under the neck
- Secure and tighten all SCBA harness straps
- Tighten facepiece head straps
- Activate back support brace prior to lifting

Donning medical response PPE will usually occur when you arrive at the incident. However, the decision to don respiratory protection should be made prior to approaching a patient or entering the room containing the patient. If the patient is known to have a communicable disease, such as tuberculosis, then respiratory protection will be required. In most cases, the use of eye, nose and mouth, and hand protection will be necessary when working with the patient.

Figure 8.22 Properly donning personal protective clothing includes ensuring a complete seal between the flashover hood and facepiece and the collar of the coat.

Removing or *doffing* the PPE is usually a simple matter. Disposable PPE such as particle masks and medical gloves should be removed and placed in a waste container. Remove contaminated medical gloves by pulling the wrist edge down over the glove trapping the contamination inside. When removing fire fighting PPE, be careful to keep the contaminated area away from exposed skin. All contaminated PPE must be cleaned and decontaminated as soon as possible.

Finally, when selecting the PPE to don, it is best to err on the side of safety. Select the highest level of protection based on the potential created by the hazard. Entering a structure in a work uniform to investigate a reported smoke odor is asking for trouble.

Another consideration in both structural and wildland fire fighting is equipment distribution and load-carrying ability. Your PPE can weigh as much as 80 pounds (36.28 kg). Additional items like hose bundles, forcible entry tools, and extinguishers can increase that load considerably. Distribute the load evenly across your body or share the load with another firefighter or emergency responder.

The weight and design of the PPE can also add to your physical stress. Structural PPE is designed with a moisture barrier as part of the inner liner. This barrier is intended to provide protection from external water and heat. However, it also helps to contain your own body heat and perspiration. The barrier can create an artificial environment within the PPE that can increase your body temperature and pull moisture out of you, causing you to become dehydrated. Heat, dehydration, and stress will cause you to tire rapidly. You must be aware of the potential and the symptoms and respond appropriately. Even though wildland PPE is lighter, similar effects can occur when fighting a wildland fire.

Inspections

Like all apparatus, tools, and equipment, PPE must be inspected and tested in accordance with NFPA® 1851 or the appropriate NFPA® standard based on the type of PPE. There are four occasions on which PPE is inspected: when accepted by the department, on a regular basis, on a formal schedule, and following repairs.

When new PPE is purchased, it is inspected when it is delivered to the department. This initial inspection is necessary to ensure that the item is complete and meets the original specifications. Some items, like SCBA and PASS devices should be tested to ensure that they perform properly. New equipment is assigned an inventory control number at this time and an individual record is started on the equipment.

In-service PPE is inspected at the beginning of each work shift. Regular inspections ensure that the PPE is ready for immediate use, that all items work properly, and that any required repairs are noted and reported. PASS devices and SCBA are tested according to manufacturer's directions. Following the inspection, the PPE is placed on or next to the apparatus.

Formal inspections are more thorough than the regular ones and are usually performed by the department's health and safety officer **(Figure 8.23)**. Formal inspections should take place annually and coincide with a periodic cleaning and preventive maintenance of the PPE. A sample inspection form in located in **Appendix D** of this manual. The health and safety officer's inspection should include the following:

- Checking the date the PPE was manufactured
- Checking the condition of the PPE:
  — stitching
  — seams
  — buttons
  — clasps
  — hook and pile fasteners
  — zippers
  — retroreflective trim
  — leather/vinyl trim
- Looking for unauthorized alterations or modifications
- Looking for cleanliness
- Checking for correct size on the wearer

SCBAs are inspected by the respiratory protection technician who performs an operational test on the unit. Facepiece fit testing also occurs annually to ensure that the facepiece seals correctly to the wearer's face.

Care and Maintenance

The importance of proper care and maintenance of PPE cannot be over stressed. NFPA® 1851 provides a guideline for care and maintenance of PPE. From a safety standpoint, equipment that is clean and in serviceable condition continues to perform the function it was intended to by protecting you. From an economical standpoint, proper care and maintenance ensures that the original cost/benefit ratio used to specify the equipment is met. Care and maintenance is the responsibility of both the administration and you. The administration provides the materials, establishes the procedures for care and maintenance, and may contract with vendors to provide maintenance. You clean and report any maintenance needs for the PPE that is assigned to you.

Figure 8.23 The organization's health and safety officer is responsible for inspecting personal protective equipment (PPE) on a regular basis.

Figure 8.24 Fire fighters are responsible for cleaning and maintaining their own PPE and reporting repair needs when they occur.

Care

In the past, soot-covered PPE was considered a mark of a *real firefighter*. The appearance was a *Red Badge of Courage* and was worn with pride: the dirtier, the better. Today firefighters are more aware that the by-products of combustion (smoke, soot, and other contaminates) are hazardous to their health. PPE that is covered in soot increases the potential for both short- and long-term illnesses. Therefore, every effort must be made to clean and decontaminate your PPE after each use and on a regular basis.

You should always follow the cleaning and care instructions of the PPE manufacturer and your department's policies. Your PPE should be cleaned following each exposure to smoke or contaminated atmospheres **(Figure 8.24)**. Water and a mild detergent are used to clean the outer shell of the coat and trousers, including the cuffs and collar, boots, hood, gloves, helmet shell, and neck shroud. The helmet headband and chin strap should be cleaned as needed in a similar fashion. Special cleaning will be required to decontaminate PPE exposed to hazardous materials such as chemical or biological waste. You must adhere to the following cleaning and care procedures:

- Never use bleach or a product containing bleach on PPE
- Clean PPE in the designated cleaning area
- Remove liners from shells and wash separately
- Clean contaminated PPE in the designated Decon Sink
- Keep PPE in the assigned storage area or in the apparatus bay
- Wash PPE in a designated washer and not in the washer used for sheets and towels
- Do not clean PPE at home
- Dry PPE in a designated dryer or by hanging on a drying rack
- Do not hang PPE in direct sunlight to dry (ultra violet light will degrade the fabric)

To follow these procedures properly, it is obvious that each fire and emergency responder would have to have two sets of PPE. For some departments, this is not economically possible. Therefore, it may be necessary to clean equipment at the end of a work shift or only after multiple exposures. It is not good policy to wear another person's PPE because this will cause cross-contamination of the clothing and the wearers.

All PPE must be stored properly in well-ventilated, secure lockers in or adjacent to the apparatus room. PPE should never be stored or taken into the living quarters of the station. Ventilation is necessary to keep clothing that is damp from mildewing and molding and to allow any contamination to be diluted by the atmosphere.

Maintenance

Simple maintenance such as the replacement of a helmet chin strap can easily be performed by the wearer. Most maintenance and repairs, however, must be performed by repair technicians with the proper tools and repair parts. This is especially true of respiratory protection equipment. Technicians who are trained and certified in the repair of SCBAs and SARs are the only ones who should repair them. These units must be tested following maintenance and prior to being place back in service.

Some departments and personnel have a tendency to make modifications to PPE to fit their own local needs or personal preferences. You must remember that each piece of PPE was tested and certified by the manufacturer to perform in its original condition. Modifications will nullify any warranty and can alter the performance of the item. In short, never modify any piece of PPE. If additional pockets or attachment points are needed, then they should be specified when new PPE is ordered. Even the use of decals on helmets can have an effect on how the helmet reacts in a fire situation.

Disposal

At some point, all PPE must be properly disposed of. The reasons for disposal include equipment that is contaminated and cannot be cleaned, equipment that is so badly damaged that it cannot be repaired or the cost of repair exceeds its value, or equipment that has reached its end of service life (ESL). State/provincial or local laws will determine the proper means of disposing of the equipment.

Contaminated Equipment

PPE may become contaminated slowly, over time through repeated exposures or through a single exposure to chemical, radiological, or biological contaminates. If the equipment cannot be thoroughly cleaned and decontaminated, then it must be disposed of properly.

Equipment that is contaminated by hazardous materials, either chemical, radiological, or biological, must be disposed of in accordance with the federal OSHA and EPA hazardous materials procedures. While wearing the proper PPE, place the contaminated items in a plastic trash bag, secure the opening, and place the bag in an overpack drum. The drum is then sealed **(Figure 8.25)**. A company that is approved for handling hazardous materials waste is then given the drum for disposal.

Damaged Equipment

The decision to dispose of equipment that has been damaged is usually based on a cost/benefit analysis. If the cost to repair the item exceeds its value or if the cumulative cost of maintenance has reached a similar amount, then the item should be disposed of. Damaged

Figure 8.25 Personal protective equipment that is contaminated with chemical, radiological, or biological materials must be disposed of properly.

equipment may be stripped of usable parts and then destroyed or placed in a trash container. Due to liability issues, it is not a good policy to donate damaged equipment to other organizations.

End of Service Life (ESL)

There are no established guidelines for determining the service life of a piece of PPE. Nor do manufacturers publish any guidelines to help determine this time. Some criteria that can be used include the following:

- Amount of use the item has received
- Cumulative maintenance costs
- Changes in regulations
- Obsolescence
- Changes in technology
- Changes in function or mission

As a general guideline, some departments keep equipment in first-line service for 7 years. At the end of this time, the item is inspected and placed in reserve, used for training, or disposed of. Records for each item must be maintained to monitor condition, assignment, and eventual disposal.

Records

Records must be kept on all PPE. The records serve many purposes including inventory control, budget justification, and liability mitigation. The types of PPE records include purchasing, inspection, maintenance, and disposal information.

Types of PPE Records

Modern record-keeping systems permit various types of files to be cross-referenced so that all information can be easily accessed. In some cases, the information is located in remote sites that may even be under the control of a different division or department within the jurisdiction. Records may be divided into purchasing, inspection, maintenance, and disposal and may be linked by a common identification number, or bar code, for each individual item of PPE.

Purchase Information

Purchasing information is usually maintained by the purchasing department of the jurisdiction, although it may also be retained by the fire and emergency services department. The information is used to ensure that products have been received and that they meet the original specifications. In addition, the information provides a background for future purchases. This information includes the following:

- Product specifications
- Purchase order
- Purchase contract
- Warranty
- Shipping/delivery papers
- Manufacturer/vendor information

Inspection Information

Local jurisdictions usually establish periodic inspection requirements for PPE. Some of these requirements are based on legal mandates, such as the inspection cycle for SCBA. Inspection information provides a basis for other types of records including the need for maintenance and disposal, to track inventory, and to determine liability in the event of a PPE-related accident. Inspection information **(Figure 8.26, p. 304)** usually includes the following:

- Inventory tracking number
- Inspector's name
- Item assigned to (person or location)
- Condition
- Age
- Repairs needed
- Remove from service reason

Maintenance Information

Maintenance records include information about cleaning, preventive maintenance, and corrective maintenance. Cleaning records would be maintained on protective clothing, ropes, and facepieces. Preventive maintenance would be scheduled repairs such as periodic battery or parts replacements, calibrations, and any manufacturer's recommended repairs. Corrective maintenance is the result of the inspection findings or of any reported equipment failure or damage. The PPE is removed from service and either a permanent or a temporary replacement is issued. Maintenance records generally include the following:

- Inventory tracking number
- Name of person requesting maintenance
- Name of person performing maintenance
- Type of maintenance
- Reason for maintenance
- Parts used
- Labor (in hours/minutes)
- Cost of parts
- Cost of labor
- Recommendation to dispose of item

Disposal Information

All state/provincial and local governments have laws concerning the disposal of property. When PPE reaches its normal ESL or repairs cannot be justified, the item must be disposed of according to those laws. Generally, there are only two ways to dispose of PPE: surplus sale or destruction.

Records must indicate how the PPE was disposed of for many reasons. First is accountability: Property that was purchased with tax money must be tracked all the way from receipt until disposal. Any revenue received from its sale must be accounted for.

Pennsylvania State Fire Academy
Personal Protective Equipment Inspection Checklist

This equipment is assigned to:	Inspection Date:

Instructions:

1. This checklist shall be used for a structural fire fighting ensemble only.
2. Soiled or contaminated gear shall be cleaned prior to inspection.
3. Use universal precautions during the inspection process.
4. Any item identified in *italics* and marked **yes (Y)** will require further inspection by either a trained individual or an Independent Service Provider (ISP).
5. The following inspection elements should not be considered all inclusive. In the event that you find something that requires further inspection, do not hesitate to have it evaluated further.
6. This inspection should in no way be interpreted as complying with the Advanced Inspection requirements established by NFPA 1851 *Selection, Care, and Maintenance of Structural Fire Fighting Protective Ensemble* 2007 Edition.
7. Place an X in the appropriate box.

HELMET			Serial Number:		
	P	F		P	F
Evaluation for fit			Suspension System		
Soiling			Damaged or missing reflective trim		
Damage to shell: Cracks, dents, abrasions				N	Y
Damage to liner: rips, tears, or thermal damage			*Visible damage to impact cap?*		

EYE PROTECTION					
	P	F		N	Y
Damaged or missing components to face-shield or goggle system			*The face shield or goggle system does not function as designed?*		

HOOD					
	P	F		P	F
Evaluation for fit			Rips, tears, cuts, or thermal damage		
Soiling				N	Y
Contamination from hazardous materials or biological agents			*Is the seam integrity compromised?*		
Loss of face opening adjustments / shrinkage			*The elastic does not rebound?*		

FOOTWEAR					
	P	F		P	F
Evaluation for fit			Damaged or deformed safety toe, mid-sole and shank		
Soiling				N	Y
Contamination from hazardous materials or biological agents			*Is seam integrity or liner compromised?*		
Rips, tears, and thermal damage			*Is the sole or heel excessively worn?*		
Loss of water resistance			*The closure system does not function properly?*		
Closure Systems Damage			*Is the liner showing excessive wear?*		

Figure 8.26 Sample personal protective equipment (PPE) inspection form. *Courtesy of Pennsylvania State Fire Academy.*

The second method of disposing of PPE is to destroy it. This is sometimes mandated if the jurisdiction continues to be liable for the equipment after it is transferred to another owner. Remember that protective clothing that is no longer acceptable for use in your department is not acceptable for use in any other department.

Disposal records include the inventory tracking number, reason for disposal, and final means and date of disposal. If the final disposal is the result of damage or loss that can be reimbursed, then additional information may be included to indicate how and when the item was replaced.

Inventory Control

PPE is purchased with revenue from taxes and other sources. Keeping track of these items is an essential function of government as stewards of the public funds and services. The process generally involves assigning an inventory control number, sometimes in the form of a bar code, to the item. As indicated in the previous paragraphs, this number is the basis of the record-keeping system and is used to determine the location, condition, maintenance, and disposal of the item throughout its service life.

Budget Justification

Another use for PPE records is to develop and justify the periodic (annual) budget requests of the department. Records provide the initial cost of equipment, its expected service life, the cost of repairs in both parts and labor, how well various types of equipment have worked, and the cost/benefit the PPE has provided. These records can be used to justify purchasing replacement equipment, changing to another type of equipment, or requesting an increase in budget allotment.

Liability Mitigation

The final reason to keep records on PPE is to reduce the liability that may result from an injury or fatality that involves the equipment. Maintaining detailed records, including the cleaning, maintenance, and original specifications, may provide a clearer picture of how the PPE may or may not have contributed to an injury or fatality.

Summary

Your PPE is designed to protect you while you perform your duties. Selecting the proper PPE is the responsibility of your organization when the equipment is purchased and yours when you don it. The selection is based on the type of hazard you will be exposed to and the level of protection required. Not only is it essential that you select the correct type of PPE but also that you wear and care for it correctly. Cleaning, maintaining, inspecting, and disposing of PPE are important safety activities that will ensure your personal safety and that of others.

No single combination of PPE can protect you against all hazards. Therefore, you must be careful when selecting and using PPE. Several types of PPE are available and listed as follows:

- Structural fire fighting PPE (see Structural section)
- High-temperature PPE (see Proximity section)

- Chemical-protective clothing (CPC) (See Hazardous Materials Section)
  — Liquid-splash protective clothing
  — Vapor-protective clothing

Review Questions

1. What are the differences in the PPE used in a medical versus a structural response?

2. What is the difference between Level A and Level D PPE used at hazardous materials incidents?

3. What are the differences between approach, proximity, and entry PPE?

4. What accessories can be used to supplement PPE and when is it appropriate to use these?

5. What criteria are use to select the appropriate PPE for a response?

6. How is PPE inspected?

7. What steps are taken during the care and maintenance of PPE?

8. What types of records are required to be kept for PPE?

Emergency Operations

Chapter Contents

Key Terms

Divider page photo courtesy of Chief Chris Mickal, New Orleans (LA) Fire Department.

chapter 9

Job Performance Requirements

This chapter provides information that addresses the following job performance requirements of NFPA® 1500, *Standard on Fire Department Occupational Safety and Health Program:*

NFPA® References

8.1	8.2.2	8.3.12	8.4.14	8.5.1	8.7.2
8.1.1	8.2.3	8.4	8.4.15	8.5.2	8.7.2.1
8.1.2	8.2.3.1	8.4.1	8.4.16	8.5.2.1	8.7.3
8.1.3	8.2.4	8.4.1.1	8.4.17	8.5.3	8.7.4
8.1.4	8.2.5	8.4.2	8.4.17.1	8.5.4	8.7.5
8.1.5	8.2.6	8.4.3	8.4.17.2	8.5.5	8.7.6
8.1.6	8.2.7	8.4.3.1	8.4.18	8.5.5.1	8.7.7
8.1.7	8.2.8	8.4.4	8.4.19	8.5.6	8.7.8
8.1.8	8.3	8.4.5	8.4.20	8.5.6.1	8.7.9
8.1.9	8.3.1	8.4.6	8.4.21	8.5.7	8.7.10
8.1.10	8.3.2	8.4.7	8.4.22	8.6	8.7.11
8.1.10.1	8.3.3	8.4.8	8.4.23	8.6.1	8.7.11.1
8.1.10.2	8.3.4	8.4.9	8.4.24	8.6.2	8.8
8.1.11	8.3.5	8.4.10	8.4.25	8.6.3	8.8.1
8.1.11.1	8.3.6	8.4.11	8.4.23	8.6.4	8.8.2
8.1.12	8.3.7	8.4.12	8.4.27	8.6.5	8.8.3
8.1.12.1	8.3.8	8.4.12.1	8.4.28	8.6.5.1	8.8.4
8.1.12.2	8.3.9	8.4.13	8.4.29	8.7	8.8.5
8.2	8.3.10	8.4.13.1	8.5	8.7.1	8.8.51
8.2.1	8.3.11				

FESHE Learning Objectives

Occupational Safety and Health for Emergency Services

Apply the knowledge of an effective safety plan to pre-incident planning, response, and training activities.

Explain the components of an accountability system in emergency services operations. Discuss the need for and the process used for post-incident analysis.

Emergency Operations

Learning Objectives

After reading this chapter, students will be able to:

1. Summarize the causes of emergency scene casualties.

2. Summarize the individual's, supervisor's, and administrator's responsibilities for safety during emergency incidents.

3. Describe the components for physical safety.

4. Describe the differences between the cold, warm, and hot control zones.

5. Describe the procedure for calculating a collapse zone for a building.

6. Define size-up.

7. Describe the information to be gathered during a preincident survey.

8. Summarize the information contained in the arrival report.

9. Describe the roles of the command and general staffs in an ICS.

10. Summarize the IAFC 10 Rules of Engagement for Structural Fire Fighting.

11. Describe the three incident priorities.

12. Summarize the general procedures for tracking resources during an emergency incident.

13. Describe the general procedures for assigning multiunit resources during an incident operation.

14. Describe the general procedures for implementing an incident action plan.

15. Describe behavioral considerations for incident safety.

16. Summarize the aspects of an incident that are analyzed during a postincident analysis.

Chapter 9
Emergency Operations

Case History

At a residential fire, two firefighters were working in the attic after most of the fire had been located and extinguished. One firefighter told his partner that he was getting low on air and that they would have to leave the attic. At the same time, the second firefighter commented about the poor condition of the roof above them and how it looked close to failure, so they should not return to the attic. Just then, a large piece of the roof system separated and fell, landing on the first firefighter's back, hitting his air tank. A Mayday was quickly called by his partner and rescue was initiated successfully. The injured firefighter walked to the ambulance and was taken to the hospital for evaluation. He had received some injury to the muscles and tendons in his back. However, if his air tank had not broken the fall of the piece of large timber beam and prevented direct contact with the firefighter, the results could have been much worse.

This incident emphasizes a number of emergency scene safety lessons. First, you must always wear the correct personal protective equipment when working at a structure fire. Second, it is important to stay in close contact with your partner and to have radio contact with the incident commander. Third, establishing and maintaining a rapid intervention crew can hasten the response to a firefighter emergency. Finally, you must constantly monitor and evaluate the changing conditions in your environment, even after the majority of the fire is extinguished.

Source: *National Fire Fighter Near-Miss Reporting System*

Fire and emergency services organizations respond to a wide range of emergency situations and incidents. Each incident presents its own hazards, placing responders' health and safety at risk. Statistics collected and analyzed by the National Fire Protection Association® (NFPA®) and the National Institute for Occupational Safety and Health (NIOSH) indicate that the majority of fatalities that occur at emergency incidents involve cardiac arrest. Suggestions for eliminating or reducing fatalities and injuries caused by cardiac arrest has been addressed in the Chapter 4, Health and Fitness Program Requirements.

Each type of emergency incident creates potential risks to you and your fellow responders. Many incidents can create multiple risks. You must be prepared for the unexpected and protect yourself by using proper personal protective equipment (PPE), adhering to your department's standard operating procedures or guidelines (SOPs/SOGs), and following the orders that are given to you by your supervisor or the officer in charge.

This chapter discusses the current trend in fatalities and injuries at emergency scenes and outlines the responsibilities that you, your supervisor, and the administration have to reduce these statistics. Various physical safety components for the incident scene will be described along with the operational considerations posed by the use of an incident command system. Behavioral considerations that can contribute to emergency scene casualties as well as the importance of a postincident analysis will also be discussed.

The act of being safe in a hazardous situation takes training, awareness, and preparedness. An emergency responder who becomes injured or incapacitated during an incident increases the risk to fellow firefighters, drains resources from the incident, and reduces the efficiency and effectiveness of the responding units. It is essential that you, your supervisor, and your department's administration do everything possible to prevent situations that will result in emergency scene casualties.

Emergency Scene Casualties

In 2008, firefighter fatalities in the United States amounted to 108 individuals. Of these, 29, or approximately 28 percent, occurred on the fireground. Thirteen of these were trapped or caught within a structure either by fire extension or collapse.

In 2009, the NFPA® issued a report on fireground injuries that covered the years 2003 to 2006. During that period, injuries that resulted from physical contact with objects, that is, crushing, cutting, falling, etc., amounted to 21 percent of all emergency incident injuries. This chapter provides suggested activities and actions that can help reduce these statistics.

Injuries

According to an NFPA® report issued in May 2009, firefighter injuries reported in the previous year were approximately 79,700, a decrease from 2007. Injuries received during emergency operations were 36,595, 45.9 percent of the total. Types of injuries included:

- Muscular pain (48.8%)
- Wound, cut, bleeding, or bruise (15.6%)
- Smoke or gas inhalation (6.2%)
- Other categories with lower percentages (29.4%)

The leading causes of injuries at emergency incidents included falls, slips, and jumps (23.5%), as well as overexertion and strains (23.1%). Being struck by (coming in contact with an object) amounted to 13 percent and burns were 12.7 percent.

Finally, this report indicates that 19.1 percent or 15,250 injuries resulted in lost time, an additional cost to the department as well as the individual.

Exposures

For reporting purposes, exposures are divided between hazardous materials and medical. There were approximately 20,650 reported hazardous materials exposures to such conditions as asbestos (found in older buildings), radioactive materials, chemicals, and vapors.

Medical exposures resulted in 10,380 reported incidents including exposure to meningitis, HIV, hepatitis, and other fluid and airborne pathogens **(Figure 9.1)**.

Fatalities

In a provisional report issued in the fall of 2009, the United States Fire Administration (USFA) provided statistics on firefighter fatalities between January 1, 2009, and September 30, 2009. During that period, 70 line-of-duty deaths (LODDs) had been documented. Of those deaths, 67.1 percent occurred while performing emergency activities. The report listed the types of incidents and the percentage of fatalities as:

- 24 Not Incident Related (34.2%)
- 20 Structure Fire (28.5%)
- 15 Wildland Fire (21.4%)
- 5 Motor Vehicle Accident (MVA) (7.14%)
- 3 Other (4.28%)
- 2 Emergency Medical Services (EMS) (2.85%)
- 1 Vehicle Fire (1.42%)

The report listed the causes of death as:

- 38 Stress/Overexertion (54.2%)
- 15 Vehicle Collision (21.4%)
- 5 Fall (7.14%)
- 3 Caught/Trapped (4.28%)
- 3 Struck by (4.28%)
- 2 Collapse (2.85%)
- 2 Lost (2.85%)
- 1 Other (1.42%)
- 1 Contact with (1.42%)

The types of activities involved when the fatality occurred included:

- 27 Not On Scene (38.5%)
- 11 Advancing Hoselines (15.7%)
- 10 Responding (14.2%)
- 6 Unknown (8.57%)
- 3 Search and Rescue (4.28%)
- 3 Other (4.28%)
- 2 Pump Operations (2.85%)
- 2 Water Supply (2.85%)
- 2 Scene Safety (2.85%)
- 2 Unknown (2.85%)
- 1 Incident Command (1.42%)
- 1 Support (1.42%)

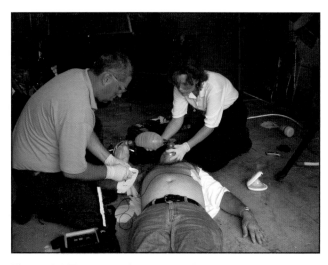

Figure 9.1 To prevent the spread of air and bloodborne pathogens, EMTs must wear personal protective equipment when treating a patient.

Victims were primarily volunteer firefighters. The distribution through types of departments was:

- 35 Volunteer (50%)
- 27 Career (38.5%)
- 4 Wildland Contract (5.71%)
- 3 Wildland Full-Time (4.28%)
- 1 Part-Time (Paid) (1.42%)

Emergency Incident Responsibilities

As in all other situations involving safety, department administrators, supervisors, and responders all have a responsibility for safety during an emergency incident. Regardless of the part that you play, you should be aware of the responsibility of the other two groups.

Administration's Responsibility

The administration is ultimately responsible for the safety of all fire department employees at all activities, at all times, and in all locations. However, every member of the department should be aware of the need for safety and appropriately communicate unsafe situations or behavior to their supervisors to avoid injury. This responsibility is especially important during emergency incidents. The administration's responsibility for emergency scene safety includes, but is not limited to:

- Developing, adopting, implementing, and managing an incident management system (IMS)
- Developing standard operating procedures or guidelines (SOPs/SOGs) and preincident plans for all types of hazards the fire department may respond to
- Training department personnel in safe operating procedures for all types of emergencies **(Figure 9.2)**
- Creating and filling the position of health and safety officer (HSO)
- Training personnel to fill the position of incident safety officer (ISO)
- Purchasing, issuing, and maintaining the appropriate types of personal protective equipment (PPE) for each type of hazard
- Being aware of all hazardous sites, processes, and conditions within the jurisdiction
- Developing automatic and mutual aid agreements with other jurisdictions and agencies
- Working with other agencies to ensure that controllable hazards are mitigated through codes and ordinances
- Reducing hazards through public life safety education, fire prevention, and code enforcement

Supervisor's Responsibility

As a supervisor, you are responsible for implementing the adopted rules, regulations, and SOPs/SOGs of the department. Your first consideration at any emergency incident is your safety and the safety of the personnel assigned to you, either in your company or in other units. Supervisors at emergency incidents are responsible for the following safety-related actions:

- Implementing or adhering to the locally adopted Incident Command System (ICS) within the IMS
- Enforcing all safety-related SOPs/SOGs during emergency incidents
- Establishing an accountability system during emergency incidents
- Undertaking or assigning the duties of ISO at the emergency incident
- Monitoring the health and safety of personnel assigned to them
- Using crew resource management (CRM) guidelines
- Implementing appropriate size-up techniques **(Figure 9.3)**
- Requesting additional resources when the situation requires it
- Adopting the correct strategy and tactics based on the analysis of the size-up information
- Establishing communication with all members of the unit and other units

Figure 9.2 Basic to all firefighter training is the safe way to handle, carry, and use all types of equipment.

Figure 9.3 Company-level supervisors must perform a size-up of the incident to determine the hazards and risks that the crew will be exposed to and make appropriate decisions to control the incident.

Your Responsibilities

Safety begins with you. If you are injured, trapped, or out of communication with your supervisor or the rest of your crew, you will increase the risk to yourself and others at the incident. You will not be able to perform your duties and will actually draw off resources to assist you. Specifically, your responsibilities are:

- Adhere to all safety SOPs/SOGs
- Operate with a partner at all times
- Apply proper CRM procedures
- Use the accountability system
- Adhere to the ICS
- Wear the appropriate PPE

- Activate your personal alert safety system (PASS) device before entering the hot zone (**Figure 9.4**)

- Use air management effectively

- Adhere to the rehabilitation requirements

- Do not freelance

- Apply situational awareness at all times

- Know firefighter self survival techniques

- Know the process for firefighter Mayday actions

Figure 9.4 You are responsible for your own safety. Always activate your PASS device before entering a hazardous area.

- Do not leave the incident scene without signing out with the incident commander (IC) or accountability officer

- Follow orders within the limits of proper CRM

Physical Safety Components

The physical safety components that are used at emergency incidents are the responsibility of the fire department administration. The administration must develop, fund, provide training, implement, and monitor the application of these components. These components are based on SOPs/SOGs, resources, personnel positions or functions, and training that must be in place prior to the emergency incident. Components include:

- Staffing
- Control zones
- Rehabilitation
- On-scene medical facilities
- On-scene lighting
- Visibility
- Exposure
- Environmental issues

Staffing

Staffing levels have been a topic of conversation within the fire service for decades. The need for adequate staffing at emergency incidents is not questioned. The debate arises over what is adequate staffing or the minimum level of personnel to perform functions such as rescue, extinguishment, and spill containment among others. In short, staffing is everyone's responsibility.

In career and combination departments, the administration is responsible because they establish how personnel will be utilized and deployed. The political leadership of the jurisdiction, based on local needs and mandates, determines the number of people that will be hired by the department. They also establish the work shift, the number of individuals that will be assigned to each apparatus during a normal work shift and the minimum number that must be present to place the unit in service. Volunteer departments determine the minimum number of positions to be filled and the minimum required to respond to an emergency before requesting outside assistance. Staffing levels are driven by national standards, such as NFPA® 1500, 1710, and 1720, Occupational Safety and Health Administration (OSHA) requirements, and insurance industry mandates. The level may also be mandated by the local labor-management agreement.

Supervisors are responsible for staffing because they must base their decision making on the number of personnel available to them during a response. For instance, an interior fire attack may not be possible if personnel are not available to provide an initial rapid intervention crew or team (IRIC/IRIT). Supervisors in volunteer departments are particularly affected by the lack of adequate staffing for the initial response **(Figure 9.5)**. Critical decisions on the strategies or tactics to implement may have to be delayed until additional personnel arrive.

Finally, firefighters, like yourself are also responsible for emergency scene staffing. You must realize that some actions may not be taken without adequate personnel to safely perform assigned tasks. You will have to balance the desire to take action with the need to wait until enough personnel are on the scene. You must also be healthy and physically fit to respond. If you are ill, you must notify your supervisor so a replacement can be assigned to your position. As a volunteer, you must be prepared to respond rapidly and safely at anytime with your PPE accessible and in good condition.

Figure 9.5 Volunteer and combination fire departments are limited in the resources that are immediately available at an incident. The minimum number of personnel may not be available for some time.

Control Zones

There are numerous reasons for establishing control zones at an emergency incident scene. First, controlling the perimeter facilitates the use of a personnel accountability system. Second, it helps in accounting for occupants or victims. Finally, it helps to keep the scene free of curious spectators.

Establishing three operating *control zones* (commonly labeled *hot, warm,* and *cold*) is the most common and effective way to control the perimeter of an incident scene **(Figure 9.6, p. 318)**. Control zones should be established as soon as possible when resources are available for this assignment. Although not always practical, zones can be cordoned off or indicated with utility rope or fireline tape.

There is no specific distance or area that should be cordoned off for each zone or from the total incident scene. The zone boundaries should be established by taking into consideration the amount of area needed by emergency personnel to work, degree of hazard presented by elements involved in the incident, wind and weather conditions, and general topography of the area. The three zones can be described as follows:

- **Hot zone** — Area where resolving the problem takes place — fires are suppressed, hazardous materials (hazmat) releases are controlled and contained, vehicle extrication is performed, etc. Only personnel who are directly involved in disposing of the problem are allowed, which limits crowds and confusion at the most critical area of the scene. The size and shape of the zone may vary greatly, depending upon the nature and extent of the problem. *Personnel requirements:*

  — Trained appropriately to manage the situation

  — Attired in complete PPE designed for the specific hazard

  — Participated in a personnel accountability system implemented by the incident commander (IC) or incident safety officer (ISO)

- **Warm zone** — Area immediately outside the hot zone for personnel who are directly supporting the work being performed by those in the hot zone — limited to personnel who are operating hydraulic tool power plants, providing emergency lighting, and providing fire protection. These personnel are in full PPE and ready to enter the hot zone. In hazardous materials incidents, this zone is where a decontamination station is normally assembled.

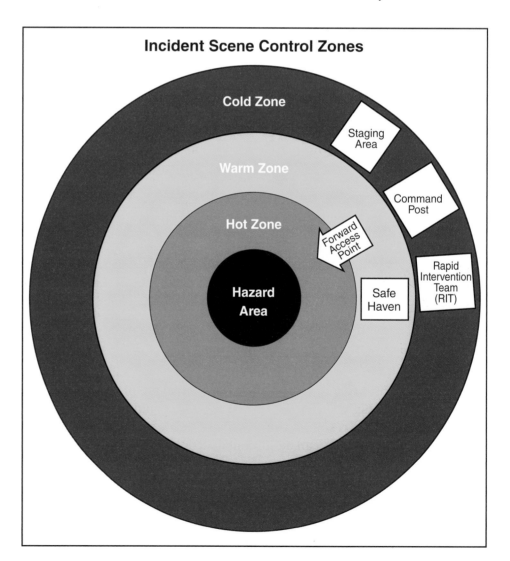

Figure 9.6 Incident Scene Control Zones.

- *Cold zone* — Area immediately surrounding the hot and warm zones — may include the incident command post (ICP) with a rapid intervention crew/team (RIC/RIT) nearby, public information officer's (PIO) location, rehabilitation area, and staging areas for personnel and portable equipment **(Figure 9.7)**. The outer boundary of this area would be the control line for the general public (crowd-control line). *Examples:*

  — Backup personnel available to enter warm or hot zones

  — Witnesses and family members of victims

  — News media accompanied by the PIO or organization representative

In addition to the control zones, structural fires and structural collapse incidents may require the establishment of a *collapse zone*. Collapse zones are established when there is evidence that a structure or portion may collapse. Strategy will shift from offensive to defensive and apparatus, personnel, and master steams must be relocated to a point of safety outside the designated collapse zone. The collapse zone is generally a space $1\frac{1}{2}$ times the height of the structure and extending the full length of the structure.

Collapse zones, sometimes referred to as safety zones, should also be established around structures that have large glass partition walls. These walls can be weakened or broken by internal fires or earthquake damage presenting a hazard of flying glass around the structure or some distance from the structure if there is a wind current.

Figure 9.7 The command post is located in the cold zone of the incident.

Calculating Collapse Zones

Traditionally, collapse or safety zones have been estimated by taking the height of the structure and multiplying it by a factor of $1\frac{1}{2}$. For example, a 3-story structure that is 30 feet tall (9.14 m) would require a collapse zone no less than 45 feet (13.72 m) from the base of the structure. As structures increase in height, this becomes impractical by creating a space that limits defensive operations. Therefore, you should consider the following guidelines when determining a collapse zone:

- Type 1 construction high-rise buildings are not as likely to collapse making the primary concern the hazard of flying glass from windows or curtain walls. In Type 1 construction, it is the contents of the building that is burning, not the structure itself. Collapse zones must be determined considering the direction and velocity of wind currents that can carry the glass shards. Structural collapse, if it does occur, will be localized and not structure wide.

- Type 2 construction consists of unprotected steel or noncombustible supports such as I beams. When exposed to temperatures above 1000°F (537.78°C), unprotected steel will expand and twist, pushing out walls, and when cooled will slightly constrict. These movements will cause floors and walls to collapse. Any type of construction that includes brick and block walls supporting unprotected steel bar joists and I beams are involved in a large number of these collapses.

Continued on page 320

Calculating Collapse Zones (concluded)

- Type 3 construction multi-story buildings should have a collapse zone of $1\frac{1}{2}$ times the height of the structure. An example would be that a building 7 stories (70 feet [21.34 m]) will require a collapse zone of approximately 105 feet (32 m) during defensive operations. In Type 3 ordinary construction, exterior load bearing walls are made of concrete, brick, or masonry while interior loads are carried by wood, masonry, or unprotected steel. Masonry construction walls can collapse in one piece or crumble in many parts. When the debris strikes the ground it can travel a distance and even cause the collapse of other structures or objects.

- Type 4 heavy timber or mill construction is one of the least likely to collapse. The weight-bearing capacity of the large dimension wood members will resist collapse unless they have been effected by a large volume of fire. A collapse zone should be established if the fire is intense or the structure has been weakened by repeated fires over time.

- Type 5 construction collapses are influenced by the style of construction. That is, a multi-story platform structure will generally burn through and collapse inward while a balloon structure can have the full walls fall outward in a single piece. Exterior masonry and veneer walls are not load bearing and are placed over load bearing wood walls. Brick veneer attached to the frame can curtain collapse into a pile or fall as a unit straight out as the ties and supports fail. Although it is rare for a Type 5 to collapse outward there is a great danger to firefighters due to interior collapses. Lightweight trusses will fail within five minutes when exposed to direct heat.

In North America, examples of structural collapse involving high rise buildings or Type 1 construction buildings are very limited. Strict building codes have ensured that structural members exposed to fire and high temperatures will remain sound until the fire is extinguished. Structural collapses due to earthquakes generally involve smaller buildings such as the one to four story buildings in the Marina Section of San Francisco in 1989. Some of these buildings simply fell over or against another building.

Collapse zones should be established when there is an indication that the structure has been weakened by prolonged exposure to fire or heat, when a defensive strategy has been adopted, or when interior operations cannot be justified. The size of the collapse zone must take into consideration the type of building construction, other exposures, and the safest location for apparatus and personnel. Remember too that church steeples, water tanks, chimneys, and false facades that extend above the top of the structure must be viewed as a potential collapse hazard even if the structure is not. It should also be noted that most collapses usually involve brick or masonry block and may be structural components or veneer. At the same time, remember that structural collapses are not limited to the actual emergency and can occur well after the fire is extinguished. Fire inspectors must ensure the structural stability of the site before entering it.

Because the collapse zone is the full length of all of the effected walls, the safest location for defensive operations is at the corner of the building. In this 90 degree area, masterstreams and apparatus can be located as long as they are far enough away that flying debris will not strike them.

Rehabilitation

Emergency scene operations place a great deal of physical stress on responders. Therefore, rehabilitation areas are necessary to provide a place where personnel can rest, relax, rehydrate, take nourishment, and have their vital signs monitored by medical personnel. Rehabilitation areas, also known as *rehab*, are located in the cold zone and are generally incorporated with the on-scene medical facilities. It is the responsibility of the administration to establish procedures for establishing a rehab at emergency incidents and to fund the items needed for the site. Water or fluid containers, folding cots, warming tents, and water mist systems are just a few of the items generally associated with rehab areas **(Figure 9.8)**.

Figure 9.8 The rehabilitation (rehab) section provides water, rest, and monitoring for emergency personnel.

The level of rehabilitation will be consistent with the size and scope of the incident. It is not unheard of to need cots or warming tents at some incidents. For example, a 4-hour hazmat incident in Massachusetts in January during a snow storm would require it. Also, even a 2-hour fire may require cots for people who have overexerted themselves for 20 minutes of intense physical exertion.

The average working fire with all responders working is going to be at a minimum 2 hours of intense operations and a couple hours of breakdown and reconstitution. Rehab does not just occur during the suppression phase, it occurs during the overhaul phase as well.

Extended operations will almost always require food. You cannot expect firefighters to burn the amount of calories they are burning during this rigorous work and not need dietary nourishment.

Rehabilitation is not just a site, however. It is also the awareness of the need for rehab that is the responsibility of supervisors and responders. Supervisors must monitor the physical condition of their subordinates and send them to rehab before they become too fatigued. Responders must monitor their own condition and report the need for rehab to their supervisors. In practice, rehab may take two forms. The first form involves resting, removing wet or damp PPE, cooling down or warming up as needed, and drinking fluids. The second form involves medical rehab that includes medical monitoring, treatment of injuries, rehydration, and/or transportation.

A general rule of thumb for structure fires, or incidents that require the use of self-contained breathing apparatus (SCBA), is to send personnel to rehab when they have used two (2) 30-minute air bottles or one (1) 60-minute air bottle. Other factors, such as weather conditions, type of work, or fire behavior will affect this estimate. For a detailed look at rehabilitation, see *Emergency Incident Rehabilitation* by Dr. Edward Dickinson and Michael Wieder.

On-Scene Medical Facilities

Locating the rehab function within close proximity to on-scene medical support is a logical approach to meeting both forms of rehab mentioned above. Trained medical personnel, along with medical supplies, and transportation are then located in one area. Medical personnel can monitor personnel in rehab for core body temperature, blood pressure and rate, hydration, and other vital signs **(Figure 9.9)**. They can provide immediate assistance and transportation if a firefighter shows symptoms of heat stress, cardiovascular problems, or other ailments. If the emergency incident is too far from a medical facility by ground transport, then an emergency helipad can be established at or near the on-scene medical facility.

On-Scene Lighting

Injuries and sometimes fatalities can be the result of obscured visibility. Visibility may be obscured by darkness, fog, snow, rain, or smoke. Fire apparatus are often equipped with spotlights mounted on the four corners of the body and telescoping lights mounted near the pump panel. Changes in apparatus design and visibility requirements have led to improved on-scene lighting. New types of apparatus lighting provide sufficient outdoor lighting for most emergency scenes. A portable generator or apparatus powered lights can illuminate structure interiors where the power has been shut off.

Care must be taken in placing portable lights, deploying electrical cords, and locating the power source. Frayed or damaged wires are an electrocution hazard. Lightbulbs can overheat and explode causing injury due to broken glass. The lights as well as the wiring can pose tripping hazards. Power plants can also provide an ignition source for flammable vapors or fire gasses, some of which are heavier than air.

On-scene lighting can actually create a hazard if it is improperly set up. Lights should not be directed toward oncoming traffic if possible. At night, the lights can blind drivers and prevent them from seeing apparatus and personnel working along the roadway **(Figure 9.10).**

Visibility

One of the leading causes of firefighter fatalities and injuries is being struck by motor vehicles while working on or near roadways. As mentioned in Chapter 6, Fire Apparatus Safety, new visibility studies have recommended ways to mark apparatus with warning lights and reflective tape. At the same time, NFPA®

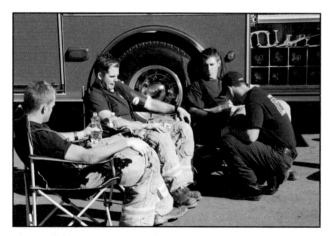

Figure 9.9 Medical personnel monitor the vital signs of firefighters and provide medical assistance or transportation to a medical facility as needed.

Figure 9.10 On-scene lighting improves visibility for firefighters but also creates a blinding hazard for nearby traffic. *Courtesy of Chief Chris Mickal, New Orleans (LA) Fire Department.*

has established the requirement for minimum reflective trim on PPE and the U.S. Department of Transportation (DOT) has established requirements for emergency scene warning signs, traffic cones, and the use of reflective vests during highway emergencies **(Figure 9.11)**.

While the administration must supply the reflective trim, apparatus markings, and signage, it is your responsibility to wear and use them. As a supervisor, you must ensure that your personnel don the reflective vests, place signage as required, and keep the reflective trim on their PPE clean and repaired.

Figure 9.11 Traffic cones, signs, reflective clothing, and flashing lights are all used at roadway incidents to reduce the risk of being struck by a passing motorist.

Exposure

Exposure to hazardous conditions is basic to the fire and emergency services. We must continually protect ourselves from a variety of hazardous conditions through the use of PPE including respiratory and eye protection. Remember that exposure is not just limited to smoke and unburned particles of combustion. It also includes the vapors from fluorescent ballasts (older ones may contain polychlorinated biphenyls [PCBs]), toxic chemical or gas leaks, communicable diseases and infection, and engine exhaust vapors. Constantly exercising situational awareness is the first step in protecting yourself from exposure to hazardous conditions. Until a responsible officer determines that the atmosphere is safe, you must always wear respiratory protection during the overhaul of fire scenes.

Environmental Issues

Protecting the environment is also a physical safety component. Containing and controlling hazardous chemicals, in any quantity, protects the soil and prevents freshwater supplies from being contaminated. Most states/provinces have restrictions on flushing contaminates, including Class A and Class B foams, into storm drains. Gasoline, oil, radiator fluid, and contaminated extinguishing agents must be contained and properly removed from motor vehicle accidents (MVAs). PPE must also be decontaminated following any incident involving hazardous materials. The water used to clean PPE and equipment must also be properly disposed of, based on the jurisdictions requirements. NFPA® 1500 requires that PPE be decontaminated in areas in fire stations and facilities especially designed for this process.

Operational Size-Up

As an emergency responder, your job is to arrive safely on the incident scene, contain, and control the hazard, and ensure the incident is properly terminated. As a fire officer and supervisor, your job further expands to include establishing command, assessing the situation, developing an incident action plan (IAP), and allocating your resources to mitigate the incident. To do this effectively and efficiently, you must collect information on the incident, evaluate it, and determine the correct strategy and tactics to use. The process you use is known as *size-up* **(Figure 9.12, p. 324)**.

Size-up — Ongoing mental evaluation process performed by the operational officer in charge of an incident that enables him or her to determine and evaluate all existing influencing factors that are used to develop objectives, strategy, and tactics for fire suppression before committing personnel and equipment to a course of action. Size-up results in a plan of action that may be adjusted as the situation changes. It includes such factors as time, location, nature of occupancy, life hazard, exposures, property involved, nature and extent of fire, weather, and fire fighting facilities.

Figure 9.12 The on-scene portion of size-up begins with an arrival report.

Size-Up

To perform an effective and thorough size-up, you consider information that is gathered before the incident occurs, during dispatch, upon arrival, and throughout the incident until all resources are placed back in service. This section provides you with a process for gathering information. It also provides important considerations that you must have when you perform a size-up. Remember that size-up is used at all types of incidents, not just at structure fires. The more practice you gain sizing up incidents, the better you will become at reading the clues and making informed decisions.

Preincident

The preincident portion of the size-up process consists of the knowledge you have gained from training, study, and experience plus the site-specific information that you gain from preincident surveys or prefire planning. Your knowledge includes fire behavior, resource capabilities, building construction, and hazardous conditions as well as the ICS, SOPs/SOGs, strategy/tactics, and fireground procedures. Specialized training in vehicle extrication, rescue techniques, confined-space entry, emergency medical care, disaster response, and hazardous materials mitigation are also part of your preincident knowledge.

A preincident survey is a familiarization tour of a site or structure performed periodically, depending on local requirements **(Figure 9.13)**. The survey provides basic information that can be used to develop a preincident plan. The preincident survey provides you with information about:

Preincident survey — The act of collecting information on a site prior to the occurrence of an incident.

Preincident plan — Plan developed from the information gathered during the preincident survey and used during emergency operations and training.

- site access
- structure
  - construction type
  - occupancy type
  - fire protection system(s)
  - content fuel load
  - unique characteristics or hazards
  - access
- available water supply
- resource needs

If time and resources allow, it is good practice to perform preincident surveys periodically on all types of structures with the exception of single-family dwellings. Preincident surveys are rarely performed on single-family dwellings. However, residential structures may have common floor plans and characteristics that can be studied. Single-family dwellings may be informally surveyed when doing in-service home inspections and/or EMS responses. Crews should familiarize themselves with home styles, designs, and layouts common to their community or response area. Every department should have SOPs/SOGs for the type of strategy and tactics used in single-family dwellings.

You should follow your local SOP/SOG for determining which occupancies to survey and how often. The criteria for which structures to inspect are usually set by the potential fire and life safety hazard of the site. Referred to as *target hazards*, these sites contain life safety concerns for firefighters as well as occupants, hazardous processes or storage that may have a high frequency or severity of fires, government buildings, prominent structures, and/or those with high contents or structure value. A few examples are:

Figure 9.13 Preincident surveys of target hazards and other structures within the company's response area are the foundation for incident size-up.

- Life safety concerns
  - Schools, colleges, and universities
  - Auditoriums, theaters, and restaurants
  - Places of worship
  - Hospitals, nursing homes, and daycare centers
  - Multi family dwellings, hotels, motels, and dormitories
  - Institutions, jails, and prisons
- Hazardous processes or storage
  - Spray paint operations
  - Metal plating
  - Chemical plants
  - Automobile fueling stations
  - Compressed-gas storage areas
  - Paint storage warehouses
- High contents/structure value
  - Mercantile occupancies
  - High-rise structures
  - Office buildings
  - Warehouses

Generally, surveys are conducted and updated annually or biannually when there has been a change of occupancy type and whenever renovations or alterations have been made to a structure. Surveys should also be made at construction sites as a means of learning about how a building is being built. Remember that construction sites are very vulnerable to fires, and your survey is a way of preparing for that possibility as well.

Your department may already have a survey checklist of information you need to gather or conditions you need to be aware of **(Figure 9.14, p. 326)**. The information you gather will be used to develop a preincident or operational plan. During the survey, you should concentrate on gathering the following information:

- Life safety concerns for firefighters and occupants
- Building construction type to determine resistance to fire spread

City of LaFosse Fire Department
Preincident Site Survey

Date _____

Company _____ Officer-in-charge _____

Site Address _____

Owner _____

Occupant _____

Emergency Contact Name _____

Phone Number _____

Number of site occupants: Day _____ Night _____ Weekend _____

Occupancy Classification _____

Construction Type _____

Roof Construction Style _____ Material _____

Number of stories: Above grade _____ Below Grade _____

Stairways _____ Hallways _____ Elevators _____

Roof access _____ Vertical openings _____

Site access _____

Portable fire extinguishers _____

Automatic fire suppression system _____

Fire pumps _____

Special agent suppression system _____

Water supply source _____

Fire department connection, location _____

Standpipe system classification _____

Fire detection and alarm system _____

Water flow alarm _____

Monitoring company _____

Fire control panel or room location _____

Location of main electrical shut-off switch _____

Location of main gas shut off _____

Location of water shut off _____

Fire doors, automatic _____ Manual _____

Hydrant locations _____

Fire flow, estimated _____

Hydrant pressure _____

Structure Contents _____

Exposures type, distance _____

_____

Additional comments or observations _____

_____

Figure 9.14 Sample Preincident Survey Checklist.

- Building services, including utility shut off, elevators, and heating, ventilating and air conditioning (HVAC)
- Building access and egress
- Building age
- Building area and height
- Construction material to determine resistance to fire spread
- Contents to estimate fire load
- Building use to determine life safety and fire load
- Exposures
- Collapse zone
- Location and capacity of available water supply
- Location of fire control and protection system control valves and connections
- Presence of existing fire control and protection systems
- Hazardous materials or processes, including flammable/combustible liquids and gases
- Location of Safety Data sheets (SDS) (previously known as Material Safety Data Sheets [MSDS])
- High-voltage equipment
- Unprotected openings
- Overhead power lines or obstructions
- Occupancy load at all hours
- Names and telephone numbers of contacts or responsible persons for owner/occupant
- Estimated quantity of water required to extinguish a fire in the structure or a portion of it (fire flow)
- Emergency evacuation plan

While Responding

The size-up process continues when the alarm is sounded or your pager is activated. Depending on the dispatch system that your department uses, you must listen carefully to all information provided through the radio broadcast. If information is provided in a printed message or on a computer terminal, read it carefully. This information contains a description of the current condition as reported by witnesses at the scene or from the automatic detection and alarm system at the site.

You must remember that not all of the information given when the assignment is dispatched may be factual or complete and could be highly inaccurate. Prior to the arrival of fire personnel, the only thing you can verify immediately is time of day, weather conditions, and capabilities of your department's response. You must be prepared for any type of situation that you may face. Even law enforcement officers, not trained as firefighters, have been known to report a building to be fully involved that later proved to be only a small kitchen fire. Conversely, reported vehicle fires have been found to be well-involved vehicles, parked in garages under occupied buildings. Local SOPs/SOGs will generally establish the minimum information that is given during a dispatch broadcast, including:

- Time — exact time of dispatch, not the time the fire started
- Situation — the type of emergency reported to the dispatch center
- Location — the address of the emergency situation
- Resources dispatched — units responding

Time is an important factor in the size-up process. Although you do not initially know the time the fire started and how long it has been burning, being able to calculate your response time and the visual clues of the fire upon arrival will help you determine the stage of the fire.

On Arrival

The most intense part of the size-up process occurs when you arrive at the emergency incident scene. You may be greeted with a scene of utter chaos or one with no visible clues that an emergency exists. In addition to the emergency situation, and those individuals directly involved in it, numerous spectators may have gathered at the scene, making it difficult to distinguish them from occupants or victims. These bystanders may be hysterical or irrational and screaming for responders to do something **(Figure 9.15)**. Some may be attempting to extinguish the fire, assist victims, or perform a rescue — actions that may place them in direct danger as well.

In the midst of this scene, you must assess and communicate critical information to other responding units and the dispatch center. Your scene assessment begins before the apparatus comes to a stop and you step from the apparatus. The incident may be categorized by the severity, extent, and dynamics of the incident. Your size-up activities depend on the type of emergency incident such as:

- structure fire
- nonstructure fire
- medical response
- hazardous materials incident
- motor vehicle incident
- rescue

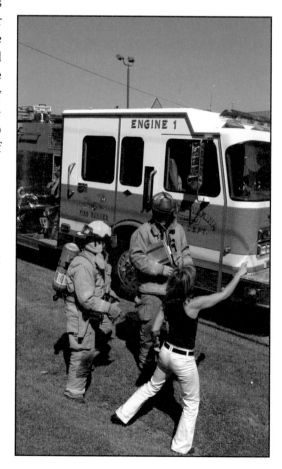

Figure 9.15 Bystanders can provide valuable information. However, their anxiety and excitement can also influence the decision-making process in a negative way.

Structure fire. At structure fires, there are two general situations that you will encounter upon arrival. First, there may be nothing evident, which will prompt an investigation of the structure. Second, smoke or fire may be visible, prompting the deployment of hoselines and a fire attack.

At structure fires, these two situations are the core of the *arrival* or *condition report (conditions on arrival* or *size-up report* in some jurisdictions) that you broadcast over the radio. Many fire and emergency services organizations mandate this type of report in their operations SOP/SOG. This arrival report is an essential communication that will inform the other responders and establish a foundation upon which to build the incident. These reports also apply to all other types of nonfire emergency incidents. In brief, an arrival report simply states:

- What we have (incident description)
- What we need (additional resources)
- What we are going to do (deployment of resources and actions to be taken)

Following the arrival report, you perform a physical size-up by looking at the scene from all sides. If you cannot physically complete a full 360-degree walk-around or survey the structure, then you must use other means to learn conditions on all sides of the incident. While checking all sides of the structure you are looking for:

- Any indication of occupants or the need for rescue
- Any indicators of fire location
- Lowest floor of fire involvement
- Smoke conditions
- Forcible entry requirements
- Special hazards: propane tanks, chemicals, electrical lines, or elevation changes **(Figures 9.16 a-d, p. 330)**
- Building construction

When the structure is small, such as a single-family dwelling, you may be able to make a 360-degree walk-around size-up. That is, walk all the way around the structure to view it from four sides. In the case of large structures, strip malls, row houses, structures with limited access, or big box retail stores, personally performing a 360 degree walk-around may not be possible. In these instances, it will be necessary to order other arriving units to provide information from the other sides. In commercial buildings, the roof is often included in this request as aerials apparatus or ground ladders are used to view the roof area. Other incoming units may be assigned to various sides of the structure and supply a report back to the IC.

You must focus on the situation and answer the question, *Can the resources at the scene and en route handle this situation?* If the answer is *no*, or even *maybe*, then additional resources must be requested *immediately.* Addressing this critical question is why the initial size-up sets the tone for the balance of the incident.

If there are not enough resources initially available and the arrival of additional ones may be delayed, personnel are likely to start the incident *behind the curve.* This type of situation will take immense work and organization to

Figures 9.16 a-d During the size-up, the officer must look for special hazards including a] propane tank, b] chemical storage, c] overhead electrical power lines, and d] elevation changes.

overcome. Therefore, it is critically important for the initial size-up to be done with an attention to detail and focus on the desired outcome and the activities and resources needed to accomplish that outcome. Considerations must be given to the level of acceptable personnel risk and which operational strategy to initiate. Properly interpreting condition indicators can help you make the correct decisions and increase the safety of you and your crew.

Once the walk-around is performed, you determine the appropriate course of action, develop an IAP, and assign duties to your crew and the remainder of your resources. Keep in mind that additional units may not arrive for some time, making it necessary for you to prioritize the activities your plan requires. You must also keep in mind that while you are setting up operations, the fire is increasing in size and extension. A risk assessment is necessary to ensure that you are making the right choice and doing it safely. Remember that while your crew may assist with information collection and have input, it is the IC who is responsible to make all decisions.

Before you commit any resources to an interior fire attack, you must have the resources to adhere to the 2-in, 2-out rule. For the first two firefighters to enter the immediately dangerous to life or health (IDLH) area, you must have two on the outside prepared to respond to a firefighter rescue. The two personnel on the exterior must be prepared to enter the structure in full PPE including SCBA and be fully trained. These individuals may be the pump operator, yourself, or another firefighter. If you cannot implement 2-in, 2-out, then you must not send personnel into the structure unless there is a known rescue, incipient stage fire, and/or fire can be extinguished with a Class 1 fire extinguisher.

NOTE: Remember that the presence of four firefighters at a structure fire does not automatically ensure that the 2-in, 2-out rule has been met. If it is necessary for one or both of the outside members to leave their primary duty area, then they may not be considered available for rescue. In other words, they may not leave their post if it will further endanger the incident scene. A member performing a task critical to the safety of other members on the scene should not be considered available for the initial rapid intervention team if by leaving an assigned task to make a rescue, that puts other on-scene members at greater risk. Your decisions must be driven by your local SOP/ SOG and by industry best practices if national, state/provincial, or local laws/ ordinances do not apply.

OSHA 2-In, 2-Out Rule

The following is the applicable portion of 29CFR 1910.134(g)(4) used to establish the 2-In, 2-Out rule:

Procedures for interior structural firefighting. In addition to the requirements set forth under paragraph (g)(3), in interior structural fires, the employer shall ensure that:

1910.134(g)(4)(i)
At least two employees enter the IDLH atmosphere and remain in visual or voice contact with one another at all times;

1910.134(g)(4)(ii)
At least two employees are located outside the IDLH atmosphere; and

1910.134(g)(4)(iii)
All employees engaged in interior structural fire fighting use SCBAs.

Note 1 to paragraph (g): One of the two individuals located outside the IDLH atmosphere may be assigned to an additional role, such as incident commander in charge of the emergency or safety officer, so long as this individual is able to perform assistance or rescue activities without jeopardizing the safety or health of any firefighter working at the incident.

Note 2 to paragraph (g): Nothing in this section is meant to preclude firefighters from performing emergency rescue activities before an entire team has assembled.

Nonstructure fire. Nonstructure fires generally involve wildlands or ground cover, materials stored out of doors, rubbish fires, sheds, or Dumpster® fires. The size of the incident may determine whether you can or need to do a 360-degree walk-around. Your size-up and condition report should consider the:

- type and quantity of burning materials
- exposures
- weather conditions
  - wind direction and velocity
  - humidity
- topography

Medical response. Departments that normally respond to medical incidents have an established protocol of situation size-up and patient evaluation. However, your size-up should include the overall situation and other potential hazards. For instance, you must consider the location of the patient, the need for extrication or removal to a safe site, bystanders that might pose a threat to you or your personnel, or physical hazards including ice, rising or rapid water, hazardous vapors, or structural collapse.

Hazardous materials incident. Like the medical incident, your organization must have an established protocol for size-up and reporting of known hazardous materials (haz mat) incidents. However, not all haz mat incidents are reported as such. You must look for clues that the situation is more dangerous than the dispatch information indicates. Visible clues may include:

- Occupancy type
- Product containers
- Location
- Time of day
- Weather
  - Wind direction
  - Wind velocity
- Unusual signs
  - Vapor clouds
  - Spilled material
- Behavior of occupants or employees
- Topography

Motor vehicle incident. Incidents involving motor vehicles may range from a minor collision to a multivehicle incident (**Figure 9.17**). Your size-up will determine the need for:

- Additional PPE
- Specialized extrication equipment
- Water supply or appropriate fire-suppression agent
- Medical assistance
- Hazardous materials specialists

Rescue. Sizing up a rescue situation depends on numerous factors including:

- Type of incident reported
- Location and condition of victims

- Visibility
- Weather
- Topography
- Condition of structure
- Access to incident site
- Bystanders or witnesses

Your arrival report will include all of these factors plus additional information such as:

- Need for specialized rescue equipment
  - Water craft
  - Earth movers
  - Structural braces
- Additional personnel

Figure 9.17 Every motor vehicle accident (MVA) creates its own risks to emergency responders.

During the Incident

After the IAP is implemented, emergency responders will be busy performing their assignments and making progress toward resolving the incident. This phase (between arrival and termination) can be relatively short or last for a considerable length of time. When the incident is relatively small and/or you make good decisions, the problem may be resolved in a few minutes. If not, additional resources and assistance from other officers may be needed to resolve the situation.

During this phase, the situation changes, either improving or worsening. In either case, initial decisions that were based on the initial size-up may or may not remain valid. Therefore, constant reassessment or ongoing size-up of incident conditions and the effect that operations have on the problem is critical. You must continue to size-up the situation, to validate or adjust the incident objectives, strategy, and tactics, making changes to the IAP as needed. By getting continuous feedback from crews, you can follow the progress of your IAP and continue to size-up the scene. An easily remembered acronym for this part of the size-up of structure fires is *CARA* which stands for:

- **C**onditions — an assessment of the interior conditions and the visible indicators on each side and roof of the structure
- **A**ctions — specific activities being performed by each crew inside and outside the structure that are engaged in fire-fighting operations
- **R**esources —includes additional personnel, apparatus, equipment, ventilation, a second hoseline, or an exchange of personnel
- **A**ir — refers to the breathing air supply available for interior fire attack as well as salvage and overhaul operations

Obviously, *CARA*, or parts of it, will apply to all types of incidents. The only part that may be optional for some is the need for breathing air.

Plan of Operation

Depending upon the nature and scope of the incident, the plan of operation or IAP may be simple or complex. As mentioned earlier, the plan does not need to be in writing on relatively small, routine incidents involving only an initial assignment. However, the use of a tactical worksheet can be very useful in maintaining control of even initial assignment incidents **(Figure 9.18)**. There should be a written IAP for long-duration or complex incidents. A written IAP is very useful when command changes or multijurisdictions are involved. How the plan is implemented is the operational phase. By assigning tactical objectives to company officers in divisions or groups, the plan is started. Then selecting the highest priority assigned first and working down your list as more crews arrive or on-scene crews complete their original tasks.

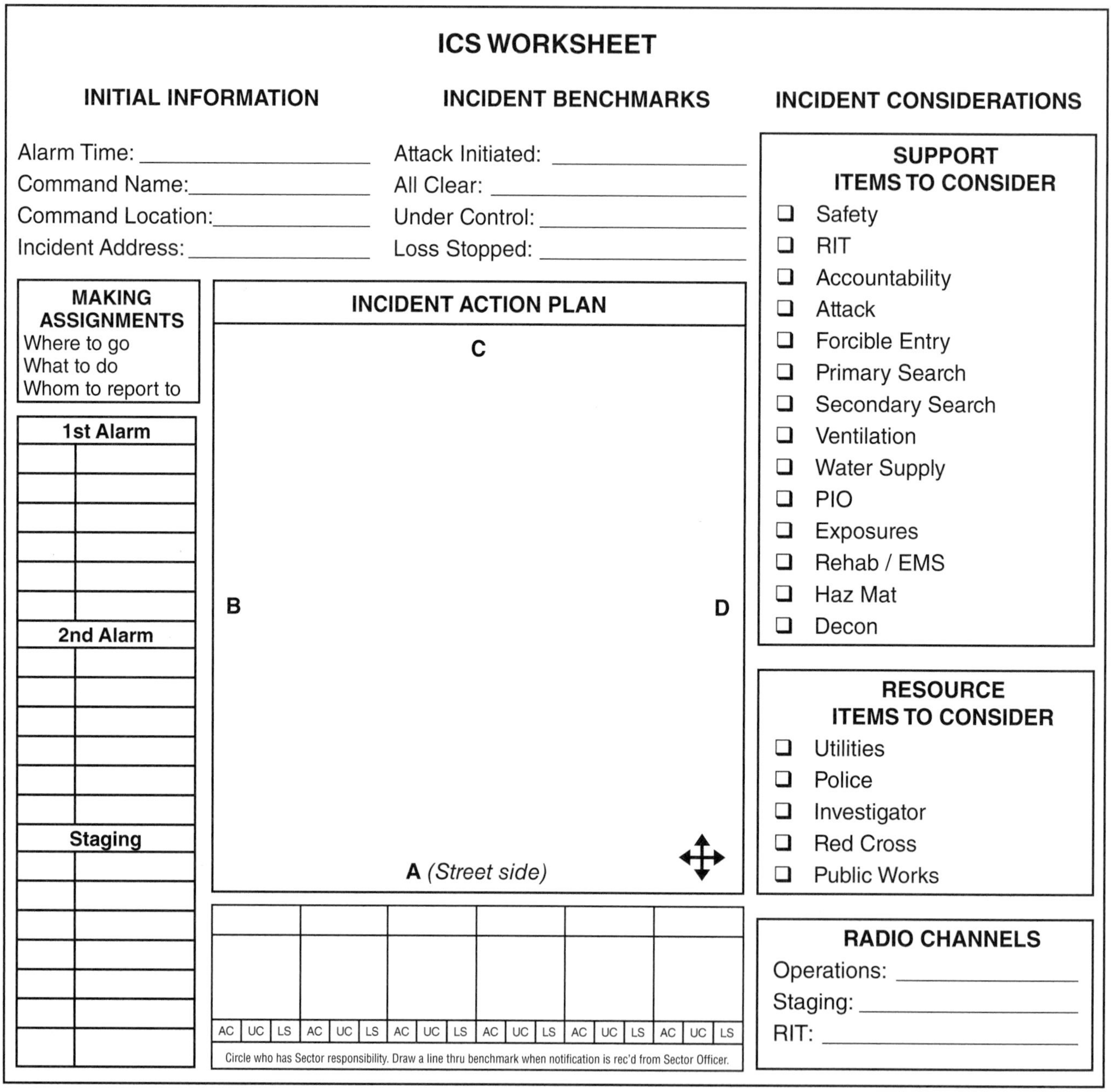

ICS WORKSHEET

INITIAL INFORMATION

Alarm Time: _____
Command Name: _____
Command Location: _____
Incident Address: _____

INCIDENT BENCHMARKS

Attack Initiated: _____
All Clear: _____
Under Control: _____
Loss Stopped: _____

INCIDENT CONSIDERATIONS

SUPPORT ITEMS TO CONSIDER
❑ Safety
❑ RIT
❑ Accountability
❑ Attack
❑ Forcible Entry
❑ Primary Search
❑ Secondary Search
❑ Ventilation
❑ Water Supply
❑ PIO
❑ Exposures
❑ Rehab / EMS
❑ Haz Mat
❑ Decon

MAKING ASSIGNMENTS
Where to go
What to do
Whom to report to

1st Alarm

2nd Alarm

Staging

INCIDENT ACTION PLAN
C

B   D

A *(Street side)*

AC	UC	LS	AC	UC	LS	AC	UC	LS	AC	UC	LS	AC	UC	LS	AC	UC	LS

Circle who has Sector responsibility. Draw a line thru benchmark when notification is rec'd from Sector Officer.

RESOURCE ITEMS TO CONSIDER
❑ Utilities
❑ Police
❑ Investigator
❑ Red Cross
❑ Public Works

RADIO CHANNELS
Operations: _____
Staging: _____
RIT: _____

Figure 9.18 Sample Tactical Worksheet.

Incident Command System

You must also have complete control over all your resources and their activities. To accomplish this, you must implement the locally adopted Incident Command System (ICS) based on the National Incident Management System-Incident Command System (NIMS-ICS). The ICS should be used, in varying forms, at all incidents. This section will provide a general overview of the NIMS-ICS in its current form.

The ICS model adopted by many jurisdictions in North America is based on NFPA® 1561, *Standard on Emergency Services Incident Management System*. The ICS provides guidance and direction for the management and control of all types of emergency incidents, ranging from single company responses to multiple agency and jurisdiction incidents. In March 2004, the U.S. government officially adopted ICS as part of the NIMS, and all federal agencies or state and local agencies that receive federal funding must use it.

Additional information on the NIMS-ICS model and its application may be found in the U.S. Department of Homeland Security (DHS) National Incident Management System document and the *Model Procedures Guide* series developed by the National Incident Management Consortium and published by Fire Protection Publications.

NIMS Compliance

Many firefighters and emergency services organizations have confused NIMS compliance with just having to take ICS courses. This is not true. ICS is part of the requirements within NIMS. An individual's responsibility in the organization determines the level of ICS training that individual must receive. Chief Officers should be at the ICS 400 level while company officers can be at the ICS 300 level. However, to be fully compliant with NIMS, organizations need to ensure that they have met the six components of NIMS. They are as follows:

- Command and management

- Preparedness

- Resource management

- Communications and information management

- Supporting technologies

- Ongoing management and maintenance

If company officers and their crews have not taken the IS-700 program they are not compliant according to NIMS.

NIMS-ICS Training

All emergency response personnel should receive NIMS-ICS training as part of their entry-level training, recurring proficiency training, and professional development. NIMS-ICS courses are offered through the National Fire Academy (NFA), the Federal Emergency Management Agency (FEMA), many state/provincial, tribal, and local agencies, as well as online.

Company officers who have completed the Fire Officer I certification training should be familiar with the basics of the ICS in use by their jurisdiction for use at all emergency incidents. As a Fire Officer II, knowledge of the ICS evolves as the responsibility for larger and more complex incidents increases. This section includes a review of some of the basic information. The ICS has a number of interactive components that provide the basis for clear communication and effective operations:

- Common terminology
- Modular organization
- Integrated communications
- Unified command structure
- Consolidated action plans
- Manageable span of control
- Predesignated incident facilities
- Comprehensive resource management

Figure 9.19 The incident commander (IC) is in overall command of the incident.

Operational Positions

To understand the application of ICS, you should know the major operational position descriptions within the ICS structure. These include Command, Command Staff, and the General Staff Sections. You may be assigned to any of these positions based on your knowledge, skills, and ability.

Command

The fire officer in overall command of an incident is the IC **(Figure 9.19)**. The IC is ultimately responsible for all incident activities, including the development and implementation of a strategic plan. This process may include making a number of critical decisions and being responsible for the results of those decisions. The IC has the authority to request resources to the incident and to release them from it. If the size and complexity of the incident requires it, the IC may delegate authority to others, who together with the IC form the Command and General Staffs. At single-unit incidents, the IC will perform the functions normally assigned to the Command Staff.

Command Staff

The Command Staff reports directly to the IC by providing management support in functions that are not directly involved with the operational tasks. Command staff positions include the Safety Officer, Liaison Officer (LO), and Public Information Officer (PIO).

- **Safety Officer** — Assesses hazardous conditions and unsafe situations and develops an incident safety plan that ensures the safety of personnel at the incident. The Incident Safety Officer (ISO) has the authority to alter, suspend, or terminate activities that are imminently dangerous to the life of an occupant or responder.

- **Liaison Officer** — Acts as a point of contact for other agencies and jurisdictions and coordinates the activities with them.
- **Public Information Officer** —Provides accurate and complete information on the incident to the media. During larger incidents, the PIO provides information to other governmental agencies that need information concerning the incident.

General Staff

As an incident becomes larger and more complex, additional tasks must be performed. This requires the addition of more personnel who will form the General Staff. The General Staff represent the major functional sections, including Planning, Operations, Logistics, and Finance/Administration **(Figure 9.20)**.

NIMS also provides for an Intelligence Officer, Section, or Unit which may be part of the Command Staff, the Planning Section, or a completely separate section.

Planning — Planning is responsible for the collection, evaluation, dissemination, and use of information concerning the development of the incident. Planning is also responsible for tracking the status of all resources assigned to the incident as well as the situation status. Command uses the information compiled by Planning to develop strategic goals and contingency plans.

Operations — Operations may be divided into divisions, or groups. Divisions have area responsibilities, while groups have functional responsibilities. NIMS-ICS recommends limiting subordinates to five. If an Operations Chief has more divisions/groups than he/she considers reasonable, branches may be organized with divisions/groups assigned to them.

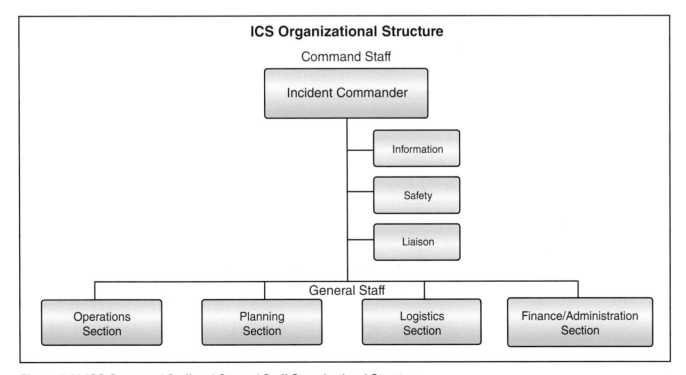

Figure 9.20 ICS Command Staff and General Staff Organizational Structure.

Logistics — Logistics is responsible for providing the facilities, services, and materials necessary to support the incident. There are two branches within Logistics: Support Branch and Service Branch.

Finance/Administration — Finance/Administration has the responsibility for tracking and documenting all costs and financial aspects of the incident. Generally, Finance/Administration will be activated only on large-scale, long-term incidents. Day-to-day mutual aid and automatic aid responses are usually considered to be reciprocal and do not require interagency reimbursement. Administration is also responsible for the legal aspects of the incident such as monitoring contracts with vendors, agreements with other agencies, and compliance with state/provincial, and national laws.

Risk Management Plan

Company officers are responsible for the safety of all personnel assigned to the unit or under their command at an emergency incident. This responsibility exists throughout the work shift and includes emergency and nonemergency situations. The officer uses a risk management model to determine the appropriate responses to the life safety risks faced by the unit and develop a risk management model. The components of the risk management plan are:

- **Risk Identification** — To identify the risks normally encountered at a structure fire, the fire officer compiles a list of all the operations and duties that occur at such an event. Ideally, the officer should take into consideration the worst possible conditions or potential events, including the need for rescue, rapid fire development, or structural collapse.

- **Risk Evaluation** — Once the risks are identified, they can be evaluated from both frequency and severity standpoints. *Frequency,* referred to by OSHA as *incidence rate,* addresses the likelihood of occurrence. *Severity* addresses the degree of seriousness of the incident and can be measured in a variety of ways such as lost time away from work, cost of damage, cost of and time for repair or replacement of equipment, disruption of service, or legal costs. Incidents of high frequency and high severity are given the highest priority in the risk analysis, while those of low frequency and low severity receive the lowest priority. The method for calculating the risk may vary from one organization to another.

- **Risk Prioritization** — Taken in combination, the results of the frequency and severity assessments help to establish priorities for determining action. Any risk that has both a high probability of occurrence and serious consequences deserves immediate action and is considered a high-priority item. Nonserious incidents with a low likelihood of occurrence are lower priorities and can be placed near the bottom of your incident action plan (IAP).

- **Risk-Control Techniques** — When the risks have been prioritized, it is time to apply risk-control measures. Several approaches can be taken in risk control, including the following:
  - **Risk avoidance** — The best risk-control choice is risk avoidance. Simply put, avoid the activity that creates the risk. At a structure fire, this may be the choice.
  - **Risk transfer** — This approach can be accomplished in one of two primary ways: physically transferring the risk to someone else and through the

purchase of insurance. Transfer of risk may be difficult if not impossible at a structure fire.

- **Implementation** — Implementation of the risk management plan requires coordination with crew members and other units as well as communication of the plan to them.

- **Monitoring** — The effectiveness of the plan becomes evident through monitoring. Any problems that occur in the process have to be revised or modified. The intent of the risk management plan is to develop a strategy for reducing the inherent risks associated with fire and emergency services operations. For instance, if the plan begins with an offensive interior attack, changes in the location and intensity of the fire may require a rapid redeployment of resources and a transition to a defensive strategy.

Risk vs. Benefit

As fire and emergency service responders, we are expected to take calculated risks to provide for life safety, incident stabilization, and property conservation. Calculated risks mean that we do not blindly go into a situation, but rather we gather information through size-up and determine what level of risk is acceptable for the given situation. Because the fire service has continued for the past two decades to have the same number and types of line-of- duty deaths (LODDs), in 2001 the International Association of Fire Chiefs (IAFC) developed a model policy called *The 10 Rules of Engagement for Structural Fire Fighting*. The policy was developed to help ensure that we can all return home safely.

10 Rules of Engagement for Structural Fire Fighting

Acceptability of Risk

1. No building or property is worth the life of a firefighter.

2. All interior fire fighting involves an inherent risk.

3. Some risk is acceptable, in a measured and controlled manner.

4. No level of risk is acceptable where there is no potential to save lives or savable property.

5. Firefighters shall not be committed to interior offensive fire fighting operations in abandoned or derelict buildings.

Risk Assessment

6. All feasible measures shall be taken to limit or avoid risks through risk assessment by a qualified officer.

7. It is the responsibility of the incident commander to evaluate the level of risk in every situation.

8. Risk assessment is a continuous process for the entire duration of each incident.

9. If conditions change, and risk increases, change strategy and tactics.

10. No building or property is worth the life of a firefighter.

In the Ten Rules of Engagement, the prohibition for committing an interior attack in an abandoned or derelict building can generate a great deal of discussion. For instance, because a structure is unoccupied, vacant, abandoned, or derelict does not mean that there is no life safety hazard present. These types of structures may contain members of the homeless population, demolition or construction workers, or security personnel. One approach is to consider all structures as occupied until proven otherwise. Of course, once you have committed personnel to an interior attack, the structure is very much occupied.

Based on the initial assessment of the incident scene, you make decisions based on acceptable risk. This concept is clearly stated in a decision-making model developed by the Phoenix (AZ) Fire Department (PFD). The model is a departmental SOP that is used to help PFD officers make reliable emergency response decisions. The essence of the model is as follows:

- Each emergency response is begun with the assumption that *responders can protect lives and property.*

- Responders will *risk their lives a lot, if necessary, to save savable lives.*

- Responders will *risk their lives a little, and in a calculated manner, to save savable property.*

- Responders will *NOT risk their lives at all to save lives and property that have already been lost.*

Phoenix Model Simplified
- Risk a lot to save a lot.
- Risk a little to save a little.
- Risk nothing to save nothing.

Firefighter safety is your primary concern. Your decisions must be based on this model, which should be applied to your initial size-up.

When dealing with an unoccupied, vacant, abandoned, or derelict structure, the most efficient and safest way to control the fire may be to make a quick attack on the seat of the fire. Doing this can decrease the property damage, reduce the exposure hazard, and reduce the risk to your fire fighters. However, you must assess the condition of the structure and the stage of the fire before committing those resources to the interior.

Implementing the System

The first fire officer arriving at the emergency incident establishes the ICS and remains in command until relieved. This officer begins by establishing the incident priorities which include:

- **Life Safety –**
  - What hazards are present and what risks do they pose to my personnel?
  - What hazards are occupants, witnesses, and other people exposed to?
  - Are any occupants trapped or injured?

- **Incident Stabilization –**
  - What has occurred?
  - What is the current status of the incident?
  - What is likely to occur?
  - Can the incident be confined or controlled with the resources I have at my disposal?
  - What specialized resources or technical assistance do I need to mitigate the incident?
- **Property Conservation –**
  - What actions must be taken to protect the property?
  - Is the property beyond saving or is the risk to my personnel too great to attempt to save it?

Incident Action Plan

If no life-threatening situation demands immediate action, the IC should begin to formulate an IAP. The plan should reflect the following priorities:

1. Ensuring personnel safety and survival
2. Rescuing or evacuating endangered occupants
3. Eliminating the hazard
4. Conducting loss control
5. Cleaning up and protecting the environment

The IC gathers enough resources, including personnel and materials, to handle the incident and organizes information to ensure that orders can be carried out promptly, safely, and efficiently. Having sufficient resources on scene will help to ensure the safety of all involved personnel.

All incident personnel must function according to the IAP. Company officers, who may be designated as crew leader, division or group supervisor, or branch director, should follow SOPs, and incident personnel should direct every action toward achieving the goals and objectives specified in the plan. When all members (from the IC to the lowest-ranking member of the team) understand their positions, roles, and functions in the ICS, the system can serve to safely, effectively, and efficiently use resources to accomplish the plan.

Incident Safety Plan

ISO or the IC develops an incident safety plan that outlines the hazard control strategies required by the IAP and the specific incident **(Figure 9.21)**. The plan is used during the incident and also as part of the postincident analysis.

Figure 9.21 An incident safety officer (ISO) designated by the IC has the authority to stop unsafe acts.

Chain of Command

Whenever the ICS is implemented, there shall be only **ONE** IC, except in a multijurisdictional incident when a unified command structure is appropriate. A multijurisdictional incident involves agencies beyond the jurisdiction of one organization or agency. Even when a unified command structure is used, the chain of command must be clearly defined. One person should issue all orders through the chain of command to avoid confusion caused by conflicting orders.

Based on your local ICS, you may retain command until the termination of the incident, or you may transfer command to a higher ranking officer or an officer with greater experience.

The initial IC provides the relieving officer with all of the information that has been gathered on the situation and the actions that have been taken to this point. This can be accomplished by giving a situation status report, which is an updated version of the incident evaluation performed on arrival. The officer assuming command should acknowledge receipt of the information by repeating it back to the current IC. If the reiteration is accurate, the recipient is ready to accept control of and responsibility for the management of the incident. The former IC can then be reassigned to an operating unit or retained at the command post (CP) as an aide or as a member of the Command or General Staff.

Situation Status Report

The situation status report should include the following information:

- Description of what happened
- Whether anyone was/is injured or trapped
- What has been done so far
- Whether the problem has stabilized or is getting worse
- What resources are on scene or en route
- Whether it appears that current resources are adequate for the situation or that more resources need to be called
- What functions of the ICS have been implemented and what additional functions should be considered

The new IC should assess the situation and determine if the operation is reaching the objectives that have been established in the IAP. If the initial CP is located within the hot zone, the new IC should consider relocating the CP to a location that provides a better view of the overall incident. That location should be in the warm zone and in an area that does not interfere with suppression or mitigation activities. It should also have space for the support functions of the command and general staffs. Finally, it should be safe from contaminated atmospheres and runoff.

Tracking Resources

One of the most important functions of an ICS is to provide a means of tracking all personnel and equipment assigned to the incident. In career departments, most units responding to an incident arrive fully staffed and ready to be assigned an operational objective; other personnel may have to be formed into units as they arrive at the scene, which may be the case in volunteer or combination departments. In order to manage these resources, the IAP must contain a tracking and accountability system with the following elements:

- Procedure for checking in at the scene

- Way of identifying the location of each unit and all personnel on scene

- Procedure for releasing units no longer needed

The IC must be able to locate, contact, deploy, and reassign the resources that are assigned to the emergency incident. At smaller incidents, the use of a tactical worksheet is recommended, while at larger incidents this is a function of the Planning Section or Resource Unit if implemented. As companies and crews arrive at the scene, the IC or Operations Chief assigns them to the part of the incident where they are most needed. The companies/crews may be held in a staging area until needed or until they are released from the incident **(Figure 9.22)**. If staging has not been implemented, company officers must check in with the IC and wait for an assignment for the system to work.

Figure 9.22 Apparatus and personnel can be retained at the staging area until needed.

Communications between resources will be through the jurisdiction's emergency radio communication system or through direct face-to-face communication. As resources are assigned to the incident, the central communications center announces the command frequency in use or automatically places all radios on that frequency. Resources that have been assigned to the incident must contact the IC to ensure that they have complete communication with the CP. Face-to-face communication occurs as personnel without radios arrive at the incident and are formed into crews.

The IC can use a number of visual aids to help manage the arriving resources. The visual aid can be as simple as a preprinted form to sketch the incident scene and the location of companies/crews as they arrive, or it can be an elaborate tracking board with magnetic symbols identifying each of the units. Regardless of the tracking device, it should be simple to read and contain as much information as necessary about the activities of all the resources on scene. The visual aid may contain the following information:

- Assigned radio frequencies

- Assigned companies/crews

- Activated ICS functions

- Site plan

- Staging areas

- Logistics location

- Control zones

Personnel Accountability System

Each organization must adopt an existing accountability system or develop its own system of accountability that identifies and tracks all personnel working at an incident. The organization should standardize the system so that it is used at every incident. All personnel must be familiar with the system and participate in the system when operating at an emergency incident. The system must also account for those individuals who respond to the scene in vehicles other than emergency response apparatus, including staff vehicles and personally owned vehicles (POVs).

Accountability is vital in the event of a change in the status of the emergency incident. At a structure fire, that change might be the extension of the fire through a concealed space, the rapid increase in the volume of fire due to the ignition of a flammable or combustible liquid, or a rapid fire development situation. The IC must know who is at the incident and where each person is located. For example, SCBAs can malfunction or run out of air; firefighters can get lost in mazes of rooms and corridors. Without having an accountability system, it is impossible to determine who and how many may be trapped inside or injured. Too many firefighters have died because they were not discovered missing until it was too late.

Wildland and urban interface fires have also claimed many lives when emergency responders were trapped by a change of direction of the fire. To effectively manage the various resources at the incident, the IC must know the status and location of all resources. Changes in the direction, magnitude, and behavior of the fire can be transmitted to the resources giving them time to relocate or take cover. At the same time, individual emergency responders have suffered heart attacks that were fatal because they could not be located in time to provide medical treatment.

Company officers are responsible for keeping track of the members of their units. When operating in the hot zone, the unit should be within visible range of each other. When the atmosphere is obscured, they should be in close proximity of each other.

Company officers must be aware that the ISO does not perform the same function. While both report to the IC, they serve two different purposes. The company officer remains in a single location that funnels personnel in and out of the operational area. The ISO must be able to move freely around the incident scene and monitor operations to ensure the safety of all personnel.

The IC is responsible for managing the personnel accountability system employed by the organization. The system may be assigned to the accountability officer, if one is available as part of the Planning Section. The system should indicate the individuals assigned to each apparatus crew or staff position, the names of people responding individually, such as staff personnel and volunteers, the time of arrival, the assigned duty or unit, and the time of release from the scene. Various systems, such as the tag system, SCBA tag system, and bar code readers, are available for tracking individuals at the emergency incident.

Tag system. A simple tag system can aid in accounting for personnel within the incident scene perimeter. Personnel can be equipped with a personal identification tag **(Figure 9.23)**. Upon entering the outer perimeter, emergency responders leave their tags at a given location or with a designated person (CP,

CAUTION

Personnel who arrive individually at an incident must always check in at the command post or staging if it has been implemented. Personnel must not freelance or assign themselves to a task. Freelancing is unsafe and can and has resulted in fatalities or injuries.

Figure 9.23 The accountability system is used to determine who is working at the incident.

apparatus compartment, company officer, accountability officer, or ISO). Tags can be attached to a control board or personnel identification (ID) chart for quick reference. Upon leaving the incident perimeter, the emergency responders collect their tags. This system enables officers to know exactly who is operating at the incident scene.

SCBA tag system. An SCBA tag system provides closer accountability for personnel inside a structure. All personnel entering a hazardous atmosphere are required to wear full personal protective clothing with SCBA. Each SCBA is provided with a tag containing the name of the user and the air pressure in the SCBA air tank. Upon entering a building, personnel give their tags to a designated supervisor. The company officer records time of entry and expected time of exit. This officer also does a brief check to ensure that all protective equipment is properly worn, used, and activated, including the PASS device. This provides complete accountability for those inside the structure and ensures that they are wearing the proper gear. Responders leaving the danger area retrieve their tags so that the control officer knows who is safely outside and who is still inside the structure or danger area.

Bar code readers. Some organizations have adopted bar code readers for tracking personnel at emergency incident scenes. Similar to the inventory control devices used by retail outlets, the system involves the assignment of a unique bar code number to each member of the organization. When that person arrives at the incident, the bar code is scanned into a computer that registers the presence of the individual at the scene. When the person leaves the scene, the bar code is scanned a second time and the person's name is removed from the site. Depending on local policies, the bar code may be attached to the helmet, protective clothing, or other equipment that is permanently assigned to each member of the organization. It may also be on a personal identification card.

Air Management

Based on the organization's air management policy, it is the responsibility of both the IC and the personnel entering an IDLH atsmophere to implement the policy. At structure fires or hazardous vapor leaks, the IC must evaluate the size and complexity of the structure's interior to determine the amount of time required to exit the structure from the point of no return. The point of no return is the distance between the point work is to be performed and the nearest safe atmosphere. This calculation will determine the maximum amount of time that can be spent working in the IDLH atmosphere.

The ISO or HSO should monitor personnel entering the IDLH area, recording their starting air supply as a means of determining when they should be replaced and begin their withdrawal. A general rule of thumb for calculating air supply is the capacity of the air cylinder in cubic feet divided by 2 equals the allowable working time. This calculation allows for a safety margin for exiting the structure.

Individual firefighters who are assigned to work in the IDLH area are responsible for:

- Performing an inspection of their SCBA when reporting to work
- Checking the air pressure on their cylinder prior to entering the IDLH area
- Reporting this pressure to their team leader or the ISO
- Monitoring air consumption and pressure continually while in the IDLH area:
  — At regular intervals
  — At 10-minute intervals when notified by the communications office
  — At a change-of-work area
  — When passing major landmarks within the structure
  — At the completion of one assignment and before another assignment
  — As the situation dictates

Because there are a wide variety of SCBA types currently in use in the fire service, the local jurisdiction must establish an air management protocol in its operational SOP/SOG. Traditionally, the activation of the low-pressure alarm on the SCBA has been used as an indicator that it is time to exit the IDLH area. In small structures, this may be sufficient time to locate an exit or a source of fresh air. In high-rise or large-area structures, the low-pressure setting may not provide enough breathing air to leave the IDLH area. Other factors that can limit the breathing air supply are the individual firefighter's breathing rate (air consumption) and the intensity of the work being performed.

One suggested air management protocol requires that interior teams must give an automatic air status report when the first team member's air supply reaches 50 percent. At that time all members of the team must begin to leave the IDLH area BEFORE any low-air alarm is activated. Some SOPs/SOGs mandate that the activation of the low-air alarm requires an immediate MAYDAY response. A sample protocol when the low-air alarm activates is to transmit a radio signal to Command stating who, where, and what your status is. You must immediately notify Command when you exit the structure.

The IC may allow team members to continue working until the ¼ full cylinder indicator light activates under the following conditions:

- The fire is declared under control.

- A structural evaluation is made by the IC or ISO to determine that there is no structural collapse or entanglement hazard, the crew's visibility is not impaired, and a personnel accountability report (PAR) indicates that the crew is intact and able to continue.

Emergency Broadcast Protocol

Fire departments throughout North America have a variety of protocols for transmitting a radio signal for an emergency situation during an incident. Although the terminology may differ, the terms can be divided into two categories:

1. Mayday protocol – The broadcast of a MAYDAY indicates an emergency situation in which a firefighter is in trouble:

   - Firefighter in low-air-alarm situation and is disoriented or unsure of location

   - SCBA failure

   - Firefighter trapped, entangled, or unable to free himself or herself within approximately one minute

   - Finding a firefighter in distress

   - At the discretion of Command

2. Emergency Traffic protocol – The broadcast of emergency traffic is used when there is an increased hazard to all personnel such as a potential structural collapse, indication of an extreme fire condition (flashover), or a shift to defensive strategy and the need to evacuate the structure or IDLH area.

   In both cases, all nonessential radio traffic ceases or is shifted to another radio channel. In the case of a Mayday, RIT/RIC resources are committed to the incident while other personnel continue to perform their duties.

Releasing Resources

As control is gained over the emergency situation, personnel and units can be released from the incident. If the personnel or units have been committed to mitigating the incident, they should be sent to the rehabilitation unit prior to being released. This helps to ensure that all personnel have an opportunity to rest, receive nourishment, and be examined by a medical professional. This can help prevent heart attacks by allowing a monitored cool down time for each person who has been actively involved in the incident.

Before leaving the scene, units that have been engaged in the incident must be inspected by the company officer to ensure that they are ready to return to service. Missing equipment must be located, and damaged equipment must be replaced or repaired. If critical equipment is not available, the unit may have to remain out of service until the equipment can be replaced.

Decontamination may be required before placing the personnel or units back into service. This can take place at the incident site or when the unit returns to the station. Small tools, medical equipment, and PPE must be decontaminated in accordance with local policy and NFPA®1581, *Standard on Fire Department Infection Control Program*. Contaminated clothing and equipment must be cleaned in a designated decontamination area at the station and may not be cleaned in kitchen or bathroom sinks or showers.

Figure 9.24 The Incident Command System (ICS) is used to control and allocate multiple resources at large incidents. *Courtesy of Chief Chris Mickal, New Orleans (LA) Fire Department.*

Assigning Multiunit Resources

The ICS is designed to permit the control and allocation of any number of units or agencies at an emergency incident. The IC may be responsible for an entire initial response that may consist of multiple engines, ladders/trucks, rescue, and ambulance units **(Figure 9.24)**. If the initial assignment is insufficient to control the incident, then the IC may be responsible for transferring command to a superior officer and assuming one of the ICS Command or General Staff positions or commanding a single unit. The company officer must be familiar with the general configuration of a multiunit incident and the terms used by the ICS to describe duties and functions.

Command Organization

The Command organization must develop at a pace that stays ahead of the tactical deployment of personnel and resources. The IC must be able to direct, control, and track the locations and functions of all operating units in order to efficiently manage the incident. Building a Command organization is the best support mechanism the IC can use to achieve the harmonious balance between managing personnel and incident needs. The Command organization can be characterized by three levels: strategic, tactical, and task.

Strategic Level — The strategic level involves the overall command of the incident. The IC is responsible for the strategic level of the Command structure. The IAP should cover all strategic responsibilities, tactical objectives, and support activities needed during the entire operational period. The IAP may include the department's SOP used for initial activities at specific incidents, such as High-Rise SOP, HazMat SOP, etc. The IAP is the basis for developing a Command organization, assigning resources, and establishing tactical objectives.

Strategic level responsibilities include the following:

- Determination of the appropriate strategy
- Establishment of overall incident objectives
- Setting of priorities
- Development of an Incident Action Plan (IAP)
- Obtainment and assignment of resources
- Planning
- Prediction of outcomes
- Assignment of specific objectives to groups or divisions

Tactical Level — Tactical-level supervisors direct operational activities toward specific objectives. Tactical-level supervisors supervise grouped resources and are responsible for specific geographic areas or functions. A tactical-level assignment comes with the authority to make decisions and assignments within the boundaries of the overall plan and safety conditions. The accumulated achievements of tactical objectives should accomplish the strategy as outlined in the IAP.

Task Level — The task level refers to those activities normally accomplished by individual companies, crews, or specific personnel. The task level is where the work is actually done. Task-level activities are routinely supervised by company officers. The accumulated achievements of task-level activities should accomplish tactical objectives.

Basic Organization

The most basic organization combines all three levels of the Command structure. For example, the company officer on a single-engine response to a dumpster fire determines the strategy and tactics and supervises the crew doing the task. The basic structure for an initial-response incident involving a small number of units requires only two levels of the Command structure. In this situation, the IC directly handles strategic and tactical levels. Emergency response units report directly to Command and operate at the task level.

The NIMS/ICS terms Divisions and Groups are tactical-level management components that assemble units and/or resources for a common purpose. *Divisions* represent geographic operations and *Groups* represent functional operations. The following examples illustrate the use of these terms.

NOTE: For ease of reading in this document, the abbreviation *DG* will be used when referencing a Division/Group hereafter. The term *DG Supervisor* will refer to the person in charge of a Division or Group.

Divisions are the organizational level having responsibility for operations within a defined geographic area. The Division level is organizationally between Single Resources, Task Forces, or Strike Teams and the Branch. For situations where the incident has an odd geographical layout (no obvious North, South, East, or West), the front of the building is designated *Division A*, and the remaining sides are given a designation of B, C, or D in a clockwise manner **(Figure 9.25)**. For clarity of purpose during radio communications, the phonetic designations of *Alpha, Bravo, Charlie*, and *Delta* are suggested. For example, *Command from Division Delta*.

In multistory occupancies, Divisions will usually be indicated by floor number (Division 6 indicates 6th floor). When operating in levels below grade, Divisions will usually be indicated by a descriptive term such as Basement Division 2 or Parking Garage Division 4.

Groups are an organizational level responsible for a specific functional assignment. The Group level is also organizationally between Single Resources, Task Forces, or Strike Teams and the Branch. Examples are Salvage Group, Search Group, Rescue Group, Haz-Mat Group, and Medical Group.

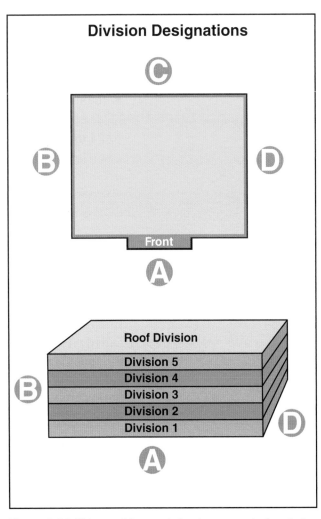

Figure 9.25 Sides and floors of structures are designated by Division as a means to assign companies and communicate with them.

Establishing Divisions/Groups (DG)

A major incident will initially have more tasks that need to be completed than the available resources can accomplish. There is a tendency to start performing these tasks immediately upon arrival, thereby postponing the establishment of ICS. *This is a major error.* The lack of direction will result in confusion and lack of coordination. This increases the risks to emergency personnel and decreases the likelihood of a successful operation.

The IC should begin to assign DGs based on the following factors:

- Situations that involve a number of units or functions beyond the IC's span of control. As additional chief officers become available, they may relieve the company officer of responsibility for the area or function.

- When units are involved in complex operations (large interior or geographic areas, hazardous materials (haz mat) operations, high-rise fires, technical rescues, etc.).

- When units are operating from tactical positions that Command has little or no direct control over (e.g., they are out of the IC's sight).

- When the situation presents special hazards and close control is required over operating units (e.g., unstable structural conditions, heavy fire load, marginal offensive situations, etc.).

When establishing DGs, the IC will assign/advise each unit:

- Tactical objectives
- Radio designation (Division/Group)
- Identity of resources assigned to that DG

Span of Control

Complex emergency situations often exceed the capability of one officer to effectively manage the entire operation. The span of control must be reduced by creating organizational tactical-level components to direct operations in specific geographic areas or to manage incident-related functions. This is accomplished by establishing DGs. Generally, the best span of control consists of three to seven subordinates with five being the preferred number **(Figure 9.26)**.

Implementing an Incident Action Plan

In most fire and emergency services organizations, operational plans (also called *prefire plans*, *preincident plans*, or *incident action plans [IAPs]*) identify the specific resources needed to successfully deal with a variety of hypothetical incidents at a particular location or type of occupancy. The incident scenarios analyzed are those that are considered to be the most likely to occur at the location or occupancy in question. For example, if an occupancy uses or stores large quantities of toxic or highly flammable materials, scenarios involving these materials are created and analyzed. Likewise, if an occupancy houses a large number of elderly or infirm residents, scenarios that require the evacuation of these residents or moving them to areas of safe refuge are created and analyzed. This process is applied to as many target hazards as time and staffing allow.

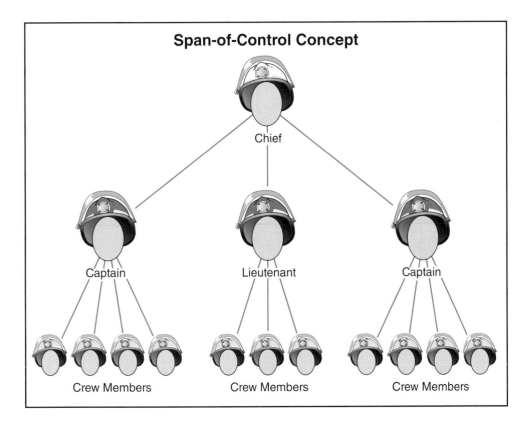

Figure 9.26 Span of Control Concept.

Span-of-Control Concept

Chief

Captain

Lieutenant

Captain

Crew Members

Crew Members

Crew Members

Operational Plans

Operational plans often include possible resource deployments. In the most likely scenarios at a given location, possible options for deploying the initial alarm resources are studied, as well as options for the deployment of those resources that would respond only if called. Scenarios based on increasingly larger and more complex hypothetical incidents help planners to identify the resources that may be needed and to identify how they can by deployed to the best advantage.

Operational plans often include provisions for a number of possible contingencies. Some possible variations include:

• Operational plans might include a variety of responses based on unusually severe weather conditions that could greatly increase the potential for wildland fires starting and increase the likelihood of these fires spreading at a greater-than-normal rate. An operational plan might specify an increase in the initial alarm resources at certain points on the daily burning index.

• Operational plans might specify different initial alarm resource levels for a given occupancy if the number of people normally in the building is significantly different at different times of the day or night. For example, the operational plan for an elementary school might specify that the initial alarm assignment be doubled if an alarm is received from the school during the hours when it is normally in operation. However, it might also specify that a lesser response be dispatched at other hours.

• The operational plans for occupancies that store large quantities of ignitable liquids might specify that crash/rescue vehicles from the local airport be a part of the initial alarm assignment so that their foam-making capabilities are immediately available if needed.

- The operational plans for large industrial complexes might specify that the first-arriving unit is to respond directly to the fire alarm panel or the plant security station but that other responding units are to stage at the facility gates and await directions from the first-in unit.

In essence, the data gathered at any particular target hazard is used to project the strategic and tactical possibilities and probabilities at that location. Based on these scenarios, the resources needed are compared to the resources available in the organization. If resources indicated are more than or different from what the organization has, planners can then recommend the purchase of the needed resources or the development of mutual aid or automatic aid agreements with nearby organizations that do have the needed resources.

Operational Plan Components

The operational plan consists of information gathered from a number of sources. The initial component is the preincident survey combined with any fire and life safety inspection data. Combined, these two sources provide a picture of the actual conditions that exist at the location or occupancy. This information must be current and accurate to be of value in the planning process.

The next component is the analysis of the fire and emergency services organization's assets and abilities. This is an inventory of such items as:

- Staffing levels
- Apparatus types and quantity
- Station locations
- Response times
- Training and certification of personnel
- Water supply
- Types and quantity of specialized extinguishing agents
- Specialized response units
- Automatic and mutual aid agreements
- Dispatch and communications system

The final component is the jurisdiction's ICS. This is the organizational framework for all single and multiunit emergency incidents. It establishes functions and assigns duties based on the complexity of the incident. You must be familiar with the system used locally and how it fits into the state/provincial and national system, NIMS/ICS.

Operational Period Safety Concerns

The majority of emergency responses can be mitigated in a short period of time. Some, however, may require more resources over longer periods of time. For this purpose, NIMS-ICS establishes operational periods as a means of setting goals and objectives that must be accomplished within that time. The NIMS-ICS does not set a specific time for an operational period. In general, operational periods may be as short as two hours but no longer than 24 hours. When an incident extends beyond one operational period, a planning meeting is held to define the goals and objectives for that period.

Safety considerations that are created by incidents that last longer than one operational period include:

- Personnel rotation
- Personnel rehabilitation
- Structural stability

Evacuation Commands

Firefighters know that an emergency incident can change rapidly and without warning. A change of wind direction and velocity can force a wildland fire to change direction and trap crews. The introduction of fresh air into a closed compartment with a smoldering fire can create a flashover or backdraft as the air and heated gases mix.

Because hazardous situations are unpredictable, fire departments must have SOP/SOGs that include protocols for evacuation of personnel. 29 CFR 1910.120, *Hazardous Waste Operations and Emergency Response*, includes the requirement for evacuation plans that can apply to fire department operations. Evacuation protocols, depending on the type of incident, must be written/and, communicated to all personnel. Then training is developed and presented. Some evacuation components include:

- Specific radio commands, referred to as *emergency traffic*
- Methods for withdrawal from the hot zone
- Accountability
- Air management
- Self-rescue techniques
- Mandatory buddy system
- RIC/RIT activities during evacuation

NOTE: **When withdrawing from a structure fire, the charged hoseline typically provides a path for withdrawal and protection from the fire.**

NIMS-ICS Forms

NIMS–ICS provides approximately 26 forms to assist in the establishment of an incident command. Examples of completed forms are included in **Appendix E**. The primary forms that are contained in most IAPs are the following:

- *Incident Objectives (ICS 202)* — Clearly stated and measurable objectives to be achieved in the specific time interval
- *Organization Assignment List (ICS 203)* — Description of the ICS table of organization, including the units and agencies that are involved
- *Assignment List (ICS 204)* — Specific unit tactical assignments divided by Branch, Division, and Group
- *Incident Radio Communications Plan (ICS 205)* — Lists the basic radio channel assignments for use during the incident
- *Medical Plan (ICS 206)* — Provides information on the location and staffing of the incident medical aid station, the types of ambulance resources available, the location of on-site ambulances, and the contact information for hospitals that are available to the IC

Behavioral Considerations

From the sound of the alarm, through the response, emergency responders are driven by a rush of adrenaline. Our heart rate increases and our minds start focusing on the potential situations that we might encounter based on the limited information that we have. This condition is compounded by our desire to save a person or structure from eminent danger.

Unfortunately, this rush can cause us to act unsafely either intentionally or unintentionally. At the same time, we may be motivated by a personality trait that drives us to be seen as being heroic or self-sacrificing, placing ourselves in harm's way to protect or save a victim. In any case, we may act in an unsafe manner that can result in our becoming a victim who is unable to help the real victim.

Actions that can result from these conditions include an attitude referred to as the Moth to the Flame or freelancing. Additionally, we, as unit or incident commanders, may improperly assign tasks to unqualified personnel, thereby placing them at risk.

Moth-to-the-Flame Attitude

The attitude referred to as Moth to the Flame is an old allusion based on a fact of nature. In nature, the moth is attracted to the a bright light such as the flame of a candle. The moth is so focused on the light that it flies too close and burns its wings. Humans have a similar tendency when they are attracted to something or someone that is dangerous to them.

For firefighters, it is the specific hazard they are responding to and the desire to control or overcome it. For example, in a structure fire, it is the desire to advance a hoseline and extinguish the fire; at a haz mat spill, it is the desire to stop the leak and control the spill; at a medical emergency, it is the desire to reach the patient and administer treatment. When this desire to act is strong enough, it can override your situational awareness, causing you to ignore indicators of personal danger to you. Stopping long enough to properly size-up the incident will give you a sense of what you need to do to protect yourself and your crew.

Freelancing

Freelancing — Operating independently of the IC's command and control.

Freelancing, defined as operating independently of the IC's command and control, is an consequence of the Moth-to-the-Flame attitude. It can affect an individual or a unit and can result in injury or worse and does not help to contain or control the incident.

To prevent freelancing, the department must establish an accountability process as part of the ICS. All units and individuals must report to the IC, ISO, or accountability officer when arriving at the incident. The IC will then assign them to function as a unit where they are most needed or hold them in reserve. A major component of accountability is allowing the IC to know where all the resources are and what they are doing to reach the situation's goals and objectives.

In most cases, freelancing occurs at the individual level. It is imperative that the integrity of assigned companies or crews be maintained at the incident, which are composed of no less than two people. Assignments, tasks as simple as going to get a tool should require two people, not simply one member of the crew, to accomplish.

Lightweight Construction

Lightweight construction is a type of construction consisting of vertical and horizontal structural members formed by a system of wood or light gauge steel components that are lighter in weight than conventional solid wood components. These lightweight components include trusses, glue-laminated beams, I-joists, structural composite lumber, structural insulated panels, and wood structural panels. They are replacing dimensional lumber in many building applications and are used to span wide areas without the need for vertical supports. Although many types of trusses exist, three typical truss construction methods are most commonly used:

Figure 9.27 Lightweight construction includes laminated wood beams.

- Heavy timber roof and floor truss systems

- Lightweight wooden roof and floor truss systems **(Figure 9.27)**

- Steel roof and floor truss systems

It is estimated that more than 60 percent of roof systems are constructed using a truss system. Lightweight and composite wood truss systems can contribute to increased structural fuel loads, create concealed spaces, and contribute to early collapse. When exposed to high temperatures in a compartment fire, lightweight steel trusses can also fail rapidly. Truss systems are designed to meet the minimum engineering requirements of the building code and are tested under controlled conditions in a laboratory setting. Because actual fires are unpredictable and uncontrolled, the failure time of truss systems cannot be predicted. In addition, lightweight and composite wood truss systems can deteriorate from prolonged exposure to moisture in the environment, a condition that may not be readily apparent during visual inspections.

Firefighters must learn about trusses and their performance under fire conditions. According to NIOSH, fire-fighting tactics must be altered to meet the fire characteristics of lightweight construction including:

- Know how to identify roof and floor truss construction.

- Report immediately the presence of truss construction and fire involvement to the incident commander (IC).

- Use a thermal imaging camera as part of the size-up process to help locate fires in concealed spaces.

- Use extreme caution and follow standard operating procedures (SOPs) when operating on or under truss systems.

- Open ceilings and other concealed spaces immediately whenever a fire is suspected of being in a truss system. *Steps:*
  — Use extreme caution because opening concealed spaces can result in backdraft conditions.
  — Always have a charged hoseline available.
  — Position between the nearest exit and the concealed space to be opened.
  — Be aware of the location of other firefighters in the area.

- Understand that fire ratings may not be truly representative of real-time fire conditions and the performance of truss systems may be affected by fire severity.

- Before emergency incidents, fire departments should take the following steps to protect firefighters:
  — Conduct preincident planning and inspections to identify structures that contain truss construction.
  — Ensure that firefighters are trained to identify roof and floor truss systems and use extreme caution when operating on or under truss systems.
  — Develop and implement SOPs to safely combat fires in buildings with truss construction.

- At the emergency incident, use the following procedures to protect firefighters:
  — Ensure that the IC conducts an initial size-up and risk assessment of the incident scene before beginning interior fire-fighting operations.
  — Evacuate firefighters performing operations under or above trusses as soon as it is determined that the trusses are exposed to fire, and move to a defensive mode.
  — Use defensive overhauling procedures after extinguishing a fire in a building containing truss construction.
  — Use outside master streams to soak smoldering trusses and prevent rekindles.
  — Report any damaged sagging floors or roofs to command

Source: National Institute for Occupational Safety and Health (NIOSH): "Preventing Injuries and Deaths of Fire Fighters Due to Truss System Failures," NIOSH Publication No. 2005-123.

Proper Use of Nonemergency Personnel

At many emergency incidents, additional personnel will be available to assist or work with you and other fire department units. These additional resources may come from a variety of sources including, among others:

- Law enforcement
  — Federal, including Military
  — State/Provincial/Tribal
  — County/Parish
  — Municipal
  — Fire Police
- Emergency medical organizations
  — Public
  — Private
- Volunteer organizations
  — Auxiliary
  — Fire Corps
  — Explorer Scouts
  — Red Cross
  — Salvation Army

How these resources are used must be defined in the fire department's SOP/SOG and in agreements with each organization. In particular, the relationship with law enforcement agencies must be determined before an incident occurs where there is joint jurisdiction such as a MVA.

Simply stated from a safety aspect, these resources must not be permitted to enter the hot zone or the collapse zone at a structure fire. The same is true at a haz mat spill including areas downwind of the incident where the vapors may have collected. Unless they have been certified and fitted for SCBA, they must not be allowed to don respiratory protection to enter an IDLH atmosphere.

Outside the hot zone, nonemergency resources may be used for a variety of functions including, but not limited to:

- Traffic control **(Figure 9.28)**
- Deploying supply hose
- Loading hose and equipment
- Establishing rehabilitation sites
- Crowd control
- Evacuation of non-hot-zone areas
- Air management
- Logistics
- Record keeping

If the use of safety equipment, such as head, eye, hand, and hearing protection, is required by the assigned function, that equipment must be provided and used. Anyone performing a specialized task, such as refilling SCBA bottles, must be thoroughly trained in the function.

Figure 9.28 Some jurisdictions use fire police for traffic and crowd control.

Postincident Analysis

The postincident analysis (PIA), evaluation (PIE), or review (PIR) is essential to the successful and safe operations of the fire and emergency services. Properly developed and written, the PIA determines the strengths and weaknesses of the organization's response to the emergency. It provides a training tool as well as the basis for future planning for emergency responses. It also motivates change in policies and procedures that may be outdated or ineffectual in meeting the current needs of the response area. This section looks at the PIA, defines it, provides a process for developing it, and describes how to use it in a postincident critique.

Postincident Analysis Defined

The PIA is a written document that is compiled by the IC or a designated member of the incident command or general staff such as the ISO. The PIA should be written for all incidents whether small or large. If the results of the PIA indicate that a postincident critique should be held, all participants should participate in the critique.

The objectives of the PIA are:

- To provide an opportunity for participants to objectively review operations in a constructive manner

- To identify effective procedures (strengths) for future emergency operations

- To identify areas needing improvement (weaknesses) and recommend changes to improve effectiveness

The analysis is intended to focus on the activities of the participants, the elements of the emergency, and the decisions made that were intended to control the incident. The PIA is *not* intended to place blame or find fault with the participants, their decisions, or actions. It must not be used to punish any of the participants and must not have the perception of a fault-finding process.

The term postincident analysis is sometimes confused with the term postincident critique. Even NFPA® 1021 infers that the analysis is both the written record and the activity that makes up the critique. For better understanding, this manual will use the term PIA to mean only the written report. The postincident critique will be the meeting of participants that may be required based on the results of the analysis.

Data Collection

The PIA is a critical and objective assessment of the emergency incident based on all available information about the incident. Information sources include:

- **Preincident survey or fire and life safety inspection report** – provides background information on the structure, facility, or occupancy (if available).

- **Size-up** – provides information on the structure, facility, occupancy, or location based on the impressions of the first-arriving company officer and crew.

- **Incident Action Plan** – provides information on the actions that were taken to mitigate the incident. These actions may or may not have actually occurred.

- **Command and General Staff records** – provides a detailed view of the activities of each of the ICS functions during the incident.

- **Outside agency reports** – provides information on the assistance provided by other agencies.

- **Interviews** – provides a narrative of the actions of each of the participants as well as the impressions of witnesses; some of this information may be provided by the fire investigator.

- **Site plan** – provides a visual aid in placing the responding units and personnel as well as casualties and occupants.

- **Incident safety plan** – provides information gathered by the incident safety officer (ISO) on all safety- and health-related issues.

- **Personnel accountability system** – provides an accurate record of the participants and their times of arrival and release from the incident.

- **Weather reports** – provides information on weather conditions that may have contributed to fire spread or hampered unit access, which is particularly important when analyzing wildland and urban interface incidents.

- **Communications records** – provides accurate dispatch times and transcripts of radio communications at the incident.
- **Miscellaneous reports** – provides additional information that may be of value such as apparatus or equipment maintenance records of vehicles or equipment that may have broken down during the incident.

The use of the NIMS-ICS forms can assist in the collection of much of this data. These forms are designed for use in all types of emergency incidents and can be acquired from the FEMA web site or other similar sources.

The information from each of these sources is then compiled into a chronological report of the incident. Photographs and sketches can help to illustrate the situation both before and after the incident. Appendices are used to provide copies of the essential reports.

Analysis

With all the information compiled into one report, you must then look for the strengths and weaknesses in the response. A checklist of questions can be developed to help you focus on each aspect of the incident. Areas that should be considered and addressed include:

Dispatch –

- Was the response time within the organization's minimum/maximum time criteria?
- How can the response time be reduced?
- Were there any extenuating circumstances that resulted in an increased response time (weather, traffic, road construction, etc.)?
- Was all available information communicated to the responding units?

Initial attack –

- Was the initial assignment adequate to mitigate the incident?
- What additional resources were required to mitigate the incident?
- Was the ICS properly implemented?
- Was the ICS adequate for the incident?
- Was all available information communicated from the incident scene to other responding units?
- What improvements could be made to the initial attack?
- What weaknesses should be corrected?
- What procedures need to be updated or changed?
- What strengths can be used as examples for similar situations?

Subsequent resource assignments –

- Were additional resources requested in time to be effective?
- Was there a staging area established and a staging area manager (SAM) assigned?
- Were IC duties delegated according to the ICS?
- Were communications with automatic mutual aid units and other support resources adequate?

- Were there any difficulties in dealing with outside agencies or jurisdictions?
- What can be improved when dealing with outside agencies or jurisdictions?

Private fire protection systems –

- How effective were the fire protection systems?
- How well did the responding units use these systems?
- Could the systems have been used to better advantage?

Health and Safety –

- Did the activity assign a Incident Safety Officer?
- How was personnel and accountability conducted?
- Did the activity necessitate for a rehab area, if so, was the location adequate?
- Did the operation identify operational periods and work/rest cycles
- Were the 2 In/2 Out programs placed in operation?
- Was there an assigned rapid intervention team (RIT) and/or a back up if needed?

Postfire activity –

- Was the property properly secured or turned over to the Fire Investigator after suppression forces vacated the area?
- Did the Fire Investigations Unit respond in a timely manner?

Once these and other questions are answered, you develop a set of recommendations based on the results of the analysis. The recommendations should be included in the PIA and in an executive summary that is provided to the organization's administration. The executive summary is a brief overview of the incident along with the recommendations. If a critique is warranted, you should recommend that one be held.

Postincident Critique

Postincident critiques can be informal or formal. The informal critique can be held by the company officer and members of the unit following any single unit incident **(Figure 9.29)**. Weather permitting, this critique occurs immediately after termination of the incident and prior to leaving the incident scene. By discussing the incident while at the scene, the unit members can provide their impressions of the initial attack, suggest alternative approaches to the attack, and learn methods to improve their actions at similar incidents. In inclement weather, the critique can take place at the station immediately after returning from the incident. In essence, an informal critique can be used as a training exercise for the unit.

The formal critique is held if the PIA indicates that it is necessary or if the incident involves an interagency or interjurisdictional response. If the incident involved only units from one jurisdiction, then all participants should participate in the critique. If the incident was interjurisdictional, representatives of each agency or jurisdiction should be in attendance.

Copies of the analysis should be provided to the participants prior to the critique to allow them time to read the report. The critique should be held in a classroom setting that provides comfortable seating, work space, audiovisual aids, and privacy. A set agenda and sufficient time are also important to a productive critique.

> **CAUTION**
>
> Postincident critiques must always be positive in nature. Personnel, units, agencies, or organizations must not be criticized for their actions, procedures, or policies. Facts should be simply stated and recommendations must be focused on improving emergency responses and interagency relationships.

Following an overview of the incident, each unit commander or agency representative should present a brief report of their participation in the incident in order of arrival. The PIA recommendations can be reviewed, and procedures for implementing them can be developed. Any additional recommendations can also be discussed and addressed.

The greatest failure of both the PIA and critique is the failure to learn from and apply the results and recommendations. Whether the critique is informal or formal, the participants must be assured that any necessary changes will be made and that successes will be celebrated. Celebrations are in order when the incident was mitigated according to ICS and local protocol and without injuries or fatalities to responders.

Figure 9.29 A postincident critique is used to improve operations and safety at future incidents by determining what occurred at the last one.

Summary

Making the emergency incident scene safer is a challenge to every member of the fire service. Each type of incident has multiple hazards, each requiring a specific approach to safety. All incidents can be made safer with administrative policies and procedures that limit the exposures members have to various hazards. The selection, purchase, use, and maintenance of PPE is another means of protecting emergency responders from potential injuries. Supervisors can help to protect their subordinates by applying approved size-up techniques to the incident and assigning resources as needed. Finally, each individual at the emergency incident can take responsibility for their own safety and act in accordance with the safety rules and regulations of the organization. Taken together, these efforts can help to reduce the injuries and fatalities that continue to plague the fire and emergency services.

Review Questions

1. What are some causes of emergency scene casualties?

2. What are the differences between individual and supervisory responsibilities for safety during emergency incidents?

3. What are the differences between the cold, warm, and hot control zones?

4. How is the collapse zone for a building calculated?

5. What is size-up and how does it relate to the preincident survey?

6. What are the IAFC 10 Rules of Engagement for Structural Fire Fighting?

7. What are the duties of the command and general staffs in an ICS?

8. How do the general procedures for tracking resources and assigning multiunit resources relate to implementing an incident action plan?

9. What are the three incident priorities?

10. What aspects of an incident are analyzed during a postincident analysis?

Facility Safety

Chapter Contents

Job Performance Requirements

This chapter provides information that addresses the following job performance requirements of NFPA® 1500, *Standard on Fire Department Occupational Safety and Health Program:*

NFPA® References

9.1.1	9.1.7	9.2.3
9.1.2	9.1.8	9.3
9.1.3	9.1.9	
9.1.4	9.2.1	
9.1.5	9.2.2	
9.1.6		

FESHE Learning Objectives

Occupational Safety and Health for Emergency Services

Describe the responsibilities of individual responders, supervisors, safety officers, and incident commanders, safety program managers, safety committees and fire department managers as they relate to health and safety programs.

Facility Safety

Learning Objectives

After reading this chapter, students will be able to:

1. Describe the design considerations for station facilities which enhance the safety and health of personnel.

2. Describe personal behaviors that ensure the safety of personnel in department facilities.

Chapter 10
Facility Safety

Case History

In 2003, a Seattle firefighter was on temporary assignment to a multistory fire station that he had not worked at previously. During the night, the firefighter got up to use the restroom and fell 15 feet through an unguarded fire pole, striking the concrete floor below. The firefighter was knocked unconscious and sustained severe head injuries, broken ribs, fractured pelvis, and damaged five lower back vertebrae. In addition, he sustained liver, bladder, and lung injuries. Subsequently, a personal injury lawsuit was filed against the City of Seattle alleging that his injuries were the result of the City's negligence for failing to install proper safety precautions.

As a result of the incident and changes in the National Fire Protection Association® (NFPA®) standards, fire poles are no longer allowed in newly constructed Seattle fire stations. Older stations that do have fire poles now have gates with latching mechanisms at the tops of fire poles and padding at the bottom to reduce the risk of injury.

Fire and emergency services facilities in North America range from extremely modern to many that are on the National Historic Register and are over a century old. Regardless of the age of these facilities, it is essential that they provide a safe and healthy work and living environment.

Facility safety can take two forms: The first is safety that is based on the design, maintenance, and condition of the facility. Safety can be engineered into the facility during the initial construction or during a renovation. Design safety reduces the likelihood that an accident will occur or that a condition will contribute to an illness.

The second form of facility safety involves the occupants' behaviors and use of the facility. No amount of engineered safety can protect against unsafe acts on the part of people. This chapter will address both forms of facility safety.

Facility Accidents

Fire and emergency services facilities are like most other types of structures. They provide spaces for work, leisure, sleep, and storage. Some are occupied 24 hours a day year round, while others are used 8 hours a day Monday through Friday. Still others are only occupied during weekly training sessions or apparatus maintenance. When these facilities are occupied, however, injuries, illnesses, and sometimes fatalities can and do occur.

A common type of accident in a fire department facility that gains the most media attention is a fire. Often, these fires are cooking- or kitchen-related and occur when the occupants are called to respond to an alarm. Other fires occur during apparatus or equipment maintenance when flammable/combustible liquids are in use. Finally, some fires occur when the facility is unattended and may be the result of an electrical problem. In any case, these fires can be prevented or their effects mitigated through the installation of fire-suppression and fire-detection systems, and changes in the occupant's behavior.

Injuries

Facility-related injuries include strains and sprains, broken limbs, cuts and abrasions, concussions, and exposure to airborne contaminates. The type of injury can be related to the type of facility and work activity. **Table10.1** depicts this relationship. In most cases, these injuries may be caused by unsafe behaviors rather than a condition in the facility.

Table 10.1
Facility Type and Injury Correlation

Type of Facility	Type of Injury									
	Carpel tunnel	Strains and sprains	Concussions	Broken limbs	Cuts and abasions	Electric Shock	Airborne contamination	Food poisoning	Hearing loss	Haz mat exposure
Administration	X	X			X		X			
Station	X	X	X	X	X	X	X	X	X	X
Maintenance	X	X	X	X	X	X			X	X
Training	X	X	X	X	X	X	X		X	X
Storage		X		X	X	X				X
Communication	X	X			X	X	X			
Aviation		X	X	X	X	X			X	X

Illnesses

Job-related illnesses may be the result of exposure to patients during medical responses, exposure to other emergency responders at the work site, or exposure to biological contaminates at the work site. Isolating the source or cause of the illness can be difficult and few records are maintained that can be analyzed to determine a trend. Typical illnesses that may result from the facility environment include the following:

- *Respiratory infections such as colds, flu, and viruses* — Germs are spread by exposure to others who are carrying the germs. The germs are transmitted through direct physical contact (touch), by touching surfaces where germs have accumulated, or by breathing the airborne particles emitted by sneezing.

- *Allergies* — Dust, pollen, mold, and mildew in air ducts and on surfaces can cause a reaction in the individual. Food allergies can result from eating certain foods.

- *Food poisoning* — This can result from eating food that is spoiled, improperly cooked, or from microorganisms on plates, silverware, cutting boards, counter tops, etc.

Fatalities

In 2008, nonemergency fatalities accounted for 18 percent of all annual fire and emergency responder deaths. These deaths included those that occurred during training, which were discussed in Chapter 5, Training Safety. Other fatalities that may occur at facilities are usually the result of cardiac arrest. Fatalities may also occur as a result of falls, equipment malfunctions, or maintenance activities.

Design Considerations

The construction of all new buildings must meet the requirements of the locally adopted building, electrical, mechanical, and plumbing codes or, in the absence of such codes, nationally accepted good practices. These codes, usually based on a national model code such as NFPA® 5000, *Building Construction and Safety Code®,* or the *International Building Code® (ICC®),* often refer to consensus standards for the exact design requirements.

Figure 10.1 The design and construction of fire department facilities must adhere to the locally adopted building code.

Local codes may also establish the requirements for alterations to existing structures. In some cases, local building codes will mandate that alterations bring an existing building up to the current new construction requirements. Although this can be costly, it ensures that life safety requirements meet current design requirements. It may prove more cost-effective to build new facilities rather than remodel old structures **(Figure 10.1)**.

This section provides a brief overview of the legal requirements for facility design and construction as well as a review of the types of facilities normally operated by fire and emergency service organizations. This information will provide a basis for the remainder of the chapter.

Legal Requirements

Building, electrical, mechanical, and plumbing codes are intended to ensure that architects, engineers, contractors, and others involved in building structures adhere to a consistent set of criteria. This criterion has been developed by professionals as an industry standard. The model codes and standards must be adopted by the authority having jurisdiction (AHJ) before they can be enforced.

Agencies contemplating new or remodel construction should seek legal advice to determine what requirements must be adhered to. The AHJ for building construction may not be the agency that will occupy the building. A municipality, county, or even state may have legal jurisdiction over the construction of the facility. In some cases, local or regional planning boards must authorize the location and even the appearance of fire facilities.

Local and state codes frequently refer to other standards, such as NFPA® 70, *National Electrical Code®,* NFPA® 101, *Life Safety Code®,* or NFPA® 13, *Standard for the Installation of Sprinkler Systems.* These standards may be adopted automatically when the building code references the standards and is adopted. The AHJ (and/or occupying agency) may also determine that additional safety requirements should be included in the building code; for example, some agencies protect their facilities with automatic fire sprinkler systems, even if not required by local building codes. These amendments to the code are locally developed and adopted along with the code. Once adopted the building code becomes the design standard for all future construction and renovations.

Fire codes may also be adopted by the AHJ to ensure that structures built under the jurisdictional building code are used in a safe manner. Fire codes establish guidelines for safe behaviors and are intended to prevent or mitigate the effect of fires. These codes are generally enforced by the local fire department.

The federal government also has regulations that apply to the safe use of facilities where volunteer and career personnel are employed. These are contained in the U. S. Code of Federal Regulations (CFR) and are enforced by the U.S. Labor Department's Occupational Safety and Health Administration (OSHA) or similar state agencies. Twenty-six states have OSHA-compliant programs modeled on national regulations.

Another federal regulation that applies to the design of fire and emergency services facilities is the Americans with Disabilities Act (ADA). This act mandates that all public facilities, with some exceptions, must be accessible to people with physical disabilities, including mobility, sight, and hearing. Existing structures or nonpublic areas may be exempted if the local authority can demonstrate the access is not necessary.

OSHA regulations also require the posting of certain types of signs and markings. In particular, these regulations apply to facilities that store or use hazardous materials. NFPA® 704 symbols are required to indicate the presence of hazardous materials and their flammability, health, stability, and reactivity hazards. Safety Data Sheets (SDSs) (formerly known as Material Safety Data Sheets [MSDS]) are also required to be maintained on the property for each chemical that is used or stored there **(Figure 10.2)**. Finally, all chemical containers must be properly labeled in accordance with OSHA regulations.

Other required signage includes the following:

- Warnings about wearing hearing and eye protection

- Warnings about electrical power equipment

- Prohibited smoking or designated smoking areas

- *Slippery when wet* signs

Figure 10.2 Material Safety Data Sheets must be maintained on site when hazardous materials are present.

Facility Types

Fire and emergency services organizations will have some or all of the following types of facilities. Safety requirements, either those that are mandated or those that represent good practices, will exist in each of these types of facilities. Common facilities include the following **(Figures 10.3 a-g, p. 370)**. Stations may include space designed and designated for any of the other functions listed:

- *Station facilities* — Structures that are intended to house emergency response apparatus and may include personnel living quarters; includes both land based and marine (fireboat) emergency response stations. Marine stations require special attention to mitigate the water hazards introduced, especially to station visitors.

- *Apparatus and equipment maintenance facilities* — Structures designed for the repair and maintenance of apparatus, vehicles, tools, and equipment. If connected to a station, care should be taken to assure that maintenance equipment or operations do not introduce additional hazardous conditions.

- *Training facilities* — Structures intended for training personnel. These facilities include classrooms, offices, smoke buildings, live burn buildings or props, and other items. If located adjacent to a station, care should be taken to assure that station operations do not introduce additional hazards to training operations or vice versa.

- *Administrative facilities* — Structures that contain offices intended to house administrative and staff personnel.

- *Storage facilities* — Structures designed to provide logistical support to the organization.

- *Communications centers* — Structures that house telephone, radio, and other communications equipment and fulfill the dispatch function of the organization.

- *Aviation facilities* — Generally located on airport property, these facilities provide hanger, maintenance, and storage space for helicopters and/or fixed-wing aircraft used in wildland fire fighting. If connected to a station or located adjacent to a station, care should be taken to assure that aircraft operations do not introduce additional hazardous conditions to station personnel and safe aircraft operations are not compromised by station operations.

Safety Requirements

Each of these facilities will have safety requirements that are particular to the activities that occur in them. At the same time, they will share many of the requirements listed in the sections that follow.

Indoor Air Quality

Poor indoor air quality leads to respiratory illnesses and allergic reactions, a condition known as sick building syndrome (SBS). The health and safety officer is responsible for performing periodic air-quality tests in all department facilities.

Figures 10.3 a-g Common types of fire and emergency services facilities include: a) stations, b) apparatus maintenance, c) training, d) administration, e) storage, f) communication, and g) aviation.

Building codes include some of the following safety requirements intended to maintain an acceptable level of air quality. These requirements include:

- Separating office and living quarters and apparatus areas
- Using an exhaust capture and filtration system (**Figure 10.4**)
- Prohibiting smoking within all facilities
- Providing ventilation hoods in areas that may produce toxic or flammable vapors
- Installing HVAC systems that circulate and filter the air at a given rate
- Installing passive ventilation systems in apparatus bays

Decontamination and Cleaning

Maintaining a clean facility reduces the possibility of liquid and airborne contamination and the passing of germs between occupants. OSHA, local health department, and fire department requirements are intended to reduce or eliminate the potential for work-related illnesses. Safety requirements include the following:

- Installing a designated decontamination area located in the apparatus bay
- Installing a designated area and equipment for cleaning and drying personal protective equipment (PPE)
- Requiring the issuing of individual sets of bed linens to each occupant
- Installing of emergency decontamination showers and eyewash stands
- Prohibiting PPE from the living quarters
- Prohibiting cleaning PPE or medical equipment in kitchen and bathroom sinks
- Providing labeled medical waste containers
- Providing stainless steel sinks and countertops in kitchens are desirable (**Figure 10.5**)
- Enforcing proper hygiene in kitchen and bath areas
- Installing an apparatus bay drain system to capture and filter contaminated water and fluids

Fire Protection

Depending on the local building code, fire protection systems may or may not be required in fire and emergency services facilities. However, it is good practice and good public relations to install them when possible. Fire protection systems include the following:

- Automatic and manual detection and alarm systems
- Kitchen ventilation hood fire-suppression systems

Figure 10.4 Apparatus exhaust capture systems are designed to reduce the levels of engine exhaust in the apparatus bays.

Figure 10.5 Stainless steel appliances and counter tops are designed for easy cleaning and to prevent the growth of bacteria.

- Automatic sprinkler systems for the entire facility
- Paint booth fire-suppression systems
- Portable fire extinguishers
- Fire-separation barrier between apparatus bay and office or living quarters
- Automatic fuel shutoff valve on cooking appliances (engaged whenever the station is dispatched)

Hazardous Materials

Hazardous materials found at fire and emergency services facilities usually include flammable/combustible liquids, compressed/liquefied gases, and corrosives. Most are in small quantities and are used for cleaning and maintenance. Both OSHA and local fire codes regulate the use and storage of these materials. Safety requirements for hazardous materials include the following:

- Maintaining a safety data sheet (SDS) book on-site
- Constructing a flammable/combustible storage room or the use of a similar approved storage cabinet
- Using approved hazardous materials containers for storage
- Providing the proper PPE for use with hazardous materials
- Providing hazardous materials containment systems

Ergonomic Concerns

Ergonomics is the scientific study of physical work intended to gain the greatest productivity in the safest manner. The study leads to the design of tools, workspaces, and work behaviors. Properly applied, ergonomics will result in a work environment that is both safe and efficient. Ergonomic issues related to fire and emergency services facilities include the following:

- Designing office workstations to relieve physical stress to backs, arms, and wrists
- Locating tools and equipment within easy reach
- Providing mechanical assistance for lifting heavy objects

Lighting

Proper lighting reduces eyestrain and associated headaches. General room illumination for the workplace is established by building codes and OSHA depending on the type of occupancy and work area. Illumination concerns include the following:

- Installation and location of exit and emergency lighting
- Installation of auxiliary lighting powered by batteries or generator **(Figure 10.6)**
- Proper level of overall lighting in each room or work area
- Connection of lights to emergency alert system in fire stations
- Installation of traffic warning and control lights
- Accessible work (drop) lights in apparatus and shop areas

Figure 10.6 Auxiliary power generators provide a limited amount of electrical power to stations for lighting and communications equipment.

Floor Surfaces

Floor surfaces should be specific to the type of activity that will occur on them. Traditionally, fire and emergency service facility floors had hard surfaces such as concrete or terrazzo. These floors were easy to maintain and difficult to damage. Today, new materials are in use such as vinyl or ceramic tiles and carpet. From the safety standpoint, floors should provide a nonslip walking surface; reduce noise levels; and reduce physical strain on legs, back, and lower body.

In apparatus bays, walkways should be designated by striping or a contrasting color. Floor drains should be located where apparatus are parked with the floor sloped to the drain. This prevents liquids from standing on the floor and creating a slipping hazard. Floors in physical fitness areas should be covered with rubber or vinyl tiles designed to absorb impact and moisture. Office, living, and sleeping areas benefit from tight woven carpeting to reduce noise. Kitchen and bathroom areas are best covered with ceramic tile or other similar easy-to-clean surfaces. Note that it is essential that the tile and grout be completely sealed to prevent liquids from staining them.

Stress Reduction

Studies have determined that cardiac-related injuries are the result of repeated exposure to alert tones. This is especially true when the individual is sleeping. New technology is being developed to counteract stress and reduce it as much as possible. Stress-reduction activities include the following:

* Testing for noise levels in all facilities **(Figure 10.7)**
* Installing noise-reduction materials on floors, walls, and ceilings
* Installing an alert system that uses colored lights combined with an audio alert that increases in volume from soft to loud

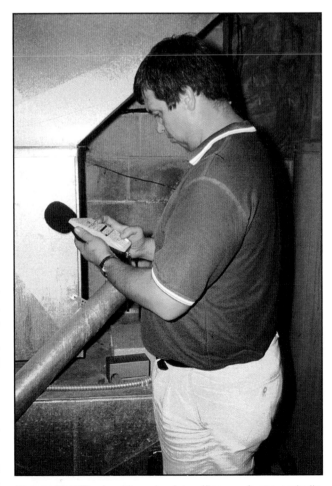

Figure 10.7 The health and safety officer performs periodic sound-level tests in all facilities.

General Requirements

There are many safety requirements that do not fit a specific category like the previous ones. They include the following:

- *Roadway access* — Station driveways should be long enough to accommodate the longest apparatus housed there. This distance provides ample time for the driver/operator to assess the traffic, activate warning devices, and prepare to enter the roadway. It also provides time for oncoming traffic to see the apparatus and come to a stop. Traffic-control lights and audio warning devices linked to the alert system should also be installed to warn motorists. Station driveways should not be obstructed by vegetation but be visible in both directions of traffic flow. Curved driveways should be avoided where possible.

- *Medical equipment* — Emergency equipment for use by facility occupants must be located where it is easily accessible by all occupants. Medical equipment includes first aid kits, defibrillators, and oxygen inhalators **(Figure 10.8)**.

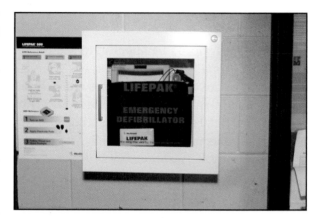

Figure 10.8 Defibrillators and other medical equipment must be accessible to occupants of all facilities.

- *Evacuation plans* — All facilities, especially those open to the public, must have evacuation plans developed and posted along routes of travel. If the structure is designed for shelter-in-place protection during a weather emergency, that space must be designated on the plan.

- *Slide poles* — Symbolic of the fire service, slide poles have been used for over a century to move firefighters from upper stories to the apparatus bay of multistory stations. Unfortunately, these same poles have resulted in foot, ankle, and leg injuries as well as fatalities when people have fallen through the opening on the upper floor. If slide poles are in use in an older existing facility, they must be equipped with the required safety accessories. These include a cushioned mat at the base of the pole, removable covering for the hole, and a guardrail with retractable gate around the hole and pole. No new installations of slide poles are permitted in some states and their use is not recommended by the NFPA®.

- *Security* — An everincreasing safety concern is security for public facilities. Property vandalism of unattended facilities, theft, and violence toward fire and emergency services personnel have led to increased efforts at facility security. Security efforts include the following:
  - Locking stations when they are unattended
  - Requiring visitors to use a single point of access
  - Requiring visitors to be escorted while on the property
  - Installing bulletproof glass in exterior doors and windows

— Installing security fences and gates around parking areas

— Installing burglar alarms

— Installing external security bars or grills on windows and doors

— Hiring personnel to provide security

— Providing a secure, locked cabinet or closet for controlled drugs or medical supplies

— Installing video surveillance camera systems

Usability

In addition to safety-related requirements, there are other design considerations that should be taken into account when designing a new facility or evaluating an existing one. How fire and emergency services facilities are used has changed drastically over the past 50 years as new services have been added and personnel requirements have changed. Usability issues include the following:

- Life expectancy of structure
- Station height
- Drive-through versus back-in station
- Privacy
- Auxiliary power
- Physical fitness equipment
- Public accommodations

Life Expectancy of Structure

Fire and emergency services facilities are usually built to last up to 50 years. However, because of the cost of replacing existing facilities, this life expectancy may increase in order to gain the greatest benefit from the capital expenditure **(Figure 10.9)**.

An increased life expectancy can pose problems. For one thing, it is difficult to predict the demographics of the service area and changes in services provided as well as personnel involved that far into the future. As land use surrounding stations changes, the apparatus and personnel needs must be reanalyzed along with the utilization of stations. Many fire agencies that assumed responsibility for ambulance service found their stations

Figure 10.9 In many parts of the country, fire stations that are over 50 years old are still in use.

were inadequate for that service. Many fire agencies hiring female firefighters found their facilities lacked female accommodations. At the same time, the station designs of today may not be able to accommodate larger apparatus or crew sizes in the future. Finally, changes in technology may make the current station designs obsolete.

In some departments, it is the practice to place temporary stations consisting of manufactured homes and large metal sheds in areas where new growth may occur. Care must be taken that these temporary facilities do not compromise personnel safety.

Station Height

The majority of fire stations that have been constructed in recent years have been single-story facilities. Departments have attempted to predict population growth and spread and purchase land that is large enough to accommodate modern single-story facilities. However, in older urban areas, multistory stations are built when property size is restricted. Restrictions on the use of slide poles are having an effect on the construction of multistory stations. There are no statistics on the accident rates of multistory stations using stairways rather than poles.

Drive-Through Versus Back-In Station

Another controversy that must be considered is the advantage of drive-through station design over back-in station design. Back-in stations require smaller lot size. However, backing up a fire apparatus is a very unsafe activity and requires two people to direct the rearward movement of the vehicle. Drive-through stations are safer **(Figure 10.10)**.

Privacy

One of the most significant changes in station design over the past 25 years has resulted from the increasing numbers of women in the fire service. Initially, it was enough to place *Occupied* signs and slide bolts on bathroom doors. Today, it is widely recognized that separate shower and toilet facilities for males and females as well as individual sleeping quarters are desirable. The change from dormitory sleeping arrangements to private cubicles or private rooms not only increases privacy but also may reduce the chance of illnesses such as colds and flu.

Auxiliary Power

Because fire and emergency services facilities depend on electricity to power more equipment than in the past, high-capacity auxiliary generators are essential. Today, generators are designed to operate with natural or propane gas or diesel and provide power for the entire facility. Not only does this increased capacity ensure that the firefighters will have uninterrupted power, but it may also provide a place of refuge for displaced citizens.

Facility Protection

Facilities that may be exposed to tornados, hurricanes, earthquakes, or other catastrophic natural disasters must be designed to protect the occupants from these hazards. Structural protection from potential terrorist attacks may also be necessary in some areas. Safe rooms may be added to existing structures as a form of protection.

Physical Fitness Equipment

With the recognition that cardiac arrest is the leading cause of firefighter fatalities, fire department administrations and labor organizations are requiring increased participation in physical fitness programs. To accommodate these programs, new facilities are being designed to include space for physical fitness equipment and areas to work out **(Figure 10.11)**. In the past, weight lifting equipment and treadmills have been located in the apparatus bay or sleeping

Figure 10.10 Fire stations with drive-through bays eliminate the possibility of accidents that occur when apparatus are backed into stations.

quarters. While this arrangement met the immediate need, it posed a number of problems not the least of which was safety.

To address safety needs, physical fitness areas should be large enough allow personnel to perform exercises, have room to access each piece of equipment, and provide a clear passage through the room. Floor coverings should be easy to clean and provide sound barriers and cushioned surfaces. The space should be well-illuminated and ventilated as well.

Figure 10.11 New station designs may incorporate dedicated physical fitness rooms in the structure.

Public Accommodations

Fire and emergency services facilities, unless privately owned, are public property. As governments have attempted to increase the cost/benefit of these facilities, they have added non-emergency-related activities to the use of the facilities. Administrative facilities, training centers, and, in particular, fire stations are being designed and opened to public use. Nongovernmental organizations (NGOs) have also collaborated with fire and emergency services organizations to provide services through these facilities. Uses include the following:

- *Community use* — These include the addition of a community meeting room or space in the facility. Uses include voting, public meetings, health examinations or vaccinations, CPR classes, or other civic events. When provided, access separate from emergency operations should be provided.

- *Place of refuge* — In times of natural emergencies or disasters, fire stations and other facilities are used as places of refuge. Location, accessibility, and space as well as the presence of auxiliary power and sanitary facilities make these facilities appropriate for this use.

- *Safe place, safe haven* — The safe-place concept allows victims of abuse or children who are lost or in danger a place to seek assistance (**Figure 10.12**). The safe-haven concept is based on legislation that addresses infant abandonment. It allows infants 72 hours or younger to be left at designated locations as a ward of the state. Station personnel are trained in the process for assisting and referring the victims to the proper authorities.

Figure 10.12 In some communities, fire stations are used as safe place or safe haven sites to provide assistance to children or victims of abuse.

Inspections and Maintenance

Facilities that are intended to last a minimum of 50 years must be inspected regularly and maintained to prevent deterioration. The responsibility for both of these duties may be spread through the fire and emergency services organization or shared with another department of local/state government. However, both duties include the participation of the administration, the supervisors, and you, the emergency responder.

Inspections

Facility inspections occur at many levels and on many time schedules. Performed properly, these inspections work together to ensure a safe and healthy work environment. We will start with the least formal and most frequent inspection, the one that every member of the organization may perform.

In a career or combination organization, a shift change provides an opportunity for the first facility inspection. Either immediately following the formal change of command or during station cleaning, personnel informally inspect the condition of the facility. Any maintenance needs, such as leaking faucets, inoperative toilets, or lightbulbs that need to be replaced are noted. Repairs are made or reported through the chain of command to the proper authority.

During the work shift, personnel repair or report any needs that they become aware of as a form of preventative maintenance, making small repairs before they become major needs. In a volunteer organization, this type of inspection may occur whenever the unit holds mandatory training or work sessions.

A more formal inspection is usually performed on a weekly basis, in all types of organizations, by the company or chief officer **(Figure 10.13)**. These inspections may coincide with a more thorough cleaning of the facility when furniture is removed and floors stripped, cleaned, and waxed. This level of cleaning and inspection allows the officer to note more significant maintenance needs, for example, repainting the facility, recaulking tile, replacement of HVAC filters, etc.

Monthly inspections are generally performed by the district/battalion chief. The intensity of these inspections is similar to the weekly inspection and may result in requests for more detailed repairs or alterations that may require greater expenditures of funds or future budget requests.

Finally, an annual inspection performed by the health and safety officer (HSO) and/or a fire prevention officer will generally occur. Compliance with the health and safety policies and the fire and life safety codes are the purposes of this inspection. The focus will be on both the condition of the facility and on safe behavior of its occupants. Two important areas that the HSO inspects are the air quality and the noise level in the facility.

To reduce the potential for illnesses from airborne germs or excessive levels of carbon monoxide, the HSO tests the air quality in the building annually or when the need arises. Testing equipment samples the air and determines the parts per million of various

Figure 10.13 Station inspections are generally performed on a weekly or monthly basis by a fire officer responsible for the facility.

types of contaminants such as dust or pollen. The inspection may result in the replacement of filters or the cleaning of the air ducts. Apparatus room contamination must also be tested and apparatus exhaust systems installed if needed.

The HSO is also responsible for testing the noise level in the facility. Noise-testing instruments determine the level in decibels (dB). Facility noise originates from the HVAC system, kitchen range ventilation hood, auxiliary power generator, and power tools and equipment. The greatest source of noise, however, is the apparatus in the apparatus bay. Although a maximum noise level in the cab of the apparatus is limited to 89 dB by federal law, the noise level outside the vehicle may be higher. If the noise level cannot be reduced, policies may be required for the use of hearing protection while the apparatus is running.

Formal inspections include the use of inspection checklists or the creation of reports that provide a record of the facility's condition. These records are generally maintained for the life of the facility or for the minimum retention time prescribed by local policy. Inspections that result in maintenance requests must be forwarded to the appropriate authority. These maintenance requests should be tracked to ensure that the work is completed in a timely fashion. The completion of safety issues should take priority. A sample facility inspection checklist form is located in **Appendix F** of this manual.

Maintenance

Like inspections, facility maintenance is the responsibility of multiple people and departments within the jurisdiction. In volunteer organizations, the majority of the work will be the responsibility of the volunteers. In large departments, the responsibility may be assigned to another department such as public works or to a private contractor.

Basic maintenance begins with continual cleaning and care of the facility. Cleaning is usually performed at the start of each work shift. Some facilities, such as administrative offices and training facilities are cleaned by private contractors during off-duty hours or on weekends. Cleaning helps reduce the potential for germs in the air and on surfaces that can cause illnesses and allergic reactions in occupants. Of particular interest is the thorough cleaning of all cooking and food-preparation areas to prevent the growth of microbial germs. These surfaces may need to be tested periodically to ensure that they are properly cleaned.

Depending on local policies, minor maintenance is generally performed by the facility personnel. Replacing lightbulbs or HVAC filters are part of this responsibility. The organization's supply or support division, the jurisdiction's public works department, or a private contractor usually performs major maintenance and repairs. Painting, roof repairs, glass replacement, plumbing and electrical repairs, and training prop repairs fall into the category of major maintenance work. In many volunteer departments, this type of work is performed by the members of the department who have the skills and abilities to perform them.

Another type of maintenance performed by emergency personnel is exterior ground maintenance, although this may also be performed by the local parks department or a private contractor. Mowing the lawn, trimming shrubs and trees, and maintaining planting beds improves the appearance of the facility and reduces the potential for small vermin from nesting in weeds and tall grass near the facility. However, the simple act of maintaining the grounds exposes personnel to injury by the use of power tools and equipment. Care must always be taken when operating lawn equipment as indicated in Chapter 7.

Personal Behavior

While a safe work and living environment can be created through the design, inspection, and maintenance of the facility, unsafe personal behavior can overcome these efforts. Safe personal behavior is founded on good common sense, training, and adhering to safety policies, procedures, and manufacturer's recommended practices.

Common sense applies to many areas of personal behavior that can ensure your safety and the safety of others. Unfortunately, it is often ignored resulting in an unsafe act that leads to property damage and injuries. The following list of good common sense behaviors, although not all inclusive, can help you focus on safe behaviors in your facility:

- Do not engage in horseplay
- Keep personal protective equipment (PPE) in the apparatus bay, designated storage areas, and cleaning areas and not in living quarters **(Figure 10.14)**
- Use approved ladders for reaching high areas
- Use mechanical aids or additional personnel to lift heavy objects
- Wear approved PPE when operating power tools and equipment
- Practice personal hygiene
- Thoroughly clean all cooking and food-preparation surfaces
- Report all facility maintenance needs immediately
- Report any illnesses or injuries that may be caused or aggravated by the facility's condition or design
- Use only approved cleaning and maintenance materials
- Update and read the SDS for all cleaning and maintenance materials
- Adhere to the safety policy and procedures manual
- Use physical fitness equipment according to instructions

Figure 10.14 Personal protective equipment (PPE) is kept in ventilated storage lockers in the apparatus bay.

- Adhere to the *No Smoking* policy
- Practice good housekeeping in all facilities and work areas

The HSO and your supervisor are responsible for providing all safety-related training to you and your unit. You are responsible for practicing the safety-related skills that you are taught. You are ultimately responsible for keeping yourself safe.

Summary

Fire and emergency services facilities exist in a multitude of types, sizes, and configurations. Their design is regulated by laws at all levels of government, including the locally adopted building and fire codes. The primary concern of all these regulations is the health and safety of the occupants. Periodic inspections and ongoing maintenance ensure the required level of health and safety and the continued longevity of the facility. Personal behavior on the part of the occupants, however, is essential to a truly safe environment.

Review Questions

1. What design considerations for station facilities can be taken to improve both the safety and health of personnel?

2. What types of personal behaviors can help ensure the safety of personnel while in the station facilities?

Safety and Health Program Administration

Chapter Contents

Job Performance Requirements

This chapter provides information that addresses the following job performance requirements of NFPA® 1500, *Standard on Fire Department Occupational Safety and Health Program:*

NFPA® References

4.2.1	4.4.4	4.5.1.1	4.6.4	7.18.3
4.2.2	4.4.5	4.5.1.2	4.6.5	11.1.1
4.2.3	4.4.5.1	4.5.1.3	4.7.1	11.1.2
4.3.1	4.4.5.2	4.5.2	4.7.2	12.1.1
4.3.2	4.4.5.3	4.5.3	4.7.3	12.1.2
4.3.3	4.4.6	4.5.3.1	4.7.4	12.1.3
4.3.3.1	4.4.7	4.5.3.2	4.7.5	12.1.4
4.4.1	4.4.8	4.6.1	4.7.6	
4.4.2	4.4.8.1	4.6.2	7.18.1	
4.4.3	4.5.1	4.6.3	7.18.2	

FESHE Learning Objectives

Occupational Safety and Health for Emergency Services

Compare the differences between standards and regulations.

Discuss the need for and the process used in post incident analysis.

Describe the components and value of critical incident management programs.

Describe the components of a wellness/fitness plan.

Learning Objectives

After reading this chapter, students will be able to:

1. Describe the legal requirements for a fire department's health and safety program.

2. Describe the procedure for researching the fire department's safety problems.

3. Summarize the reasons for implementing an occupational safety and health program.

4. Summarize the role of a department's occupational safety and health committee.

5. Summarize the role and responsibilities of a department's health and safety officer.

6. Summarize the steps for developing a department's risk management plan.

7. Explain the techniques that can be used to control risk.

8. Describe the training and policy measures that can be implemented to address hazards in the workplace.

9. Summarize the elements required to be in a respiratory protection plan.

10. Describe the process for investigating exposures and accidents.

Chapter 11
Safety and Health Program Administration

Courtesy of Cherry Hill (NJ) Fire Department.

Case History

On February 20, 2009, a 45-year-old male volunteer firefighter assisted in fighting an 800-acre wildland fire. After fighting the fire for approximately 10 hours, the firefighter was asked to assist with "mop up." While backing up a brush truck to provide water to crew members, the firefighter collapsed. A crew member yelled for help and moved the gearshift to the park position. The firefighter was pulled from the brush truck and found to be unresponsive, not breathing, with a weak pulse. Shortly thereafter, his pulse stopped, and an oral airway was placed while cardiopulmonary resuscitation (CPR) was begun. An automated external defibrillator (AED) was retrieved, a shock was delivered, and a pulse returned. An ambulance arrived and began advanced life support treatment. About 30 minutes after he collapsed, a Life Flight helicopter arrived and took over patient care. The firefighter went into cardiac arrest again and CPR continued for another 32 minutes. The Medical Control Authority was contacted and informed of the firefighter's condition. Medical Control advised the flight crew to pronounce the firefighter dead, approximately 67 minutes after his collapse. The death certificate and the autopsy, completed by the medical examiner, listed "atherosclerotic and hypertrophic cardiovascular disease" as the cause of death. NIOSH investigators conclude that the physical stress of performing fire-extinguishing activities, coupled with the firefighter's severe underlying atherosclerotic coronary artery disease (CAD), probably triggered a heart attack and his subsequent sudden cardiac death.

The NIOSH investigator offers the following recommendations to address general safety and health issues. Had these recommended measures been in place prior to the firefighter's collapse, his sudden cardiac death may have been prevented.

- Provide preplacement and annual medical evaluations to firefighters consistent with National Fire Protection Association (NFPA®) 1582, *Standard on Comprehensive Occupational Medical Program for Fire Departments.*

- Incorporate exercise stress tests following standard medical guidelines into a fire department medical evaluation program.

- Phase in a comprehensive wellness and fitness program for firefighters consistent with National Fire Protection Association (NFPA®) 1583, *Standard on Health-Related Fitness Programs For Fire Department Members.*

- Perform an annual physical performance (physical ability) evaluation consistent with NFPA® 1500, *Standard on Fire Department Occupational Safety and Health Program.*

- Provide firefighters with medical clearance to wear self-contained breathing apparatus (SCBA) as part of the fire department's medical evaluation program.

Source: NIOSH Firefighter *Fatality Report 2009-09*

One of the most critical responsibilities of fire and emergency service organizations and the jurisdictions they serve is to provide for the safety and health of their employees. The ability of the organization to provide fire-suppression, prevention, public education, emergency medical services, and other services depends on the safety, health, and wellness of their members. Therefore, it is the organization's responsibility to develop, implement, and manage a comprehensive safety and health program.

The occupational safety and health policy as required by Chapter 4 of NFPA® 1500, *Standard on Fire Department Occupational Safety and Health Program*, establishes the infrastructure for the development and implementation of the organization's occupational safety and health program. The purpose of the policy is to clarify for all members their responsibilities in the safety and health process as well as the organization's responsibility. The following paragraph is an example of a clear and concise occupational safety and health policy based on the NFPA® 1500 requirements:

> *It is the policy of the fire department to provide and to operate with the highest possible levels of safety and health for all members. The prevention and reduction of accidents, injuries, and occupational illnesses are goals of the fire department and shall be primary considerations at all times. This concern for safety and health applies to all members of the fire department and to any other persons who could be involved in fire department activities.*

Based upon the fire and emergency service organization's risk assessment, members are exposed to a variety of hazards that create a complex set of occupational safety and health issues and concerns. The objective of this policy is to provide the necessary control measures to assist members so that they may perform their assigned tasks safely, efficiently, and effectively. For the policy to be effective, it must be communicated to all members of the organization, monitored for desired results, reviewed periodically, revised as necessary, and applied fairly. It cannot just remain in a binder on a bookshelf in the organization's main office.

According to NFPA® 1500 and 1521, *Standard for Fire Department Safety Officer,* the organization's designated health and safety officer (HSO) is responsible for the safety and health program based on the policy. In this chapter, the term *health and safety officer* or *HSO* is used regardless of the rank of the person assigned to that function. This position may be held by a chief officer or other officer or a nonuniformed health and safety professional who meets the educational and skills requirements of NFPA® 1521. There are very practical methods for smaller agencies to meet this requirement. These could include using training officers as the HSO with assistance from other staff members. Organizations may also contract private or outside agencies to provide the HSO function.

The HSO develops a plan to meet the goals of the organization's occupational safety and health policy. This chapter describes the legal foundation of the policy and the creation of the overall compressive safety and health program (including the functions of occupational safety and health committees and roles and responsibilities of HSOs) **(Figure 11.1)**. Each of the elements/components that compose the overall program is also explained to provide an understanding of the importance of each to the organization and its members. The elements/components that compose the comprehensive program include the following:

- Risk management plan
- Accident, injury, and illness prevention program
- Medical exposure management program
- Employee physical fitness and wellness program
- Employee assistance program (EAP)

Legal Requirements

Besides the NFPA® standard requirement for a safety and health program, other legal mandates also exist. As mentioned previously, NFPA® standards are not mandatory unless they are adopted by the authority having jurisdiction (AHJ). However, other laws or ordinances may cite the standards, making them part of the law or ordinance. For instance, the local ordinance that establishes the fire department may adopt NFPA® 1500 by reference or through an enabling ordinance. Adoption by reference means that the local jurisdiction will follow the standard exactly as written. Adopting a standard through an enabling act allows the local jurisdiction to select portions of the standard that fit the local needs and delete others.

Figure 11.1 The organization's health and safety committee, under the leadership of the HSO, develops plans to meet the safety and health policies.

Comparison Between Standards and Regulations

Standards are a set of published criteria developed to serve as a model or example of desired performance or behaviors. The standards contain requirements and specifications outlining minimum levels of performance, protection, construction, or behavior.

A *regulation* is a collection or compilation of rules and regulations enacted by a legislative body to become law in a particular jurisdiction. The jurisdiction can be any lawfully chartered political subdivision such as a municipality, county, or state/province. When adopted the regulation can be composed partially or entirely of standards.

The major difference between a regulation and a standard is that a regulation must have administrative provisions to explain how and when the standards are to be applied. Simply stated, a regulation is a law that may be based on or incorporate a standard. A standard, however, only becomes a law when it is legally adopted by a jurisdiction or included as part of a code.

> **Regulations** — Rules or directives of administrative agencies that have authorization to issue them.

> **Standards** — Criterion documents that are developed to serve as models or examples of desired performance or behaviors and that contain requirements and specifications outlining minimum levels of performance, protection, or construction. No one is required to meet the requirements set forth in standards unless those standards are legally adopted by the authority having jurisdiction, in which case they become law.

Safety and health requirements may be the result of federal or state/provincial acts, laws, or regulations. Federal laws will usually take precedence over state/provincial or local laws unless the state/province enacts similar legislation. In the case of safety and health requirements, the federal regulations generally apply because they must be consistently applied throughout the country.

The Occupational Safety and Health Administration (OSHA) enforces worker-protection requirements for all federal employees and in states that have established approved OSHA regulations. A General Duty Clause may also be used to enforce OSHA regulations in non-OSHA states.

Federal hazardous materials laws regarding the manufacture, use, transportation, and disposal of hazardous materials are also universally enforced. As part of OSHA' ss Right-to-Know Regulations, safety data sheets (SDSs) must be provided for all potentially hazardous materials. SDSs must be available in the workplace, and employees must be trained in the mitigation of hazards caused by the materials.

Besides local, state/provincial, and federal safety and health laws, the labor/management agreement with the fire and emergency services organization (or jurisdiction) will also contain safety and health clauses. These clauses may include requirements for personal protective equipment (PPE), a wellness program, an employee assistance program (EAP), or annual medical examinations. The labor/management agreement may directly cite language in a NFPA® standard, essentially adopting that standard for the duration of the agreement.

Researching the Safety Problem

Developing an effective safety, health, and wellness program depends on accurate and current information. Information is based on raw data that must be gathered from a variety of sources. Gathering the data, also called data mining, requires time and effort. Once gathered, the data must be categorized by type and them evaluated to determine the information that it provides. To be effective at researching a topic, you must know what to look for, where to look for it, and how to determine the validity of the data that is collected.

Data Collection Process

The project topic helps determine the data to be collected. For example, if the project involves developing a safety program, the data collected would include the following:

- Local accident, injury, and line-of-duty (LODD) death reports
- Similar reports for other fire departments in the region or of similar size
- National statistics on firefighter injuries and LODDs
- Statistics on similar injuries and deaths in other high-risk occupations based on similarity of activity
- Insurance industry statistics
- Fire and emergency responder pension program statistics
- Worker's compensation data
- Data compiled by governmental agencies

Sometimes it is just as difficult to determine what to exclude as it is to decide what to include within the search parameters. If the focus of this project is injuries and fatalities, as in this example, then data on illnesses would not be needed.

The first step of the process is to write down the topic that is to be researched. List all possible topics that are similar to the main topic. Next list the various types of data that may support the topic such as internal reports, regional or national reports, legislation, product reviews, cost estimates, and so on. Then list the possible sources for each information type. Index cards can be used to record this information and the sources. The information can also be placed in a computer database for easy retrieval and sorting. The information types **(Figure 11.2)** may include the following:

- *Expert opinions* — Statements by credible experts in a particular field or someone who has analyzed or experienced a similar situation. Personal knowledge is an acceptable starting point when looking for this type of information; that is, what do you or the person doing the research know about the subject.

- *Trends* — Patterns that can be traced over time and used to forecast the future. Trends may be developed from raw data such as incident scene strains/sprains; however, raw data must be the same type for each entry on the timeline.

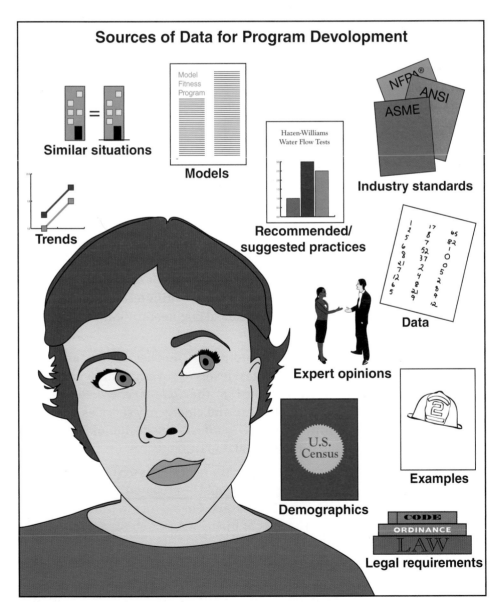

Figure 11.2 Sources of Data for Program Development.

- *Models* — Frameworks composed of accepted practices that an organization can adapt to its own needs such as a model fitness program.

- *Similar situations* — Information from interviews with other people on how they handled particular situations. Sources for this type of research include industry journals, newspaper articles, or simply networking with peers through professional organizations.

- *Data* — Raw numbers such as lost-time injury leave or cost of overtime.

- *Examples* — Representative images of situations, processes, items, or anything else that a researcher can use to illustrate concepts. Good examples can be used to create a *benchmark;* that is, something that is considered an informal standard or goal that an organization is attempting to attain. From using benchmarks, an organization can determine how well it has done in reaching or surpassing the standard.

- *Analyses* — Third-party testing laboratories, universities, and other organizations continually evaluate equipment, procedures, and trends and produce written analytical reports of the behavior of those items. The resulting reports can provide researchers with objective reviews without having to perform the testing.

- *Demographics* — Collected information on the population in a specific area or of a specific group such as the members of the fire and emergency services organization. This type of information is probably the easiest to obtain because it is readily available.

- *Recommended/suggested practices* — Suggestions for how to do something; usually based on trial-and-error experiences of other people or scientific research. The Hazen-Williams Water Flow Tests are examples of this type of information because they provide a basis for determining nozzle pressures on hoselines.

- *Industry standards* — Generally accepted methods for accomplishing some task or function (similar to recommended practices). For example, the NFPA® provides the industry standard for fire and life safety. Other types of standards are developed by engineering and scientific testing organizations such as the American Society of Mechanical Engineers (ASME) or American National Standards Institute (ANSI).

- *Legal requirements* — Laws, codes, ordinances, and decrees that are legally binding requirements created at most levels of government to ensure the safety and welfare of a society. In the U.S., the *Code of Federal Regulations (CFR)* applies to many areas of the fire and emergency services, while the Americans with Disabilities Act (ADA) affects the design of facilities to ensure access for people with physical impairments.

With the knowledge of what to look for, the next step is to start locating the sources of the information. However, you should be prepared to seek assistance from a professional research firm if necessary. The savings in time, effort, and frustration can easily offset the cost of hiring a research organization that specializes in the particular field of research.

Information Sources

Finding information sources, or where to look, depends a great deal on the topic that you are researching. In general, data may be located on the Internet; from government agencies, libraries, educational institutions, testing and standards organizations; nonprofit organizations; and vendors/manufacturers. The greatest challenge in locating reliable data is determining the credibility of the source. It is a mistake to *believe everything one reads in the paper* because most information (regardless of the source) is biased in some way. Information provided by a government agency may be written in such a way as to justify that agency's actions or existence. Information from vendors tends to show the positive results of using their products rather than providing an unbiased view like the one provided by third-party testing agencies. The Internet is the least credible of all sources because the primary source for the information may have been lost or altered. Authorship of online articles should always be questioned.

Internet

The creation and development of the Internet has greatly increased the amount of information that is easily available to researchers. Finding it, however, is still a challenge. Some information is proprietary such as the formula for Class A foam concentrate. Other information is considered vital to national security and access to it is restricted.

It is important to remember that not all of the information found on the Internet is accurate, valid, and current. One may print or save it only to discard it once it proves to be unusable. Chat rooms and message boards are the least accurate or credible sites for information because the authors of the information can remain anonymous.

Government Agencies

Canadian and U.S. Government agencies at all levels are the best sources of raw data on fire and emergency services topics. Laws, ordinances, and statutes that affect the fire service and emergency services exist at national, state/provincial, and local levels. In addition, the U.S. Government maintains data of fire and emergency responses through the National Fire Incident Reporting System (NFIRS) and on incidents involving firefighter fatalities through the National Institute for Occupational Safety and Health (NIOSH). Agencies responsible for transportation, manufacturing, and natural resources all maintain information on incidents that occur on highways, airways, railroads, and ship channels and in factories and national forests and grasslands. The Centers for Disease Control and Prevention (CDC) and the World Health Organization (WHO) can provide information on health and wellness topics. The U.S. federal government also funds research into fire and life safety hazards.

The U.S. Department of Labor as well as state departments of labor can provide data on worker's compensation claims. This data can provide a view of the length of rehabilitation time required for injuries, the cost, and the types of injuries that are more prevalent. Insurance companies can also provide injury compensation data based on an experience modifier. The experience-modification factor compares an organization's actual loss experience to an expected loss experience over the past three years. If the organization's losses are lower than expected, the experience modifier will be less than 1.00, which reduces the organization's insurance premium. If the losses are higher than expected for the fire service, the experience modifier will be greater than 1.00. This results in higher insurance premiums for the organization.

As is the case with all information collected by the government, it is subject to manipulation, biased analysis, being out-of-date and incorrect collection, and recording processes. In some cases, data that is supposed to be collected never gets to the agency responsible for it, which makes it difficult to accurately forecast or analyze a trend.

Libraries

Libraries are located at all levels of government, within educational institutions, and in some third-party organizations. Many are accessible through the Internet or through other libraries. A few are specific to the fire service for research such as the U.S. National Fire Academy Learning Resource Center or the Fire Protection Publications Library at Oklahoma State University **(Figure 11.3)**. Library reference sections can be a valuable source for information. In addition, library reference staff can assist in the research process.

Figure 11.3 Libraries like the one located at Fire Protection Publications provide excellent sources for resources when developing programs.

Because libraries are networked and have sharing agreements, materials that are not available locally can be acquired from other libraries (interlibrary loans). This service is available at most libraries. The material can be checked out for three to four weeks rather than the usual two weeks for local materials. Many libraries also have computers that are dedicated to Internet research. Fire and emergency services organizations that lack this capability may wish to locate a library that has it.

Educational Institutions

Vocational schools, community colleges, and universities are another source of information. The libraries within these institutions are usually linked to public libraries through interlibrary loan agreements. In addition, many of these institutions have research projects underway or individual professors doing research that may be of benefit when researching topics ranging from fire science to social/political issues. Educational institutions are also a source for assistance in applying for and obtaining funding grants.

Professional Organizations

Fire and emergency services professional organizations are also sources for safety-related data, information, and model programs. Among these organizations are the following:

- International Association of Fire Chiefs (IAFC)
- Canadian Association of Fire Chiefs (CAFC)
- International Association of Fire Fighters (IAFF)

- Canadian Volunteer Fire Fighters Association (CVFFA)
- National Volunteer Fire Council (NVFC)
- Fire Department Safety Officers Association (FDSOA)
- American Society of Safety Engineers (ASSE)
- Canadian Society of Safety Engineers (CSSE)
- National Fire Protection Association® (NFPA®)

State/provincial level organizations associated with these groups may also be able to assist in the gathering of data. Membership may be required to take advantage of the services offered by these organizations.

Information is also available from the Volunteer Firemen's Insurance Services® (VFIS®), an organization that provides educational material through its subsidiary, the Emergency Services Education and Consulting Group (ES-ECG). The International Fire Service Training Association (IFSTA) is another source for information through its research library and technical editors. The Congressional Fire Services Institute (CFSI) and the Home Fire Sprinkler Coalition are both sources of information.

Another source that may not be directly associated with the fire service but is still an excellent source of information is the International City/County Management Association (ICMA). It has a variety of publications and can provide information on local government, public management, human resources, and trends in laws that affect employment practices.

Testing and Standards Organizations

Third-party testing and certification organizations provide statistics on the quality of equipment and products and the ability of that equipment or product to perform a specific task. Most fire and emergency services personnel are familiar with the testing that is performed by Underwriters Laboratories Inc. (UL), Underwriters' Laboratories of Canada (ULC), Factory Mutual, and other similar organizations. Test results, specifications, and performance criteria are available from these organizations and may assist in writing specifications or in establishing testing criteria for internal equipment evaluations. NFPA® maintains data on fire-related loss, injuries, deaths, fire causes, and other fire- and safety-related topics. In addition, they create consensus standards.

Nonprofit Organizations

Nonprofit organizations that have safety-related missions are also excellent sources of information for chief officers. Most have the mission of reducing injuries and death due to accidents by educating the population in proper safety activities. Some of those groups include the Home Safety Council (HSC), the Association of State and Territorial Health Officials (ASTHO), Safe Kids, and other public safety councils at federal, state/provincial, and local levels.

Safety and Health Program

The organization's HSO is usually assigned the task of developing and implementing a safety and health program that meets the requirements, demands, needs, and concerns of the organization. Developing a safety and health program is made easier if the HSO uses current safety and health laws, codes, and standards as foundations for program development. Although these safety standards have requirements for employers, the regulations also contain significant requirements for employees. Each employee must comply with the requirements of the occupational safety and health standards.

One of the principal tasks of the HSO is to manage the organization's safety and health program. The HSO is the catalyst that initiates proven safety programs that reduce injuries and fatalities within a fire and emergency medical organization. Supervisors and members have a responsibility to support and abide by these programs in the interests of improving safety in the work environment. Essential to the success of the program is documented and public support from the top level of management.

The following are important reasons for implementing an occupational safety and health program:

- Fulfills an ethical obligation to prevent injuries, illnesses, and fatalities and reduce property loss

- Provides sensible economic investment by reducing the frequency and severity of injuries and fatalities involving vehicle crashes, personal accidents, and organizational property damage, which (in turn) reduces fire and emergency service organizations' costs and expenditures to worker's compensation and liability insurance

- Ensures compliance with applicable national, state/provincial, and local laws, codes, and standards

The safety and health of an organization's members are paramount issues for the successful operation of the organization. Many components of the occupational safety and health program ensure a successful process. This process is a comprehensive team effort that requires management support and member participation.

The fire and emergency services organization's HSO may also be called upon to administer the appropriate sections of the local jurisdiction's safety and health plan as it applies to the organization's membership. This plan may include sections on worker's compensation, workplace violence, smoking, and drug and alcohol abuse. If the local jurisdiction does not have a policy, then these issues need to be included in the fire and emergency services organization's comprehensive safety and health program.

One of the instrumental and valuable components of a fire and emergency services organization's safety and health program is the development, implementation, and operation of an occupational safety and health committee. The organization's HSO is the chair of this committee and facilitates its work. This individual is the ultimate manager of the safety and health program and reports to the chief/manager of the organization. The roles and responsibilities of this position are also discussed in this section.

Occupational Safety and Health Committee

The occupational safety and health committee is a centralized group within the organization that functions as a clearinghouse for activities, problems, and issues relating to employee safety and health. The issues that confront this committee are unlimited. Ensuring employee safety and health is a constantly changing process based upon the needs of the organization and the commitment to safety by the administration. Other factors that can affect the successful operation of this committee are the levels of activity of the committee and examining and managing current safety and health issues within the organization.

The goals of the occupational safety and health committee are to develop and recommend solutions to resolve conflicts. However, one issue that should **not** be an objective of this committee is disciplinary action. The activities and issues that are addressed must be within the scope of the committee. As part of their duties, occupational safety and health committee members may be required to perform the following tasks:

- Identify situations that may be a source of danger to members.
- Investigate complaints of violations of the organization's safety policy.
- Make recommendations to the chief or manager of the organization on matters reported to the committee and on rules and regulations promulgated by outside regulatory agencies that relate to safety issues.
- Evaluate safety rules and regulations established by the organization to ensure compliance with state/territorial/provincial mandates.
- Review the annual injury and illness report for trends.
- Review the safety and health inspection reports to assist in correcting identified unsafe conditions or practices.
- Evaluate the incident investigations conducted since the last meeting to determine if the cause of an unsafe act or unsafe condition was properly identified and corrected.
- Evaluate the accident- and injury-prevention program and make recommendations for improvement where indicated.
- Make recommendations to the administration and members for the improvement of members' safety and health.
- Recommend, maintain, and monitor safety and health programs and procedures.
- Consider forwarded reports from outside agencies addressing safety and health in order to make recommendations to management regarding these issues.
- Take and post meeting minutes that include attendance and the topics of discussion at occupational safety and health committee meetings.

NFPA® 1500 defines the criteria for establishing and using an occupational safety and health committee. The requirements are as follows:

- The fire and emergency services organization needs to establish an occupational safety and health committee that will serve in an advisory capacity to the chief or manager of the organization.
- The intent of the occupational safety and health committee is to conduct research, develop recommendations, and study and review matters pertaining to occupational safety and health within the fire and emergency services organization.
- The occupational safety and health committee must hold regularly scheduled meetings and special meetings as deemed necessary. Regular meetings must

be held at least once every 6 months. Written minutes of each meeting must be maintained and distributed to all members of the organization. Minutes of each committee meeting will be prepared and filed for a period of at least 1 year and made available for review by members of the organization.

The record-retention policies of the AHJ may vary from this criteria and should comply with all applicable statutory requirements.

Regardless of the size and type of organization, every fire and emergency services organization should establish an occupational safety and health committee. The total number of members will vary depending on size and type of organization. A career organization uses members from each shift, each district/battalion, each division, or other combination depending on the organization's size. Volunteer and combination organizations should recruit a representative cross section of the organization that will provide sufficient diversity of views, opinions, and skills. An effective occupational safety and health committee usually consists of three members from management, three labor representatives, and the HSO as a nonvoting member. Additional members may be appointed, including the fire and emergency services organization's physician. You should remember that committees that are too large become unwieldy and ineffective.

As with any organization, it is important to manage the committee's time well. With training mandates as well as other time requirements, the key to participation is efficiency and effectiveness. An organization must emphasize the importance of good time management to make the committee process work.

A checklist is helpful in assisting you to establish of a safety and health committee. A sample checklist is located in **Appendix G**.

Health and Safety Officer Roles and Responsibilities

The HSO, under the direction of the chief of the organization or other chief officer, must be able to apply analytical skills to the development and evaluation of a comprehensive safety and health program. These skills also apply to the program elements that include the risk management plan; accident, injury, and illness prevention program; medical exposure program; and employee physical fitness and wellness program). NFPA® 1521 is very specific in defining the roles and responsibilities of the HSO in the safety and health program. As a minimum standard, roles include (but are not limited to) the following tasks:

- *Ensure safety training and education* — Relate training to all types of operations and functions in which the organization is likely to be involved (including proper safety supervision of all training activities and live-burn exercises). Create educational materials for all members from information collected from accidents, injuries, occupational deaths, job-related illnesses, exposures, and postincident analyses.

- *Manage the accident or loss prevention program* — Periodically review operations, procedures, equipment, and facilities and recommend any changes in work practices and procedures to the authority having jurisdiction (AHJ). Instruct all personnel in safe work practices (in both emergency and nonemergency operations), including effective driver/operator training and testing policies that cover the safe operation of all types of vehicles within the organization's fleet. Program development, review, and supervision may be delegated.

- **Investigate accidents or incidents** — Investigate incidents that result in hazardous conditions, injuries, illnesses, exposures, and fatalities involving members of the organization. Include incidents involving apparatus, vehicles, facilities, and equipment owned and operated by the organization **(Figure 11.4)**. Review the procedures in use at the time of the incident and develop corrective procedures as necessary.

Figure 11.4 One of the many responsibilities assigned to the HSO is accident investigation and analysis.

  — *Immediate responsibility:* Ensure that transportation and medical treatment are provided for any injured personnel. Develop and implement procedures that ensure this care is provided at the most appropriate health-care facility (may include establishing appropriate contracts with health-care facilities and private ambulance services not provided by the jurisdiction).

  — *Agency cooperation:* Work closely with law enforcement agencies to ensure complete and accurate reporting in the event of litigation or liability. Include this information as part of the postincident analysis.

  — *Incident review:* Include recommendations to the chief/manager of the organization or AHJ for corrective action to prevent future injuries, fatalities, or property-loss incidents.

  — *Investigating and reporting:* Develop and review procedures periodically; ensure compliance with federal, state/territorial/provincial, and local requirements.

- **Maintain records management and data analysis** — Maintain good record management and data analysis on all accidents, occupational deaths, injuries, illnesses, and hazardous materials exposures to provide justification for corrective actions. Collect and analyze this information and recommend corrective action to the chief/manager of the organization.

  — Create safety and health standard operating procedures or guidelines (SOPs/SOGs).

  — Maintain records pertaining to periodic facility, apparatus, and equipment inspections.

  — Perform safety testing of the organization's apparatus and equipment.

  — Inspect and test in-service personal protective equipment (PPE).

  — Inspect the organization's facilities.

  — Keep records of recommended safety corrections and implementation of safety and health procedures and accident-prevention policies.

— Develop an annual safety and health report (includes records relating to accidents, occupational injuries, illnesses, deaths, and hazardous materials exposures) that is given to the chief/manager of the organization and communicated to all personnel.

— Compare the annual report with the organization's records and national fire and industry safety and health data to determine development of unfavorable trends; however, national data is not always complete and may not be dependable.

— Develop and maintain a data-collection system to comply with both NFPA® 1500 and U.S. Occupational Safety and Health Administration (OSHA) regulations where applicable. Canadian regulations have similar requirements. Collect permanent records of all job-related accidents, injuries, illnesses, exposures to infectious agents and communicable diseases, or fatalities.

● *Review equipment specifications and assist in acceptance testing* — Review all specifications for new apparatus, equipment, and personal protective clothing and equipment (in conjunction with the chief officer responsible for the design and purchase of fire apparatus and equipment). Ensure specifications comply with all OSHA, NFPA®, and U.S. Department of Transportation (DOT) standards and regulations. Canadian regulations have similar requirements. Assist in the acceptance testing of new equipment and apparatus upon receipt and make recommendations as necessary to ensure compliance with applicable codes, standards, and regulations.

— *Periodic service tests:* Assist with the organization's equipment tests, including (but not limited to) hose tests, ladder tests, safety and rescue equipment tests, apparatus pump and aerial device tests, and power extrication equipment tests **(Figure 11.5)**.

— *Protective clothing program:* Develop, implement, and maintain program; write specifications, perform acceptance tests, perform in-service inspections, implement a cleaning and repair program, and determine clothing and equipment replacement criteria.

● *Ensure compliance* — Inspect all fire and emergency services organization's facilities for building, safety, and health code violations as applicable, and report all violations, make recommendations for corrections, and ensure compliance with the recommendations. Compliance is based on NFPA®

Figure 11.5 The HSO assists in the periodic testing of the organization's hoses, ground ladders, and rescue equipment. Testing may be performed by a third-party organization.

1500; state/territorial/provincial and local building, fire, and health codes; OSHA regulations; National Electrical Code®; and life safety codes as applicable.

- *Comply with health maintenance requirements* — Ensure that the organization complies with the requirements for heath maintenance found in NFPA® 1500, Chapter 10, which covers medical requirements for hiring firefighters/emergency responders, physical performance and fitness requirements, infection control, individual medical data, and postinjury or illness rehabilitation. NFPA® 1500 also contains the requirement for a fire and emergency services organization physician. The health maintenance program must also include the components of medical surveillance, physical fitness, wellness, nutrition, and rehabilitation.

- *Serve as internal and external liaisons* — Serve as an internal liaison between the organization's administrative personnel and its members; serve as an external liaison between the organization and equipment manufacturers and outside agencies involved in safety issues. Also serve as liaison between the organization and its physician to ensure that the medical needs of the organization are met and that administrative personnel are aware of any medical trends that may be developing.

  — *Internal liaison:* Serve as a member of the occupational safety and health committee and report committee recommendations to the chief/manager of the organization (includes working with chief officers in charge of training, equipment purchase and maintenance, and policy development to ensure that unsafe practices and designs are eliminated). Also provide assistance to emergency personnel in identifying potential hazards within their response areas. This assistance may involve working with fire prevention and inspection personnel, building inspectors, plans review officers, and the hazardous materials officer and team.

  — *External liaison:* Meet and work with equipment manufacturers and vendors to ensure that equipment acquired meets all standards and specifications. Develop an alliance with members of standards committees (such as NFPA®) and regulatory agencies along with consultants and safety specialists in both the public and private sectors. Establish this alliance by attending fire and safety conferences and manufacturer-sponsored meetings and serving on standards committees and participating in and networking with local health and safety councils and risk management associations.

- *Act as infection control officer* — Ensure that the requirements of NFPA® 1581, *Standard on Fire Department Infection Control Program*, and OSHA Title 29 *CFR* 1910.1030, *Occupational Exposure to Bloodborne Pathogens*, are met. Become familiar with the reporting procedures for the Centers for Disease Control and Prevention (CDC). Maintain a close liaison with this officer if someone else holds the position.

- *Develop a critical incident stress management plan* — Ensure that this plan is developed in the form of a written policy and implemented (in accordance with Chapter 12 of NFPA® 1500). The plan is part of the organization's wellness initiative along with the member assistance program and is available to both members and their families.

- **Conduct postincident analyses** — Ensure that safety and health issues are included in these analyses, which give opportunities to identify, analyze, and correct any problems or deficiencies discovered during an incident. Include information gathered by the incident safety officer (including both the incident action plan and the incident safety officer's safety plan). Include additional information pertaining to the use of protective clothing, personnel accountability system, rehabilitation (rehab) operations, hazardous conditions, and any other issues relating to the safety of personnel at an incident in the final written analysis.

- **Address workplace violence** — Consider addressing workplace violence as a general responsibility, even though it is not directly addressed in NFPA® 1521. Perform the following tasks:

  — Include this type of situation in the safety and health program.

  — Monitor conditions and morale of all personnel.

  — Train supervisors in the various levels and symptoms of stress and appropriate intervention techniques.

  — Recommend appropriate member assistance programs to help mitigate any potential problems.

  — Consider providing confrontation management training to assist personnel in identifying and managing potentially violent situations.

- **Leadership advocacy** — The HSO must be a vocal advocate for the role of leadership in safety. No matter what procedures are written or equipment purchased, they are of minimal value without the leadership of the officers who ensure that the procedures are followed and the equipment is used. This is critical to the success of any program.

Other Safety and Wellness Issues

It is important to remember that safety and wellness involve a multitude of concepts and issues. Some may not seem like they belong in a safety and wellness program. The following are issues that are related directly or indirectly to safety and wellness.

Staffing

Constant staffing levels have a direct effect on accidents, injuries, and fatalities at emergency incidents. When a unit or group of units is forced to operate at minimum staffing levels, greater physical stress is placed upon the personnel. Units also do not operate at their maximum efficiency, causing the completion of the operation to take longer.

Drug-Free Workplace

The use or abuse of drugs and alcohol either before or during a work shift impairs judgment, increases the likelihood of mistakes, and slows reaction times. The use of some drugs can lead to cardiac arrest. Administrative policies of career and volunteer departments prohibit personnel from reporting to duty while under the influence of drugs or alcohol. It is essential that supervisors and members of the organization monitor personal behavior of their peers and prevent them from attempting to work while they are impaired. The organization must provide medical and psychological assistance to any member who uses or abuses drugs or alcohol.

Employee Assistance Program (EAP)

One form of assistance provided by organizations is the employee assistance program (EAP). This program provides counseling on topics such as smoking cessation, domestic violence, stress relief, financial planning, and other stress inducing concerns. It is mandated by NFPA® 1500.

Safety Stand Down

The national Safety Stand Down is sponsored by the IAFC and the IAFF as a means of reducing accidents and focusing attention and training on safety in the fire service. The concept has been in use by the military and private sectors for a number of years. The fire service began using it in 2005. All fire departments are encouraged to implement it during June, teaching safety-related topics during each work shift.

Taking Care of Our Own

Originally conceived by the National Volunteer Fire Council (NVFC) and the National Fallen Firefighters Foundation, the concept of taking care of our own involves training chief officers in methods for dealing with family and friends of personnel killed in the line of duty. The training includes five modules beginning with preparing for the possibility of a line-of-duty death (LODD) and continuing through providing support for a victim's family **(Figure 11.6)**.

Figure 11.6 Funerals for firefighters killed in the line of duty occur all too often. Fire departments should have a program in place to support the family and coworkers of the fallen.

Good Ol' Boy vs. Fit and Safe Workforce

Part of the ingrained culture of the fire service has been that we take care of our own. This concept has resulted in our making excuses for or covering up the unsafe behavior of our subordinates or peers. In the end, it has often resulted in unnecessary accidents, injuries, and fatalities. By covering up drug or alcohol abuse, we have allowed personnel to operate apparatus or enter hazardous conditions. This Good Ol' Boy approach to safety and wellness must be changed.

The alternative is a fit and safe workforce. This is achieved by implementing and adhering to the safety and wellness program and by helping subordinates and peers to do the same. The application of a crew resource management or similar approach can be one method for reaching the goal of a fit and safe workforce.

Public Safety Officer's Benefits Program (PSOB)

Enacted in 1976, the Public Safety Officer's Benefits (PSOB) Program provides a one-time financial payment for survivors of fire and emergency services personnel who are killed in the line of duty or die as a result of injuries sustained in the line of duty. Personnel who have been disabled by a catastrophic injury sustained in the line of duty are also eligible for disability benefits. The program is administered by the U.S. Department of Justice.

Risk Management Plan

Regardless of the size of the community or service area, every fire and emergency services organization should develop, implement, and use a risk management plan to guide its operations. Implementation of this plan is a dynamic and aggressive process that the chief officer must monitor and revise as needed but at least annually. A sample plan can be found in Annex D, Figure D.2, of the current edition of NFPA® 1500.

In everyday conversation, the terms *hazard* and *risk* are often used interchangeably; however, technically, they describe two different things. The term *hazard* usually refers to the source of a risk. A *risk*, on the other hand, is the likelihood of suffering harm from a hazard. Risk can also be thought of as the potential for failure or loss. In other words, *risk* is the exposure to a hazard. A *hazard* is a condition, substance, or device that can directly cause an injury, fatality, or property loss.

These terms can be used when discussing safety and health issues of the members of the fire and emergency services organization. The hazards are those that the individual must face while carrying out the duties of firefighter or emergency responder. Those hazards include those found at the emergency scene and that exist in the organization's facilities, apparatus, and nonemergency operations. The risks that must be minimized include the ones that are caused by the high physiological and psychological nature of the profession.

The chief of the department is responsible for the development of a risk management plan. That officer may have the option of delegating the responsibility to the organization's HSO. Regardless of whether it is the chief officer or an assigned officer, it is important that all members of the organization be familiar with the plan. Personnel who are assigned the responsibility for the risk management plan must be capable of performing multiple skills such as training, investigating, evaluating, analyzing, implementing, and communicating. Each of these skills can be applied to the tasks required of the HSO. In each task, however, the HSO is the risk manager, applying the elements of risk management to each of the various tasks.

To understand how this concept works, it is necessary to understand the risk management model incorporated in NFPA® 1500. The risk management model adopted by NFPA® has been successfully used by general industry for decades. The safety and health components of risk management were incorporated into NFPA® 1500 during the 1992 revision process. In the 2007 edition of the standard, they are outlined in Chapter 4 and applied to emergency operations in Chapter 8. The role of the HSO in the risk management process is defined in Chapter 5 of NFPA® 1521.

The risk management plan described in NFPA® 1500 is a process that incorporates several components that are applied to the operations of a fire and emergency services organization. This plan is not a stagnant document that is developed, described in a printed document, placed on a shelf and used only occasionally. Essentially, a risk management plan serves as documentation that risks have been identified, evaluated, and that a reasonable control plan has been implemented and followed. An effective risk management plan has a positive effect on the department from the operational, safety, financial, and liability standpoints.

This plan must be reviewed and revised annually if necessary. Responsibility for the review and revision may be assigned to the HSO or the occupational safety and health committee by the fire chief. The plan may be a part of the overall community/jurisdiction plan and may be the responsibility of that organization's risk manager such as the community health and safety manager or loss control manager.

In Chapter 4 of NFPA® 1500, the requirements of the risk management plan are simply stated **(Figure 11.7, p. 404)**. The fire department shall adopt an official written risk management plan that covers administration, facilities, training, vehicle operations, protective clothing and equipment, operations at emergency incidents, operations at nonemergency incidents, and other related activities. At a minimum, the plan shall include risk identification, risk prioritization, risk evaluation, risk control techniques, and risk monitoring.

An understanding of the concepts of risk management and system safety is essential to all fire officers and the personnel who fill the position of HSO for the organization. These concepts are the basis for the majority of the roles and responsibilities for those officers who plan, develop, and manage the safety and health program.

Personnel Risk Analysis

The use of risk analysis is applied to the safety and health of the internal customers of the organizations: employees and members. Because many injuries and some fatalities occur during training evolutions or physical fitness sessions, it is important to apply the risk analysis to all nonemergency activities of the organization in order to prevent or reduce the number of occurrences. The risk analysis includes identification, evaluation, prioritization, and control techniques.

Risk Identification

To identify the risks, the HSO compiles a list of all emergency and nonemergency operations and duties in which the organization participates. Ideally, the HSO should take into consideration the worst possible conditions or potential events, including major disasters and multiple events. There are many sources to assist with this identification process. The first (and possibly the most effective) is the department's loss prevention data, which consists of annual fire-loss reports by occupancy type, loss value, frequency, etc. Although most departments are too small to rely on their own database for a statistically valid trend, national averages and trends are available from NFPA® and the National Fire Academy (NFA). National data is not always complete or accurate due to collection inconsistencies, and a time lag of 1 to 2 years is required to collect, analyze, and publish it.

The HSO should seek input and ideas from department personnel, trade journals, professional associations, and other service providers to identify the potential risks. When using information provided by other fire departments or organizations, the HSO should consider local circumstances that might

II. EXECUTIVE SUMMARY

The LNU Fire Management Plan's purpose is to identify the high value, high-risk areas within the six counties, and to provide the planning basis for reducing the damaging effects of wildfire. This is accomplished through a comprehensive approach designed to minimize the costs and losses due to wildfire by a variety of means, including response and evacuation planning, cooperative fuel reduction projects, fire prevention, and education.

The Plan utilizes stakeholders' input and the best available Geographic Information System (GIS) data along with other data, to analyze fire hazards and assets at risk. These various analyses are then combined into a fire plan assessment. One purpose of this analysis is to identify the high value and high risks where the potential exists for costly and damaging wildfires. These areas can then be prioritized for the development of mitigating treatments designed to reduce future costs and losses. The four basic components of this fire plan assessments are:

- Assets at Risk
- Vegetation Fuel Hazards
- Fire History and Frequency of Severe Fire Weather
- Ignition Workload Assessment and Management Prioritization

The overall goal of these assessments is to reduce the total costs and losses from wildland fire by protecting assets at risk through focused pre-fire management prescriptions and increasing initial attack fire suppression successes. To accomplish this goal, the framework laid out in the 1996 California Fire Plan guides is utilized. This framework forms the basis of an ongoing fire planning process to monitor and assess the Unit's wildland fire environment. It consists of five strategic components:

1. **Wildland Protection Zones**. A key product of this fire plan is the development of enhanced wildlife safety and "Firewise" communities that reduce citizen and firefighter risks from future large wildfires.

2. **Initial Attack Success**. The fire plan attempts to assess the initial attack fire suppression successes and the Department's ability to provide an equal level of protection to lands of similar vegetation type. This measurement is the percentage of fires that are successfully controlled before unacceptable costs are incurred. Knowledge of the level of service will help define the risk to wildfire damage faced by Public and private assets in the wildlands.

3. **Assets Protected**. The fire plan establishes a methodology for defining assets protected and their degree of risk from wildfire. The assets addressed in the plan are citizen and firefighter safety, watersheds and water, timber, wildlife and habitat (including rare and endangered species), unique areas (scenic, cultural, and historic), recreation, range, structures, and air quality. Stakeholders for each of the assets at risk are identified. The assessment will enable the Unit and other fire service managers to set priorities for prefire management project work.

4. **Pre-fire Management**. The plan facilitates development of a wide range of management prescriptions, utilizing every program and tool available to the Department, for protecting assets at risk. These tools include every conceivable combination of fuels reduction, ignition management, fire-safe engineering activities, code development and enforcement, public education, and forest health enhancements to protect Public and private assets.

5. **Fiscal Framework**. The State Board of Forestry and CDF are developing a fiscal framework for assessing and monitoring annual and long-term changes in California's wildland fire protection systems. State, local, and Federal wildland fire protection agencies, along with the private sector, have evolved into an interdependent system of pre-fire management and suppression forces. As a result, any changes related to budgeted levels of service of any of the entities directly affects the others and the overall services delivered to the Public. Monitoring system changes through this fiscal framework will allow the Board and CDF to address public policy issues that maximize the efficiency of local, state, and federal firefighting agencies.

The ongoing implementation of the Unit's Fire Management Plan is expected to enhance the wildland fire protection system in the following ways:

- Identify for local, state, and federal officials and for the Public those areas of concentrated assets and high risk.
- Allow CDF to create a more efficient fire protection system focused on meaningful solutions for identified problem areas.
- Give citizens an opportunity to identify public and private assets to design and carry out projects to protect those as sets.
- Identify, before fires start, where cost effective pre-fire management investments can be made to reduce tax payer costs and citizen losses from wildfire.
- Encourage an integrated intergovernmental approach to reducing costs and losses.
- Enable policy makers and the public to focus on what can be done to reduce future costs and losses from wild fires.
- Integrate elements of the Fire Management Plan into the land use and safety elements of the general plans of each of the Unit's six counties.

Figure 11.7 Sample Executive Summary.

present a different set of emergency and nonemergency operations and duties. Other risk identification sources include risk management plans developed by local industry and hazardous substance sites, vulnerability analyses, and the U.S. Environmental Protection Agency (EPA) plans among others.

Risk Evaluation

Once the HSO identifies the risks, they can be evaluated from both frequency and severity standpoints **(Figure 11.8)**. Frequency, referred to by OSHA as incidence rate, addresses the likelihood of occurrence. Typically, if a particular type of incident, such as injuries related to lifting, has occurred repeatedly, it will continue to occur until a job hazard or task analysis has been performed to identify the root cause and effective control measures have been implemented. In this example, the HSO or occupational safety and health committee must develop and implement guidelines that outline proper lifting techniques and physical fitness requirements or provide mechanical aids for lifting.

Figure 11.8 Apparatus accidents may be very severe yet occur infrequently.

Severity addresses the degree of seriousness of the incident and can be measured in a variety of ways such as lost time away from work, cost of damage, cost of and time for repair or replacement of equipment, disruption of service, or legal costs. Incidents of high frequency and high severity must have the highest priority in the risk analysis, while those of low frequency and low severity receive the lowest priority. The method for calculating the risk may vary from one department to another.

Risk Prioritization

Taken in combination, the results of the frequency and severity assessments help to establish priorities for determining action. Any risk that has both a high probability of occurrence and serious consequences deserves immediate action and is considered a high-priority item. Nonserious incidents with a low likelihood of occurrence are lower priorities and can be placed near the bottom of the action-required list.

Risk Control Techniques

Once the HSO prioritizes the risks, it is now time to apply risk control measures. Once control measures have been implemented, they need to be evaluated to measure their effectiveness. Several approaches can be taken in risk control, including the following:

- *Risk avoidance* — The best risk control choice is risk avoidance. Simply put, avoid the activity that creates the risk. In a fire and emergency services organization, this approach frequently is impractical. Examples:

  — Lifting a stretcher presents a serious back-injury risk, but personnel cannot avoid this risk and still provide effective service. Training in the use of safe-lifting techniques and/or safer equipment would be solutions that are more acceptable.

— Risk avoidance could include a policy prohibiting smoking by departmental/organizational candidates when they are hired, thereby reducing the potential for lung cancer and other smoking-related illnesses among members.

- **Risk transfer** — Can be accomplished in one of two primary ways: physically transferring the risk to someone else or through the purchase of insurance. Transfer of risk may be difficult if not impossible for a fire or emergency medical services (EMS) organization. Examples:

— Contracting the cleanup and disposal of hazardous waste would transfer risks to a private contractor who accepts the liability of the risks associated with those activities.

— The purchase of insurance transfers financial risk only and does nothing to affect the likelihood of occurrence. Buying fire insurance for the station — while highly recommended to protect the assets of the department — does nothing to prevent the station from burning. Insurance is no substitute for effective control measures such as installing an automatic sprinkler system.

- **Control measures** — Effective control measures (risk reduction) are the most common method used for the management of risk. While control measures will not eliminate the risk, they can reduce the likelihood of occurrence or mitigate the severity. Effective control measures include safety, health, and wellness programs; ongoing training and education programs; and well-defined standard operating procedures or guidelines (SOPs/SOGs). Examples:

— Changes in station apparatus bay design and apparatus-backing procedures have been very practical. The risks associated with backing apparatus into station bays are well documented. The simplest solutions are improved driver/operator training, painted guidelines on the apparatus bay floors, and a policy that requires a second person to guide the backing operation from the rear of the vehicle. A more expensive solution would be the replacement of older single-door stations with new drive-through stations.

— Typical control measures instituted to control incident scene injuries include use of accountability systems, use of full-protective clothing, mandatory respiratory protection plans, training and education sessions, and health and wellness SOP/SOGs.

Plan Implementation Procedures

Once the analysis process determines appropriate control measures, the plan is implemented. The procedures for implementing the plan are the same as those mentioned in previous chapters that involved risk analysis and program development. The HSO or other assigned officer is responsible for implementing, monitoring, evaluating, reporting, and revising the plan. Administrative support is essential in this process for the plan to meet its goals.

Implementation

Implementation of the risk management plan requires communication, training, and application. The plan, produced in written form as part of the organization's SOPs/SOGs, is distributed to the membership of the organization.

The distribution includes the public acknowledgement by the administration and the leadership of the jurisdiction of the importance of the plan. Administration support of the risk management plan is essential to membership acceptance and support. Communication of this support helps to ensure a positive response from not only the internal customers but also the external customers and stakeholders.

Thorough training sessions on the plan involve communicating the plan and its importance, use, and intended results. Training is organization wide and begins in the entry-level training for new employees. Training sessions for changes in the plan as well as refresher training sessions are provided periodically to current members of the organization. The effectiveness of the program depends on proper training.

The application of the plan takes place daily as officers and members follow the prescribed policies and procedures. The risk control techniques must become second nature to all personnel, including you. Whether these techniques involve applying proper lifting techniques when picking up heavy objects or putting on respiratory protection when entering a contaminated atmosphere, they must be performed naturally without questions (**Figure 11.9**).

Figure 11.9 Back injuries can be reduced if proper lifting techniques are taught, practiced, and applied.

Monitorship

The effectiveness of the plan becomes evident through monitoring. This step ensures that the system is dynamic and facilitates periodic reviews of the entire program. Any problems that occur in the process have to be revised or modified. The intent of the risk management plan is to develop a strategy for reducing the inherent risks associated with fire and emergency service operations. Regardless of the size or type of organization, every emergency services organization should operate within the parameters of a risk management plan. Operation of this plan is a dynamic and aggressive process that the HSO must monitor and revise as needed but at least annually.

Evaluation

Applying the evaluation techniques discussed earlier, the chief or HSO compares the desired results with the actual results of the plan. Data for making the comparisons include injury and fatality reports, amount of participation in safety training, fitness testing results, preemployment physical fitness reports, and alterations that have been made to address preplan risks. The sources for the data include the target risks or hazards, policies and procedures intended to eliminate the risks, emergency incident reports, daily attendance reports, medical leave requests, training records, and physical fitness reports.

The results of the comparisons, which include an increase, decrease, or no change in the risks, determine the effectiveness of the risk management plan. If there is a decrease in the risk, indicated by a reduction in medical leave taken or lost-time injuries, for example, then the plan may be considered as being effective. If, however, the evaluation indicates that there is no change or an increase in the number of injuries, then the implemented risk control techniques must be reviewed and alternate solutions applied.

It will also be important to determine if the cost/benefit is appropriate: Is the cost of the risk control techniques less than the cost of the results of the risk? For example, if the cost of altering apparatus storage compartments is greater than the cost of lost-time injuries due to back strains resulting from removing equipment from the compartments, then it may not be worth the control cost.

Revision

Revision procedures are included in the risk management plan when it is developed. The revision process involves following the same steps that the initial risk management plan followed, although it only focuses on those risks that require revision and not the entire plan. Fire officers should be aware of the reasons why revisions may be necessary and be able to recognize these reasons when they appear in the plan evaluation.

Some reasons may include the following:

- Increase in injuries, fatalities, or loss due to the target risks
- Increase in medical leave requests
- Increase in risk-related costs
- No apparent change in the risk results
- Ineffective cost/benefit
- Changes in the target risks
- Ineffective training

Sometimes at the very beginning of a mandatory physical fitness training program, there may be a temporary increase in on-the-job injuries. This may be due to improper warm-up techniques or attempts to make the program fail.

Accident, Injury, and Illness Prevention Program

One of the elements of the comprehensive safety and health program is the accident, injury, and illness prevention program. Its purpose is to ensure the protection of members of the fire and emergency services organization by providing information and training regarding the hazards that exist in their work environments. These environments include all facilities, apparatus, and emergency incident sites and training exercises. Hazards include those chemicals normally used and stored by the organization and conditions that present a risk of physical injury, damage to the respiratory system, hearing loss, or other occupational illness.

Although the chief of the fire and emergency services organization is ultimately responsible for developing this program, the responsibility is usually delegated to the HSO and occupational safety and health committee. The HSO and committee members identify the hazards, develop the program, assist in its implementation, monitor its progress, evaluate its effectiveness, and recommend revisions as indicated. The result of their work is a written accident prevention program tailored to the particular operation and hazards that are present in the workplace.

Identify Hazards/Corrective Measures

The first step in developing an accident, injury, and illness prevention program is to determine the hazards that are present in the workplace. While the hazards associated with fire suppression may be all too familiar to members of the fire service, those that exist within facilities may not. The hazards can be divided into the following categories:

- Physical injury
- Occupational illness
- Motor vehicle related accidents
- Hearing loss
- Respiratory injury or illness
- Hazardous materials exposure

These hazards may occur at emergency incident scenes, during training, or while working in the organization's facilities. Some hazards may be present in one form or another in each of these situations. When the hazards have been identified and prioritized, then the appropriate corrective measure (whether policy, procedure, or equipment required to protect personnel) is identified.

Just as the fire and emergency services organization is responsible for providing a safe work environment, members are responsible for following safety procedures, adhering to all safety policies, and using safety equipment and clothing. Information is provided during the initial entry-level training course and in annual refresher courses to current members. The organization helps to ensure that members are aware of their responsibilities by providing training in the following areas:

- How and when to report injuries, including instruction as to the location of first-aid kits, oxygen inhalators, defibrillators, and decontamination facilities **(Figure 11.10)**
- How to report unsafe conditions and practices
- Use, care, selection and maintenance of required personal protective equipment (PPE), including respiratory, hearing, and eye protection
- Proper actions to take in event of emergencies, including the routes of exiting from areas during emergencies
- Description of the organization's comprehensive safety and health program along with the accident, injury, and illness prevention element
- On-the-job review of the practices necessary to perform initial job assignments in a safe manner

Injuries that may occur to fire and emergency service personnel include muscle strains and sprains, traumas, burns, electrocution, and abrasions. Causes include improper lifting techniques, improper use or lack of

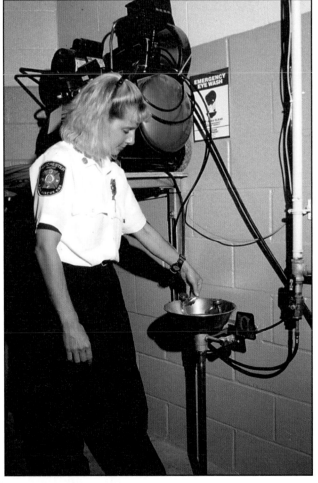

Figure 11.10 Decontamination facilities should be provided in the workplace.

protective equipment, and the lack of training. Written policies that may be established to protect personnel from physical injuries or fatalities include the following topics:

Figure 11.11 The health and safety officer (HSO) is responsible for enforcing lockout/tag-out procedures.

- Lockout/tagout procedures to prevent unexpected energization, start up, or release of stored energy **(Figure 11.11)**

- Two members or use of mechanical devices to lift objects over a predetermined weight

- Appropriate protective clothing when performing certain tasks; may include eye protection, hand protection, hearing protection, respiratory protection, or ice cleats worn on boots or shoes

- Heavy items mounted in lower compartments on apparatus

Based on an analysis of firefighter fatalities data collected over a period of 30 years provided by the U.S. Fire Administration (USFA), fatalities could be significantly reduced by putting more emphasis on the following:

- Proactive wellness/fitness programs

- Annual medical fitness evaluations

- Safe-driving practices

- Enforcement of seat-belt requirements

- Ongoing incident strategy training for fire officers

- Training designed to prevent firefighters from getting lost inside burning buildings

Training is the key to reducing all types of physical injuries and fatalities among fire and emergency services personnel. Proper procedures should be established and all personnel trained in their use; training may include the following topics:

- Proper lifting techniques

- Lockout/tagout programs

- Use of all types of personal protective clothing and equipment

- Use of first-aid kits mounted in the facilities and on apparatus

Rapid Intervention Teams (RIT)

Rapid intervention teams (RITs), crews (RICs), or firefighter assistance and search teams (FASTs) are essential to ensuring personnel safety at emergency operations. The establishment of RITs is mandated by OSHA in the United States and the provincial Workers' Compensation Boards in Canada. RITs are required as an element of any command structure for an incident where the requirement applies. The incident commander and the incident safety officer are responsible for assigning this duty and managing these assets.

Occupational Illnesses

Occupational illnesses are those diseases that are directly related to the duties and tasks of fire and emergency services personnel, including heart attacks, cancer, strokes, respiratory ailments, stress ailments, and exposure to communicable diseases. While some of these can be controlled through training and physical fitness, exposures must be controlled by the use of PPE that is specifically designed to protect wearers from airborne and bloodborne pathogens. Emergency medical responders are particularly susceptible to these types of exposures. Other exposures may occur at hazardous materials incidents or while using certain types of chemicals for maintenance or cleaning duties. Further, occupational illnesses may not become apparent for many years following exposure; some personnel have not shown symptoms until long after retirement.

Policies. Policies may also be included in the respiratory protection and hazardous materials units of the accident, injury, and illness prevention program and the employee physical fitness and wellness program. Policies that may be initiated to reduce occupational illnesses include the following:

- Give medical examinations and evaluations at the time of employment and annually during service career.

- Report medical exposures within a set period such as 24 hours following the suspected exposure.

- Include required information in the occupational illness report such as the following:

  — Location of the incident

  — Time of the incident

  — Number of fatalities or hospitalized members/employees

  — Contact person

  — Phone number

  — Brief description of incident

  — Type of symptoms or physician's report

- Investigate and report each occupational illness to the appropriate authorities.

- Keep all records relating to the suspected occupational illness or medical exposure with requirements for privacy and duration of retention.

- Maintain safety data sheets (SDSs).

- Define the types of reportable cases of occupational illnesses.

Training. Training for all personnel must be provided in the established policies, which should include recognition and immediate care of persons with symptoms of occupational illnesses. Additionally, training in the selection and use of personal protective clothing and equipment during emergency medical responses is essential for all personnel who respond to such incidents. All emergency and nonemergency personnel should be trained and certified in cardiopulmonary resuscitation (CPR) techniques in the event of a heart attack in the workplace. Oxygen inhalators should be available in all facilities and training provided in their use.

Motor Vehicle Related Accidents

Since 1984, motor vehicle related accidents involving fire and emergency service personnel have accounted for between 20 and 25 percent of fire and emergency responders' fatalities annually. One quarter of firefighters who died in these accidents were killed in private/personally owned vehicles because of the large number of volunteers who respond to incidents in their personal vehicles. According to USFA statistics, the type of apparatus most often involved in fatal collisions included mobile water tenders (tankers), engines/pumpers, and aircraft. More firefighters were killed in tender/tanker collisions than in engines and ladder apparatus combined. About 27 percent of fatalities involved in motor vehicle accidents were ejected from the vehicle at the time of the collision; only 21 percent of firefighters were reportedly wearing seat belts before the collision.

Policies. NFPA® standards and state/territorial/provincial as well as local laws and ordinances require the installation and use of seat belts in fire and emergency services apparatus. Fire and emergency service organizations can reduce these statistics through policies and procedures such as the following:

- All persons riding on fire and emergency apparatus must be seated and secured to the vehicle by seat belts or safety harnesses while the vehicle is in motion.

- Personnel are prohibited from standing while riding on apparatus.

- Personnel are prohibited from riding on tailsteps or tailboards or any other exposed position such as sidesteps or running boards while the vehicle is in motion.

- Rearward movement of vehicles that have an obstructed view to the rear is prohibited unless the vehicle has an audible reverse signal alarm distinguishable from the surrounding noise level.

- All fire and emergency apparatus must be brought to a full stop before personnel are allowed to step from the apparatus.

- Two members, one equipped with a radio in contact with the apparatus driver/operator, must assist in the rearward movement of all apparatus.

- Apparatus parked at incidents must be placed to protect personnel from traffic.

- Safety harnesses must be worn when loading hose in hosebed **(Figure 11.12)**.

- Driver/operator training and certification is required in the operation of specific types of apparatus (includes recertification criteria).

- Daily inspection, maintenance, and repair of all vehicles operated by the organization are required.

- Preventive maintenance programs (including record keeping for each vehicle) are required in order to record and track potential or ongoing problems.

Training. Training for apparatus driver/operators should meet the requirements provided in NFPA® 1002, *Standard for Fire Apparatus Driver/Operator Professional Qualifications.* Ambulance driver/operator training programs may be based on the requirements provided by the Commission on Accreditation of Medical Transport Systems (CAMTS) and the Commission on Accreditation of Ambulance Services (CAAS). All personnel, including volunteers who operate privately owned vehicles while responding to emergencies, should have defensive-driving and driver-safety training. Model training programs are available from many organizations such as the National Safety Council (NSC), the Transportation Safety Institute (TSI), Canada Safety Council (CSC), and OSHA.

Hearing Loss

Hearing loss is a continuing health problem for fire and emergency services personnel. The causes of hearing loss include the following:

- Exposure to apparatus audible warning devices (sirens and air horns)
- Exposure to power equipment noises
- Continued exposure to constant low-level noises such as those from air-handling units or engines or apparatus pumps
- Lack of hearing protection
- Viral diseases
- Aging process

Hearing conservation plan. As part of the accident, injury, and illness prevention component of the comprehensive safety and health program, a hearing conservation plan should be developed. The development of this plan begins with the determination of the types of noises employees are subjected to and then determining what types of noise attenuation may be used to reduce the effect on employee hearing. The accepted approach is to develop a written hearing conservation plan that protects members from exposure to noises that are in excess of 85 dB.

Figure 11.12 To prevent slips and falls, safety harnesses must be worn when loading hose beds.

Entry-level employees are given a baseline hearing evaluation at the time they are hired **(Figure 11.13)**. Periodic hearing examinations are then provided as part of the organization's medical examination requirements. They may also be administered upon the member's request or following any exposure to noises in excess of the established minimum level. A hearing examination should also be performed upon separation from the organization. Audiometric tests are performed by a licensed or certified audiologist, otolaryngologist, or other qualified physician or technician who is certified by the Council for Accreditation for Occupational Hearing Conservation. A technician who performs audiometric tests must be responsible to an audiologist, otolaryngologist, or other qualified physician.

Policies. Policies should be established that require the use of hearing protection when operating equipment or riding on apparatus. These types of protection usually include earplugs, earmuffs, or two-way communication systems that include earphones. Protective hoods or helmet earflaps can provide a limited amount of protection at fire-related incidents.

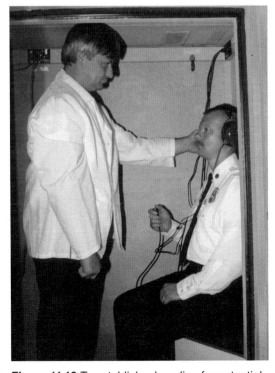

Figure 11.13 To establish a baseline for potential future hearing loss, hearing tests must be given to entry-level personnel.

The HSO is also required to perform sound-level tests in all facilities and for all types of apparatus and equipment to determine the level of noise present. Hearing protection is then provided for and used by all members when exposed to an 8-hour time-weighted average of 85 dB or greater or when exposed to noise in excess of 115 dB from power tools, vehicle engine warm-ups, drafting/ pumping, or other such activities (except in situations where the use of such protective equipment would create an additional hazard to the user such as in fire-suppression activities).

Training. A final component of the hearing conservation plan is a training session that is provided to members in the selection, use, and care of all hearing protectors. The training session should be repeated annually for each member of the organization. Information provided in the training session should be updated to be consistent with changes in protective equipment, the work environment, and policies and procedures.

Respiratory Injury or Illness

Members of the fire and emergency services are continually exposed to various types of hazards that can affect the respiratory system. Sources of these hazards include the following:

- Toxic or hazardous atmospheres found at incidents

- Concentrations of dust, paint, and other particulates produced during cleaning and maintenance

- Airborne microorganisms that may be present during medical emergencies or in the air-handling systems of the organization's facilities

Symptoms may develop rapidly (as in the case of smoke inhalation or oxygen deficiency) or slowly (as in the case of viral infections). Each type of situation requires a different level of personal protection and a different type of equipment. Self-contained breathing apparatus (SCBA) can provide protection against all of these circumstances. However, they are generally used to protect against only the toxic and hazardous atmospheres found at emergency incidents. Dust or particle masks provide the minimum level of protection and can be used in the workshop environment. High-efficiency particulate air (HEPA) filter-equipped masks can be used in the shop environment as well as at the medical emergency. All masks or respiratory protection must meet the minimum requirements for the environment that the employee is in. NFPA® provides standards that define the selection, use, and care of all types of respiratory protection, including those associated with chemical, biological, radiological, nuclear, or explosive (CBRN[E]) incidents as mentioned earlier in this manual **(Figure 11.14)**.

Figure 11.14 Respiratory protection is provided by masks designed for specific hazards and levels of protection.

Respiratory protection plan. The HSO is responsible for the development of the organization's respiratory protection plan as outlined by OSHA in Title 29 *CFR* 1910. A respiratory protection plan requires the following elements:

- Selection criteria for equipment

- Inspection criteria

- Proper-use policy
- Individual fit testing for employees
- Maintenance schedule
- Training session
- Air-quality testing schedule
- Medical certification for those wearing the equipment

Equipment selection criteria can be established by the occupational safety and health committee or any other designated body within the organization. The HSO should be part of the selection team to ensure that safety requirements are met. Considerations in the selection process are as follows:

- Types of hazards members may be exposed to
- Cost of equipment purchases
- Periodic testing and maintenance requirements
- Training requirements
- Design requirements established by NFPA®

Respiratory protection equipment used in fire-fighting activities must be third party certified for structural fire fighting. Particle masks and HEPA-filter masks must meet the minimum requirements set by National Institute for Occupational Safety and Health (NIOSH).

Policies. A written policy covering the proper use of all respiratory protection equipment is required by NFPA® 1500 and NFPA® 1404, *Standard for Fire Service Respiratory Protection Training*. This policy can be developed by the HSO or by the occupational safety and health committee. When adopted by the organization, it must be communicated to all members. Like all operating procedures, it must not become static, but be monitored constantly, reviewed periodically, and revised as necessary. Policies must also be established that define the use of all types of respiratory protection. Examples are as follows:

- Respirators will be provided for (and used by) all personnel working in areas where (1) the atmosphere is hazardous or (2) the atmosphere is suspected of being hazardous or (3) the atmosphere may rapidly become hazardous.
- SCBA or supplied-air respirators (SARs) must be worn in all oxygen-deficient atmospheres (less than 19-percent oxygen content).

Individual fit testing is required of all employees who are going to use respiratory protection equipment. This testing helps to ensure a full and complete seal of the mask to the face. Testing must be done before allowing any employee to use the mask in a hazardous atmosphere. The HSO may be assigned this testing activity.

Training. All employees must be trained in the use of respiratory protection equipment. Training includes when to wear the equipment, how to put it on and take it off, operation of the equipment, daily inspection procedures, and how to clean, maintain, and dispose of the equipment. This training begins during entry-level training sessions and is then provided annually to all members of the organization. The HSO may find it necessary to work with the training officer or staff in providing this training.

Air-quality testing. Air-quality testing is also the responsibility of the HSO and takes two forms. The first is atmospheric testing in all facilities owned or operated by the organization. This testing is intended to determine the presence of mold, mildew, or other airborne contaminants in the air-handling system of the facility. Apparatus exhaust emission must also be tested and appropriate ventilation systems or policies established. The second form of air-quality testing involves the breathing air used in SCBA and SAR systems. Whether the breathing air is generated by air-generating systems operated by the organization or purchased under contract from certified suppliers, the air must be tested for moisture content and contaminates quarterly.

Hazardous Materials Exposure

Fire and emergency service personnel may be exposed to hazardous materials in two ways. The first is through exposure to hazardous materials at emergency incidents. For example, chemical spills or fires that create hazards that can affect the human body through inhalation (airborne), absorption (contact), or ingestion (orally). The second source of exposure is through common usage of products that may contain or emit hazardous fumes or liquids.

Policies. Protection against exposures at emergency incidents includes hazard recognition, policies and procedures governing hazardous materials responses, and the use of proper personal protective clothing and equipment. Because structural fire-fighting clothing does not protect the wearer from all types of hazardous materials, it is necessary to establish a procedure that will contain the incident until personnel with the appropriate protective clothing and training arrive at the incident.

Hazardous communication plan. Protection against common hazards found in the workplace is included in the OSHA-required hazardous communication plan. Its purpose is to ensure that information on the hazards of all chemicals used by fire and emergency service personnel is communicated to all members when the products are placed into service. Members must be made aware of where hazardous chemicals are stored and used in their work areas. They also must be informed of the requirements of the Hazard Communication Standard (Title 29 *CFR* 1910.1200), availability and location of the written unit, list of hazardous chemicals, and safety data sheet (SDS) provided with each product. The SDS is used to relay chemical hazard information from the manufacturer to the user and is maintained at the workplace where the chemical is used.

Training. Training topics include the following:

- Which protective equipment to wear when using specific products
- Product-labeling system
- How to obtain and use SDSs
- Physical and health hazards of the chemicals
- How to recognize, avoid, and prevent accidental entrance of hazardous chemicals into the work environment

Workplace protection. Passive protection in the workplace includes the following items:

- Ventilation systems **(Figure 11.15)**

- NFPA® 704, *Standard System for the Identification of the Hazards of Materials for Emergency Response,* placards on storage rooms

- Warning signs

- Spill protection and controlled drainage

- Fire-protection systems

- Alarm notification systems

- Eyewash and decontamination showers

- First-aid kits with appropriate ointments or medications for treatment of injuries

- Eye and hand protection for use in the event of spills

Implement Program

The reduction of accidents and occupational illnesses, injuries, and fatalities is the responsibility of all fire and emergency service officers and members. Personal safety is a professional quality that must be adhered to in the fire and emergency services.

Implementation of the accident, injury, and illness prevention program includes communication and training. Current employees are trained in the various requirements of the program, policies that mandate it, and use of equipment and procedures that are intended to protect them from accidents, illnesses, and injuries. New employees receive a safety orientation *before* beginning work. Orientation of new employees, rehires, and those transferred from another section/unit within the organization begins the first day on the new job.

Figure 11.15 Work area ventilation hoods remove hazardous vapors from the area.

This orientation provides an introduction of the organization's policies and rules and includes a thorough safety briefing. It includes a tour of appropriate facilities to acquaint employees with their areas of responsibility and the locations of safety-related equipment. A safety orientation session describing the fire and emergency services accident, injury, and illness prevention program includes (but is not limited to) the following elements:

- How and when to report injuries, including instruction as to the location of first-aid facilities

- How to report unsafe conditions and practices

- Use and care of required personal protective equipment (PPE)

- Proper actions to take in the event of emergencies, including the routes of exiting from areas during emergencies

- Identification of the hazardous gases, chemicals, or materials involved along with instructions on the safe use and emergency action following accidental exposure

- Description of the organization's total safety and health program

- On-the-job review of the practices necessary to perform the initial job assignments in a *safe manner*
- Procedures to follow if exposed to bloodborne pathogens or any contagious respiratory illnesses
- Proper procedures for lifting and back care
- Proper handling and use of power tools
- Rules and regulations regarding the use of vehicles in both emergency and nonemergency situations
- Approved safety procedures for incidents involving trenching/shoring and confined-space rescues
- Safety procedures designed to provide fall protection
- Use of respiratory equipment and the type of systems used
- Proper care, use, selection, and maintenance of PPE along with its limitations

The HSO should also begin noise-level testing of all apparatus, equipment, and facilities as well as air- and water-quality testing in all facilities. Breathing-air quality testing should be contracted from a certified third-party testing organization. Monthly facility inspections for safety hazards should also be implemented. These inspections include (but are not limited to) tools, apparatus, extinguishers, protective equipment, and life-safety equipment.

The HSO (in coordination with the appropriate chief officer) should establish hose, apparatus, and equipment testing criteria based on NFPA® requirements. Third-party testing agencies such as Underwriters Laboratories, Inc. (UL), Underwriters Laboratories of Canada (ULC), and FM Global usually provide annual certification tests. Some testing such as the service testing of aerial devices may only be required on a 5-year basis. These are minimum recommendations, and greater safety can be provided if the testing occurs on a more frequent schedule. The department's insurance carrier may also perform safety inspections and should coordinate with the HSO. In addition, the worker's compensation carrier may also perform an audit of the recent injuries and recommend programs, reporting or training options.

Monitor, Evaluate, and Revise Program

The accident, injury, and illness prevention program is monitored, evaluated, and revised continuously. Like other programs implemented by the fire and emergency services organization, the results of the program are compared against the desired or planned goals, and deviations are corrected. Revisions are dictated by the recognition of new safety hazards (such as the use of biological agents by terrorists), failure of current equipment or procedures to meet the hazards (such as the need for greater respiratory protection), or failure of personnel to follow established procedures (such as entering a structure containing smoke without proper respiratory protection).

Fire officers and the HSO must continually monitor the respiratory protection plan to ensure that it meets the changing challenges of the work place. Organizations like NFPA®, NIOSH, International Association of Fire Fighters (IAFF), International Association of Fire Chiefs (IAFC), and CDC provide information on new respiratory protection equipment, requirements, and procedures as well as the rapidly expanding hazards that will damage the respiratory system. Fire officers must actively pursue this information and provide it to the administration and membership of the organization.

Medical Exposure Management Program

Some occupational illnesses can be the result of medical exposures to individuals or situations that may transmit a disease or other hazardous materials. To control and manage this potential threat to the safety and health of an organization's members, it is essential that a medical exposure management program be established. The HSO, working with the occupational safety and health committee and the organization's physician, is responsible for developing and managing this program.

Infection Control

The program begins with a written infection (exposure) control plan that clearly explains its intent, benefits, and purposes. The plan must cover the standards of exposure control such as the following:

- Education and training requirements
- Vaccination requirements for potential threats such as anthrax or hepatitis B virus
- Documentation and record-keeping requirements
- Cleaning, decontamination, and disinfection of personnel and equipment
- Exposure control and reporting protocols

The organization should establish the position of infection (exposure) control officer who ensures that an adequate infection control plan is developed and that all members are trained and supervised in the plan. The HSO or other officer who has the knowledge, training, and skills to fulfill the required duties may fill this position.

Exposure reports must be maintained in a record-keeping system **(Figure 11.16, p. 420)**. Because some symptoms may not become apparent for many years following exposure, these records must be maintained for 30 years following termination of an employee. These records are confidential and may only be released to the member or a designated representative. Records are also maintained for training that involves the proper use of PPE, exposure protection, postexposure protocols, and disease modes of transmission as they relate to infectious diseases. The HSO and safety and health committee members are required to annually review the infectious disease plan, updates, protocols, and equipment. Policies should also be developed that specify the types of protection that must be used to prevent an exposure such as in the following example:

> *Members will wear a particulate respirator when entering areas occupied by individuals with suspected or confirmed tuberculosis (TB), performing high-risk procedures on such individuals, or transporting individuals with suspected or confirmed TB in a closed vehicle. A NIOSH-approved 95-percent efficient particulate air respirator is the minimum acceptable level of respiratory protection.*

City of LaFosse Fire Department
Medical Exposure Report Form

Name:_____ Work Phone:_____

Age:_____   Sex: ❑ Male ❑ Female

Address:_____ Home Phone:_____

City, State, Zip:_____

Company or Assignment: _____

Supervisor: _____

SOURCE OF EXPOSURE

Date of Incident:_____ Time of Incident: _____am / pm

Incident Location:_____

Reference Number to Incident (Dispatch Number):

Type of Incident _____

Name of Source Patient or Individual: _____

Age:_____   Sex: ❑ Male ❑ Female

Address:_____ Home Phone: _____

City, State, Zip:_____

Other Phone:_____

Other identification (e.g. operators permit number, vehicle license plates, etc.) _____

_____

Receiving Facility of Source Patient or Individual (e.g., hospital, mortuary, etc.,) _____

_____

Address:_____

City, State, Zip:_____

Phone: _____

Patient's Attending Physician: _____

Work Phone: _____

Address: _____

City, State, Zip:_____

Known Infectious Disease: _____

Describe the Significant Exposure: _____

_____

Describe any action taken in response to the exposure to remove the contamination: _____

_____

What Personal Protective Equipment and Procedures were you using at the time of the exposure: ___

_____

_____

Any other information related to the incident: _____

_____

List witnesses to the exposure: _____

_____

Signature_____ Date_____

Figure 11.16 Sample Exposure Report Form.

Hazardous Materials Exposure Control

Although incidents involving exposure to bloodborne pathogens and other infectious diseases are increasing, fire and emergency services personnel are regularly exposed to hazardous materials during emergency operations. The tasks involved with hazardous materials safety may be assigned either to the HSO or to a hazardous materials branch safety officer. In either case, the skills necessary to perform these tasks are outlined in Chapter 10 of NFPA® 472, *Standard for Competence of Responders to Hazardous Materials/Weapons of Mass Destruction Incidents.*

The HSO develops exposure protocols and implements them through the organization's SOPs/SOGs. As with all other protocols, they must be reviewed and updated periodically and made a part of the organization's training sessions. During a known hazardous materials incident, either the HSO or the hazardous materials branch safety officer monitors the potential hazards to emergency personnel and advises the incident commander. Risks to consider during a hazardous materials incident (according to Title 29 *CFR* 1910.120) include the following:

- Exposures exceeding the permissible exposure limits and published exposure levels
- Immediately dangerous to life or health (IDLH) concentrations
- Potential skin absorption and irritation sources
- Potential eye-irritation sources
- Oxygen deficiencies
- Explosion sensitivity and flammability ranges

Should the presence of hazardous materials become known following an incident, Title 29 *CFR* 1910.120 requires that the fire and emergency services organization provide exposed employees with all available data on the materials to which they were exposed. Record keeping for hazardous materials exposure incidents follows the same guidelines as those given for infectious diseases.

Accident Investigation Process

Accidents are unplanned, uncontrolled events resulting from unsafe acts and/or unsafe occupational conditions, which may result in injury, death, or property damage. To reduce the potential for accidents to occur or to reduce the severity of accidents, the organization must develop and implement an accident investigation policy and procedure.

Too often, investigators only seek to identify the person, if anyone, responsible for the incident. This approach does nothing to prevent a reoccurrence or to address the *root cause* (basic reason) for the incident.

The policy should define accidents, establish the authority for investigating each type of accident, and establish a procedure for accident investigation. A fire officer, the HSO, and the safety and health committee will have the ultimate authority for accident analysis. The company officer will have the responsibility for performing the initial accident investigation based on the procedure.

Figure 11.17 Accidents involving fire apparatus may be the result of equipment defects, behavior, or environmental conditions. *Courtesy of Mike Mallory, Tulsa (OK) Fire Department.*

Accident investigations are not limited to accidents according to NFPA® 1500. Company officers are also responsible for investigating job-related injuries and illnesses, fatalities, and exposures to infectious diseases and hazardous materials or atmospheres.

Accidents

When an accident occurs, an investigation is conducted to determine the root cause or the most basic reason for the accident and its source or origin. Accident investigations should be objective, impartial, and directed toward fact-finding, not fault-finding. Investigations should never be intended as punishment for those involved. Several reasons to investigate workplace accidents are to identify and document the following conditions:

- Behavior or condition that caused an accident (root cause)
- Previously unrecognized hazards
- Apparatus/equipment defects or design flaws **(Figure 11.17)**
- Additional training needs
- Improvements needed in safety policies and procedures
- Facts that could have a legal impact on an accident case
- Historical trends

When a workplace accident investigation is conducted, all participants and witnesses should be interviewed and all relevant factors documented. Prominent among these is the human factor (personal characteristics). To conduct a thorough and comprehensive investigation, the investigators must have some knowledge of human behavior.

Human Factors

Safety research in private industry has shown that accidents happen frequently to some people and infrequently to others. This research indicates that accidents are *not* uniformly distributed throughout the workforce. Workers who fail to control the factors leading to an accident because of mental, psychological, or physical reasons will be involved in accidents more often than other workers will. These workers are described as *accident prone.*

The accident-prone phenomenon can be explained by the term *human factors*, which are an individual's attributes or personal characteristics that cause the individual to be involved in more or fewer accidents than other individuals. In most cases, an organization can mitigate negative human factors through motivation, training, or technical revision. Human factors that often contribute to accidents have been classified into the following three broad categories:

- ***Improper attitude*** — Includes willful disregard, recklessness, irresponsibility, laziness, disloyalty, uncooperativeness, fearfulness, oversensitivity, egotism, jealousy, impatience, obsession, phobia, absentmindedness, excitability, inconsideration, intolerance, or mental unsuitability in general. Readjusting any of these attitudes or personality traits through counseling, training, or discipline can lead to accident reduction.

- *Lack of knowledge or skill* — Includes insufficient knowledge, misunderstandings, indecision, inexperience, poor training, or failure to recognize potential hazards. These problems can be reduced or eliminated through training.

- *Physically unsuited* — Includes problems of hearing, sight, weight, height, illness, allergies, slow reactions, disabilities, intoxication, or physical limitations in general. Correcting these physical limitations can often reduce accident rates. If they cannot be corrected, personnel should not be assigned to tasks where their limitations might create a hazard or be potentially dangerous to themselves or others.

An organization's effectiveness in mitigating the human factors that lead to accidents often depends upon a number of other factors. Some of these factors include time and resources committed to developing and implementing safety policies and procedures, safety training, and certification on the safe use of equipment. Safety training must be documented and the policies and procedures enforced.

You should be aware that NIOSH requires that an investigation be made of all incidents involving firefighter LODD. In addition, NIOSH may require that additional training be provided within the organization to change any contributing behavioral characteristics that may have caused a fatality.

Accident Scene Control

When the incident involves serious personal injuries, the victim must first be stabilized and transported to a health care facility for medical treatment. Minor injuries may be treated at the scene. The officer in charge of the incident next ensures that the accident scene is secured and all equipment, clothing, or vehicles are left in place.

If conditions require that vehicles be moved, an attempt to photograph or at least sketch the location of each vehicle should be made **(Figure 11.18)**. The officer in charge must remember that anything that was involved in the incident is evidence first for reconstructing the accident and preventing future accidents and second for supporting potential legal cases that may result.

Figure 11.18 Accident scenes should be photographed to provide a reference for the accident investigation and analysis. *Courtesy of Mike Mallory, Tulsa (OK) Fire Department.*

Accurate Data Collection

The accident, injury, illness, and fatality statistics that are used to analyze accidents and determine trends are only as accurate as the original information gathered by investigating officers. All personnel must follow the prescribed procedures and properly complete accident investigation forms provided by the jurisdiction.

The form may be a simplified version that permits the investigating officer to enter words into blanks and complete a narrative account of the incident. Forms may also be designed so that they are similar to the National Fire Incident Reporting System (NFIRS) forms that require the use of predesignated codes to describe an incident. In either case, the form should be designed to record the essential information for accidents, injuries, illnesses, or fatalities.

In conducting an investigation, investigating officers collect basic information about the participants, event, or incident. The investigation should provide the following information:

- *General information* — Items:
  — Date and time of incident
  — Type of incident, illness, injury, or fatality
  — Location and emergency response type
  — Names of witnesses and their accounts of the situation

- *Employee characteristics (participant)* — Items:
  — Name and unit assignment (company/shift)
  — Age and gender
  — Rank/function
  — Personal protective clothing or equipment in use

- *Environmental information* — Items:
  — Weather and temperature
  — Day or night conditions
  — Noise and visibility
  — Terrain

- *Apparatus/equipment information* — Items:
  — Type of equipment involved
  — Age and condition
  — Location
  — Maintenance history
  — Distinguishing characteristics

A narrative description of the incident is the final portion of the investigation report. This narrative includes observations on the part of the officer, eyewitness reports, participant interviews, and information from other sources such as law-enforcement reports and dispatch information.

Data gathered by the investigating officer is transmitted through the system to the HSO who uses the information to analyze the accident, job-related injuries, and illnesses, fatalities, and exposures. Documentation must be retained and signed by the investigating officer.

Injuries, Illnesses, and Exposures

When a member of the unit experiences a job-related injury, illness, or exposure, the first duty of a fire officer is to ensure prompt medical treatment for the individual. When the injury, illness, or exposure is serious, the individual must be transported to the closest available medical facility. Once the officer has provided for the treatment of the member, an investigation can commence.

Investigations of job-related injuries, illnesses, or exposures generally include gathering the same information in the same way as accident investigations. Some differences may exist depending on the nature of the incident. For instance, an investigation into a claim that involves an illness that is perceived to be job-related may require more time and the services of a specialty organization such as one that samples air or water. These types of investigations may occur at some time after the event. The investigating officer must focus on the idea that information gained from the investigation may lead to the discovery of an environmental cause for the illness.

An exposure investigation may occur immediately following the incident or at some point in the future when symptoms of exposure appear in the individual. Accurate and thorough documentation of all incidents and exposures is basic to any current or future investigation.

Long-Range Investigation

An example of a long-range investigation involved the occurrence of cancer in a predominately-high number of personnel who worked at a single station in one Midwestern city. The symptoms did not appear until many years after some had retired. Attempts to determine the cause of the trend included water and air sampling, testing for asbestos, and a review of all emergency responses the members had been on. Tragically, no evidence of a single cause was ever determined.

Summary

An efficient and effective fire and emergency service organization depends on a safe, healthy, physically fit, and emotionally stable membership. It is the responsibility of the administration, fire officers, and the HSO to provide the membership with a safe working environment though training, education, equipment, policies, procedures, leadership, and supervision. This safe environment is accomplished through the development and implementation of a comprehensive safety and health program that contains individual components to address both the obvious and obscure hazards to the health and well-being of the members.

The organization must also make physical fitness a reality through the implementation of a holistic employee physical fitness and wellness program. The wellness component includes proper nutrition, back care, heart and lung disease awareness, and stress-reduction counseling. The organization should work to address all areas of health and wellness both through prevention and education.

Through these efforts, the organization can reduce injuries, reduce fatalities, reduce health and lost-time costs, and improve the morale of the membership. Fire officers of all levels, along with the HSO and other officers of the organization, must make safety and health their primary concerns for the membership. To do so, they must set the example for others to follow and ensure that all operations are performed in a way that is consistent with SOP/SOGs and safe practices.

Review Questions

1. What legal requirements affect a fire department's health and safety program?

2. When researching the fire department's safety problems, what types of sources should be consulted during the data-collection process?

3. Why is it important to implement an occupational safety and health program?

4. What are the main differences between the responsibilities of the occupational safety and health committee versus a health and safety officer?

5. What specific techniques can be used to control risk and how can these be used in developing a risk management plan?

6. What specific training and policy measures can be taken to address hazards in the workplace?

7. What elements are required for a respiratory protection plan to be complete?

8. What four main types of information need to be gathered when investigating exposures and accidents?

Appendices

Contents

Appendix A
Chapter Correlation to NFPA® and FESHE Requirements

Chapter References	NFPA® Standard and Paragraph Numbers	FESHE Learning Objectives *Occupational Safety and Health for Emergency Services*
1	NFPA® 1021, 4.7.1 NFPA® 1021, 5.7.1 NFPA® 1021, 7.7.1	List and describe the components of risk identification, risk evaluation, and incident management.
2	None	Describe the history of occupational health and safety. Identify occupational health and safety programs for industry and emergency services today.
3	NFPA® 1001, 6.3 NFPA® 1001, 6.4.1 NFPA® 1021, 4.1.1 NFPA® 1021, 4.2.1 NFPA® 1021, 4.2.2 NFPA® 1021, 4.7 NFPA® 1021, 4.7.1 NFPA® 1021, 6.7.1 NFPA® 1021, 7.6.2 NFPA® 1021, 7.7 NFPA® 1021, 7.7.1 NFPA® 1071, 6.2.3.3 NFPA® 1041, 4.4.2 NFPA® 1041, 5.4.3 NFPA® 1051, 1.3.12 NFPA® 1051, 1.3.16 NFPA® 1051, 5.5.3 NFPA® 1081, 8.2.1 NFPA® 1081, 8.2.3	Describe the responsibilities of individual responders, supervisors, safety officers, and incident commanders, safety program managers, safety committees and fire department managers as they relate to health and safety programs.
4	NFPA® 1500, 5.2 NFPA® 1500, 10.1 NFPA® 1500, 10.2 NFPA® 1500, 10.3 NFPA® 1500, 10.4 NFPA® 1500, 10.5 NFPA® 1500, 10.6 NFPA® 1500, 10.7	Describe the components and value of critical incident management programs. Describe the components of a wellness/fitness plan. Identify and analyze the major causes involved in line-of-duty firefighter deaths related to health, wellness, fitness, and vehicle operations.

Continued

Chapter References	NFPA® Standard and Paragraph Numbers	FESHE Learning Objectives *Occupational Safety and Health for Emergency Services*
5	NFPA® 1500, 5.1 NFPA® 1500, 5.2 NFPA® 1500, 5.3 NFPA® 1500, 5.4 NFPA® 1500, 5.5	Apply the knowledge of an effective safety plan to pre-incident planning, response, and training activities.
6	NFPA® 1500, 6.1 NFPA® 1500, 6.2 NFPA® 1500, 6.3 NFPA® 1500, 6.4 NFPA® 1500, 6.4. NFPA® 1001,5.3.2 NFPA® 1001,5.3.3 NFPA® 1002, 4.1 NFPA® 1002, 5.1 NFPA® 1002, 6.1 NFPA® 1002, 8.1 NFPA® 1002, 10.1	Apply the knowledge of an effective safety plan to pre-incident planning, response, and training activities.
7	NFPA® 1500, 6.5	Describe the relevance for safety in the work place including the importance of PPE.
8	NFPA® 1500, 7.1 NFPA® 1500, 7.2 NFPA® 1500, 7.3 NFPA® 1500, 7.4 NFPA® 1500, 7.5. NFPA® 1500, 7.6 NFPA® 1500, 7.7 NFPA® 1500, 7.8 NFPA® 1500, 7.9 NFPA® 1500, 7.10 NFPA® 1500, 7.11 NFPA® 1500, 7.12 NFPA® 1500, 7.13 NFPA® 1500, 7.14 NFPA® 1500, 7.15 NFPA® 1500, 7.16 NFPA® 1500, 7.17	Describe the relevance for safety in the work place including the importance of PPE.
9	NFPA® 1500, 8.1 NFPA® 1500, 8.2 NFPA® 1500, 8.3 NFPA® 1500, 8.4 NFPA® 1500, 8.5 NFPA® 1500, 8.6 NFPA® 1500, 8.7 NFPA® 1500, 8.8	Apply the knowledge of an effective safety plan to pre-incident planning, response, and training activities. Explain the components of an accountability system in emergency services operations. Discuss the need for and the process used for post-incident analysis.

Chapter References	NFPA® Standard and Paragraph Numbers	FESHE Learning Objectives *Occupational Safety and Health for Emergency Services*
10	NFPA® 1500, 9.1 NFPA® 1500, 9.2	Describe the responsibilities of individual responders, supervisors, safety officers, and incident commanders, safety program managers, safety committees and fire department managers as they relate to health and safety programs.
11	NFPA® 1500, 4.2 NFPA® 1500, 4.3 NFPA® 1500, 4.4 NFPA® 1500, 4.5 NFPA® 1500, 4.6 NFPA® 1500, 4.7 NFPA® 1500, 7.18 NFPA® 1500, 11.1 NFPA® 1500, 12.1	Compare the differences between standards and regulations. Discuss the need for and the process used in post incident analysis. Describe the components and value of critical incident management programs. Describe the components of a wellness/fitness plan.

NOTE: Complete chapters for NFPA® 1500 are listed above while specific section references for the other NFPA® standards are provided. NFPA® standards listed in this correlation:

NFPA® 1001, *Standard for Fire Fighter Professional Qualifications*

NFPA® 1002, *Standard for Fire Apparatus Driver/Operator Professional Qualifications*

NFPA® 1021, *Standard for Fire Officer Professional Qualifications*

NFPA® 1041, *Standard for Fire Service Instructor Professional Qualifications*

NFPA® 1051, *Standard for Wildland Fire Fighter Professional Qualifications*

NFPA® 1071, *Standard for Emergency Vehicle Technician Professional Qualifications*

NFPA® 1081, *Standard for Industrial Fire Brigade Member Professional Qualifications*

NFPA® 1500, *Standard on Fire Department Occupational Safety and Health Program*

Appendix B
EVERYONE GOES HOME:
Firefighter Life Safety Initiatives Program

1. Define and advocate the need for a cultural change within the fire service relating to safety; incorporating leadership, management, supervision, accountability and personal responsibility.

2. Enhance the personal and organizational accountability for health and safety throughout the fire service.

3. Focus greater attention on the integration of risk management with incident management at all levels, including strategic, tactical, and planning responsibilities.

4. All firefighters must be empowered to stop unsafe practices.

5. Develop and implement national standards for training, qualifications, and certification (including regular recertification) that are equally applicable to all firefighters based on the duties they are expected to perform.

6. Develop and implement national medical and physical fitness standards that are equally applicable to all firefighters, based on the duties they are expected to perform.

7. Create a national research agenda and data collection system that relates to the initiatives.

8. Utilize available technology wherever it can produce higher levels of health and safety.

9. Thoroughly investigate all firefighter fatalities, injuries, and near misses.

10. Grant programs should support the implementation of safe practices and/or mandate safe practices as an eligibility requirement.

11. National standards for emergency response policies and procedures should be developed and championed.

12. National protocols for response to violent incidents should be developed and championed.

13. Firefighters and their families must have access to counseling and psychological support.

14. Public education must receive more resources and be championed as a critical fire and life safety program.

15. Advocacy must be strengthened for the enforcement of codes and the installation of home fire sprinklers.

16. Safety must be a primary consideration in the design of apparatus and equipment.

Appendix C
The IAFF/IAFC Fire Service Joint Labor Management Candidate Physical Ability Test Program Summary

The IAFF/IAFC Wellness-Fitness Task Force

The IAFF and IAFC have teamed up with 10 of North America's leading fire departments and unions through the Fire Service Joint Labor Management Wellness/Fitness Initiative to develop the IAFF/IAFC Candidate Physical Ability Test (CPAT). The participating departments on the Task Force are:

- Austin, TX / IAFF Local 975

- Los Angeles Co., CA / IAFF Local 1014

- Calgary, ALB / IAFF Local 255

- Metro Dade Co., FL / IAFF Local 1403

- Charlotte, NC / IAFF Local 660

- New York City, NY / IAFF Local 94/854

- Fairfax Co., VA / IAFF Local 2068

- Phoenix, AZ / IAFF Local 493

- Indianapolis, IN / IAFF Local 416

- Seattle, WA / IAFF Local 27

The Task Force successfully developed the Fire Service Joint Labor-Management Wellness-Fitness Initiative in 1997 to address the need for a holistic and non-punitive approach to wellness and fitness in the fire service. The Task Force then discovered that municipalities were hiring people who would not be physically capable of a successful career in the fire service. The Task Force, therefore, unanimously agreed to develop a physical ability test for pre-employment testing of candidates. The Task Force has developed and validated the Candidate Physical Ability Test (CPAT), resulting in a consistent test for hiring of candidate fire fighters.

Developing the CPAT

The Task Force directed the Technical Committee to develop a performance test for the ten jurisdictions that measures the critical skills of fire fighter candidates. Committee members reviewed six of the ten jurisdictions job analysis and job task surveys. They also reviewed each of the ten jurisdiction's current candidate performance tests and job descriptions.

Additionally, a complete equipment and demographics survey was also completed by the ten jurisdictions. The equipment survey provided types and weights of all fire fighter protective clothing, protective equipment, fire department equipment and fire department tools. It also assessed the average weight of fire fighters in each jurisdiction as well as the average weight of emergency room and hospital admitted patients. Demographic profiles of each jurisdiction were also developed.

Using the data collected, the Committee then derived a list of 31 tasks to investigate. They developed survey questions directly related to these 31 tasks. The critical and physical task skills that all fire fighters should possess were validated (through questionnaires) by 1,000 fire fighters from the ten departments. Selection of the 1,000 fire fighters was random and anonymous, but consistent with the race and gender diversity of the specific department. The completed results of the surveys resulted in the development a series of physical ability tests that were selected based on the results of the data indicating the physicality and criticality of the tasks performed by fire fighters.

Props for each event were built and modified to obtain the necessary information regarding candidate's ability. The props were then placed in a sequence that would best simulate their use in a fire scene. The Technical Committee chose to move the candidates along a predetermined path from event to event in a continuous manner requiring a mandatory walk between events.

Because of the test set up, individual event times would be an ineffective measure of a candidate's ability, so an overall test time needs to be established. Numerous individuals were run through the test including: technical staff members, incumbent fire fighters, lay people and a recruit class.

The overall consensus found the test to be a good predictor of an applicant's ability to perform basic fire fighting tasks. The Technical Committee was confident that the ability test would provide the fire service with a physically competent recruit. There was also unanimous agreement that the test equaled or was superior to current tests run by each of the ten jurisdictions.

These tests were filmed at different speeds and presented to supervisors in the 10 jurisdictions (captains and battalion chiefs with experience in fire fighter training). At the completion of this process an entry test was finalized and presented to the Task Force for adoption. The Task Force unanimously adopted the full Candidate Physical Ability Test with a pass/fail time of 10 minutes and 20 seconds. The test may only be administered on a pass/fail basis, municipalities may not rank candidates based upon CPAT completion times.

Administering the CPAT

In developing the CPAT Program, the Task Force recognized that the fire department should reflect the community it serves. In today's society, communities are increasingly diverse and fire fighters are continually challenged to operate in multi-cultural environments. The goal of the CPAT is to test for those individuals physically qualified to perform the job of fire fighter without separating from the department's broader goal of attaining a properly trained and physically capable workforce whose members reflect the diversity of the community. Diversity should never come by lowering validated entry standards. Rather, it should come from actively recruiting qualified men and women candidates from all racial and ethnic backgrounds for careers in the fire service. It is the position of the IAFF/IAFC Joint Labor Management Wellness-Fitness Initiative Task Force that fire departments should increase the diversity of their workforce by actively recruiting candidates from throughout their communities rather than lowering candidate physical ability standards. The CPAT Program details avenues fire departments can pursue to raise awareness of job opportunities and to recruit and mentor qualified candidates within the fire service before the test is administered.

The CPAT goal was to develop a fair and valid evaluation tool in the selection of fire fighters to ensure that all fire fighter candidates possess the physical ability to complete critical tasks effectively and safely. This CPAT Program covers every aspect of administering the CPAT: from recruiting and mentoring programs, to providing recruits with fitness guidance to help prepare them for the CPAT, to setting up and administering the test. The entire validation process is discussed in detail, as well as the legal issues that departments might face when implementing the actual program.

After a municipality has completed all aspects of recruiting and mentoring candidates they may administer the actual eight-event test. During the entire test the candidate must wear a 50 lb. weighted vest (simulating the weight of a fire fighters protective clothing and equipment). The eight events are:

- Stair Climb (climbing stairs while carrying an additional 25 lb. simulated hose pack)
- Ladder Raise and Extension (placing a ground ladder at the fire scene and extending the ladder to the roof or a window)

- Hose Drag (stretching uncharged hoselines, advancing lines)
- Equipment Carry (removing and carrying equipment from fire apparatus to fireground)
- Forcible Entry (penetrating a locked door, breaching a wall)
- Search (crawling through dark unpredictable areas to search for victims)
- Rescue Drag (removing victim or partner from a fire building)
- Ceiling Pull (locating fire and checking for fire extension)

Appendix D
PPE Inspection Form

PPE INSPECTION FORM

EMPLOYEE NAME_____EMPLOYEE SIGNATURE_____

SUPERVISORS NAME_____ SUPERVISORS SIGNATURE_____

STATION_____SHIFT_____DATE_____ EMPLOYEE#_____PRIMARY OR RESERVE_____

COAT Shell	GOOD	MAINTENANCE REQUIRED	REPLACE	COMMENTS
Bar Code#_____				
CLEANLINESS				
RIPS, TEARS, HOLES				
THERMAL DAMAGE CHARRING, BURN HOLES, MELTING, DISCOLORATION				
FABRIC INTEGRITY-UV, CHEMICAL DAMAGE				
SEAMS-OPEN BROKEN STITCHES				
BROKEN/MISSING STITCHES				
EXCESSIVE WEAR				
CLOSURE SYSTEM-HOOKS/DEES, SNAPS, VELCRO, ZIPPER				
VELCRO CLEANLINESS, DAMAGE, STITCHING				
HARDWARE-MISSING, DAMAGED, FUNCTIONING-SNAPS, HOOKS & DEES, ZIPPERS				
WRISTLETS-ELASTIC, DAMAGE, THUMBHOLE				
TRIM-INTACT & EFFECTIVE				
NFPA LABELS				
POCKETS-USABLE, INTACT				
CUFFS-INTACT				
REINFORCING MATERIAL-LEATHER, ARASHIELD, FABRIC INTACT FUNCTIONAL				
FIT-OVERLAP				
ADD ONS-APPROVED				

PPE INSPECTION FORM

EMPLOYEE NAME_____EMPLOYEE SIGNATURE_____

SUPERVISORS NAME_____ SUPERVISORS SIGNATURE_____

STATION_____SHIFT_____DATE_____ EMPLOYEE#_____PRIMARY OR RESERVE_____

PANTS Shell	GOOD	MAINTENANCE REQUIRED	REPLACE	COMMENTS
Bar Code#_____				
CLEANLINESS				
RIPS, TEARS, HOLES				
THERMAL DAMAGE CHARRING, BURN HOLES, MELTING, DISCOLORATION				
FABRIC INTEGRITY-UV, CHEMICAL DAMAGE				
SEAMS-OPEN, BROKEN STITCHES				
BROKEN/MISSING STITCHES				
EXCESSIVE WEAR				
CLOSURE SYSTEM-HOOKS/DEES, SNAPS, VELCRO, ZIPPER				
VELCRO CLEANLINESS, DAMAGE, STITCHING				
HARDWARE-MISSING, DAMAGED, FUNCTIONING-SNAPS, HOOKS & DEES, ZIPPERS				
WRISTLETS-ELASTIC, DAMAGE, THUMBHOLE				
TRIM-INTACT & EFFECTIVE				
NFPA LABELS				
POCKETS-USABLE, INTACT				
CUFFS-INTACT				
REINFORCING MATERIAL-LEATHER, ARASHIELD, FABRIC INTACT FUNCTIONAL				
SUSPENDERS				
SUSPENDER BUTTONS				
FIT-OVERLAP				
ADD ONS-APPROVED				

Continued

PPE INSPECTION FORM

EMPLOYEE NAME_____EMPLOYEE SIGNATURE_____

SUPERVISORS NAME_____ SUPERVISORS SIGNATURE_____

STATION_____SHIFT_____DATE_____ EMPLOYEE#_____PRIMARY OR RESERVE_____

COAT-Liner	GOOD	MAINTENANCE REQUIRED	REPLACE	COMMENTS
CLEANLINESS				
LINER ATTACHMENT SYSTEM-BROKEN SNAPS, ZIPPER, CLOGGED VELCRO				
NFPA LABELS				
FIT OF LINER IN SHELL				
PANT-Liner				
CLEANLINESS				
LINER ATTACHMENT SYSTEM-BROKEN SNAPS, ZIPPER, CLOGGED VELCRO				
FIT OF LINER IN SHELL				
CROTCH				
NFPA LABELS				
MOISTURE BARRIERS				
RIPS, TEARS,CUTS, HOLES				
THERMAL DAMAGE, MELTING, SCORCHING				
SEAM INTEGRITY				
SEAM TAPE IN PLACE				
PUDDLE TEST				
THERMAL LINER				
RIPS, CUTS, TEARS, HOLES				
NFPA LABELS				
THERMAL DAMAGE				
UV DAMAGE-FACE CLOTH FADED-CHECK MOISTURE BARRIER UNDER UV DAMAGE				
FRAYED, BROKEN QUILT STITCHING				
INSULATION IN PLACE				
EXCESSIVE WEAR				

Continued

PPE INSPECTION FORM

EMPLOYEE NAME_____EMPLOYEE SIGNATURE_____

SUPERVISORS NAME_____ SUPERVISORS SIGNATURE_____

STATION_____SHIFT_____DATE_____ EMPLOYEE#_____PRIMARY OR RESERVE_____

HELMET	GOOD	MAINTENANCE REQUIRED	REPLACE	COMMENTS
SHELL				
CLEANLINESS				
THERMAL DAMAGE BUBLING, SOFT SPOTS, WARPING, DISCOLORATION				
DELIMINATION				
DENTS, MICKS, CRACKS, GOUGES, FLAKING				
REFLECTIVE TRIM IN PLACE & FUNCTIONAL				
ADD ONS APPROVED				
FACESHIELD/ GOGGLES				
FUNCTIONABILITY				
COMPONENTS WORK				
CLEAR, SCRATCHES, DISCOLORATION				
EAR FLAPS				
RIPS, CUTS, TEARS				
CHARRING, BURN HOLES, MELTINS				
CUFFS-INTACT				
ATTACHMENT SYSTEM OK				
IMPACT CAP				
DAMAGE-CRACKED OR THERMAL DAMAGE				
SYSPENSION SYSTEM FUNCTIONAL				
RETENTION SYSTEM FUNTIONAL				
CHIN STRAP OK				
NFPA LABEL				

Continued

PPE INSPECTION FORM (Concluded)

EMPLOYEE NAME_____ EMPLOYEE SIGNATURE_____

SUPERVISORS NAME_____ SUPERVISORS SIGNATURE_____

STATION_____ SHIFT_____ DATE_____ EMPLOYEE#_____ PRIMARY OR RESERVE_____

HOODS	GOOD	MAINTENANCE REQUIRED	REPLACE	COMMENTS
CLEANLINESS				
RIPS,TEARS, HOLES				
STRETCH				
WEAR				
FACE OPENING ELASTICITY				
SEAMS				
EXCESSIVE WEAR				
NFPA LABEL				
GLOVES				
CLEANLINESS				
RIPS, TEARS, CUTS				
CHAR, BURN HOLES				
LINER INTEGRITY				
SEAMS				
PLIABILITY				
NFPA LABEL				
BOOTS				
CLEANLINESS				
CUTS, TEARS, PUNCTURING, CRACKING,SPLITTING, DELAMINATION				
CHAR, BURN HOLES, MELTING				
EXPOSED, DEFORMED STEEL TOE, SHANK, MIDSOLE				
TREAD				
SEAMS				
WATER RESISTANCE				
CLOSURE FUNCTIONAL				
NFPA LABELS				

ICS Form 202

INCIDENT OBJECTIVES	1. INCIDENT NAME	2. DATE	3. TIME

4. OPERATIONAL PERIOD (DATE/TIME)

5. GENERAL CONTROL OBJECTIVES FOR THE INCIDENT (INCLUDE ALTERNATIVES)

6. WEATHER FORECAST FOR OPERATIONAL PERIOD

7. GENERAL SAFETY MESSAGE

8. Attachments (☑ if attached)

☐ Organization List (ICS 203)   ☐ Medical Plan (ICS 206)   ☐ Weather Forecast _____

☐ Assignment List (ICS 204)   ☐ Incident Map   ☐ _____

☐ Communications Plan (ICS 205)   ☐ Traffic Plan   ☐ _____

9. PREPARED BY (PLANNING SECTION CHIEF)	10. APPROVED BY (INCIDENT COMMANDER)

Organization Assignment List, ICS Form 203

ORGANIZATION ASSIGMENT LIST		1. INCIDENT NAME	2. DATE PREPARED	3. TIME PREPARED

POSITION	NAME	4. OPERATIONAL PERIOD (DATE/TIME)

5. INCIDENT COMMAND AND STAFF

INCIDENT COMMANDER	
DEPUTY	
SAFETY OFFICER	
INFORMATION OFFICER	
LIAISON OFFICER	

6. AGENCY REPRESENTATIVES

AGENCY	NAME

7. PLANNING SECTION

CHIEF	
DEPUTY	
RESOURCES UNIT	
SITUATION UNIT	
DOCUMENTATION UNIT	
DEMOBILIZATION UNIT	
TECHNICAL SPECIALISTS	

8. LOGISTICS SECTION

| CHIEF | |
| DEPUTY | |

a. SUPPORT BRANCH

DIRECTOR	
SUPPLY UNIT	
FACILITIES UNIT	
GROUND SUPPORT UNIT	

b. SERVICE BRANCH

DIRECTOR	
COMMUNICATIONS UNIT	
MEDICAL UNIT	
FOOD UNIT	

9. OPERATIONS SECTION

| CHIEF | |
| DEPUTY | |

a. BRANCH I- DIVISION/GROUPS

BRANCH DIRECTOR	
DEPUTY	
DIVISION/GROUP	
DIVISION/ GROUP	
DIVISION/ GROUP	
DIVISION/GROUP	
DIVISION /GROUP	

b. BRANCH II- DIVISIONS/GROUPS

BRANCH DIRECTOR	
DEPUTY	
DIVISION/GROUP	
DIVISION/GROUP	
DIVISION/GROUP	
DIVISION/GROUP	

c. BRANCH III- DIVISIONS/GROUPS

BRANCH DIRECTOR	
DEPUTY	
DIVISION/GROUP	
DIVISION/GROUP	
DIVISION/GROUP	

d. AIR OPERATIONS BRANCH

AIR OPERATIONS BR. DIR.	
AIR TACTICAL GROUP SUP.	
AIR SUPPORT GROUP SUP.	
HELICOPTER COORDINATOR	
AIR TANKER/FIXED WING CRD.	

10. FINANCE/ADMINISTRATION SECTION

CHIEF	
DEPUTY	
TIME UNIT	
PROCUREMENT UNIT	
COMPENSATION/CLAIMS UNIT	
COST UNIT	

PREPARED BY (RESOURCES UNIT)

Sample Assignment List, ICS Form 204

1. BRANCH	2. DIVISION/GROUP	**ASSIGNMENT LIST**

3. INCIDENT NAME	4. OPERATIONAL PERIOD DATE _____ TIME _____

5. OPERATIONAL PERSONNEL

OPERATIONS CHIEF _____	DIVISION/GROUP SUPERVISOR _____
BRANCH DIRECTOR _____	AIR TACTICAL GROUP SUPERVISOR _____

6. RESOURCES ASSIGNED TO THIS PERIOD

STRIKE TEAM/TASK FORCE/ RESOURCE DESIGNATOR	EMT	LEADER	NUMBER PERSONS	TRANS. NEEDED	PICKUP PT./TIME	DROP OFF PT./TIME

7. CONTROL OPERATIONS

8. SPECIAL INSTRUCTIONS

9. DIVISION/GROUP COMMUNICATIONS SUMMARY

FUNCTION		FREQ.	SYSTEM	CHAN.	FUNCTION		FREQ.	SYSTEM	CHAN.
COMMAND	LOCAL				SUPPORT	LOCAL			
	REPEAT					REPEAT			
DIV./GROUP TACTICAL					GROUND TO AIR				

PREPARED BY (RESOURCE UNIT LEADER)	APPROVED BY (PLANNING SECT. CH.)	DATE	TIME

Sample Incident Communications Plan, ICS Form 205

INCIDENT RADIO COMMUNICATIONS PLAN		1. Incident Name	2. Date/Time Prepared	3. Operational Period Date/Time
4. Basic Radio Channel Utilization				
System/Cache	Channel	Function	Frequency/Tone	Assignment	Remarks

5. Prepared by (Communications Unit)

Medical Plan, ICS Form 206

MEDICAL PLAN	1. Incident Name	2. Date Prepared	3. Time Prepared	4. Operational Period

5. Incident Medical Aid Station

Medical Aid Stations	Location	Paramedics Yes	No

6. Transportation

A. Ambulance Services

Name	Address	Phone	Paramedics Yes	No

B. Incident Ambulances

Name	Location	Paramedics Yes	No

7. Hospitals

Name	Address	Travel Time Air	Ground	Phone	Helipad Yes	No	Burn Center Yes	No

8. Medical Emergency Procedures

Prepared by (Medical Unit Leader)	10. Reviewed by (Safety Officer)

Appendix F
Sample Facility Inspection Checklist

Facility: _____ Time: _____ Date: _____

Compliance with each subject MUST be 100% or attach documentation of corrective action taken.

☐ **ITEM OK**   ☐ **ITEM HAS PROBLEM**

1. **Walking & Working Surfaces**

   A. Aisles
   B. Stairways
   C. Catwalks
   D. Guard Railings
   E. Work Platforms

2. **Compressed Air & Gas**

   A. Storage & Restraint
   B. Handling
   C. Labeling
   D. Proper Use

3. **Mechanical Safeguards**

   A. Point of Operation
   B. Drive Train
   C. Operating Controls
   D. Emergency Stops
   E. Start Up Alarms

4. **General Environment**

   A. Noise
   B. Lighting
   C. Ventilation

5. **Material Handling Operations**

   A. Conveyors
   B. Crane
   C. Fork Truck
   D. Slings & Rigging
   E. Attachments

6. **Housekeeping & Sanitation**

   A. Work Areas
   B. Rest Rooms
   C. Locker Rooms
   D. Break Areas
   E. Waste Disposal
   F. Leakage & Spillage
   G. Proper Storage of Materials

7. **Material Storage & Stacking**

   A. Damaged Containers
   B. Improperly Stacked
   C. Distance from Sprinklers
   D. Blind Corners
   E. Blocked Aisles
   F. Projecting Materials
   G. Safe Heights
   H. Floor Load Limits
   I. Proper Coil Steel Storage
   J. Proper Blank Steel Storage

8. **Personal Protective Equipment**

   A. Side Shields
   B. Face Protection
   C. Hearing Protection Needed
   D. Respirators Required
   E. Hard Hats Required
   F. Protective Clothing

9. **Fire & Emergency Equipment**

   A. Alarms
   B. Sprinklers
   C. Extinguishers
   D. Fire Hoses
   E. Eye Washes & Showers
   F. Stretchers
   G. Emergency Lighting

10. **Fall Hazard Control**

    A. Anchorage Points
    B. Fall Protection Equipment
    C. Ladders
    D. Scaffolds
    E. Manlifts

11. **Means of Egress**

    A. Exits Marked
    B. Clear & Unobstructed
    C. No Locked Exits
    D. Posted Evacuation Plan
    E. Posted Shelter Areas

12. **Electrical Safeguarding**

    A. Covers for Switches
    B. Covers for Outlets
    C. Doors on Panel Boxes
    D. Disconnects can Lockout
    E. Proper Labeling
    F. Posted Lockout Procedure
    G. Identify High Voltage
    H. Extension & Fixture Cords
    I. Temporary Hookups

13. **Hazardous Materials**

    A. Storage
    B. Handling
    C. Personal Protective Equipment
    D. Labeling
    E. Warning Signs
    F. All Chemicals Approved by HMCC
    G. Safe Use Instructions Reviewed with Employees
    H. AC240 Completed for New or Transferred Employees

14. **Cranes & Hoisting Equipment**

    A. Daily Inspection prior to use
    B. Inspection tag attached to every hoist
    C. Authorized/Trained user
    D. Preventive Maintenance Performed

15. **Hand & Portable Power Tools**

    A. Ergonomics
    B. Carrying & Handling
    C. Damaged Tools
    D. Guards in Place
    E. Special Tools
    F. Grounding

16. **Industrial Vehicles**

    A. Operation of Trucks
    B. Railroad Cars
    C. Utility Vehicles
    D. Manlifts
    E. Operation of Heavy Equipment

17. **Safe Working Practices**

    A. Adherence to Safety Rules and Regulations
    B. Adherence to Lockout Procedure
    C. Work Habits
    D. Unsafe Acts
    E. Corrective Instruction or Training of Employees

Based on form developed by the National Safety Council

Appendix G
Safety Committee Checklist

The following checklist represents the minimum requirements for a Safety Committee in most states:

_____ Form a safety committee.
_____ Committee shall be composed of an equal number of employer and employee representatives (number of employees may be greater).

Committee members will be provided with:
_____ Purpose and Operation of the Committee
_____ Access to Applicable State and Federal Rules
_____ Methods of Conducting Meetings
_____ Staff to Record and Distribute Agendas and Minutes

Safety Committee Operations:
_____ Elect a chairperson.
_____ Have a written agenda for conducting committee business during meetings.
_____ Hold monthly meetings.
_____ Take minutes at each meeting.
_____ Maintain minutes for three years.
_____ Post copy of minutes of each meeting and send copy to committee members.
_____ Establish reasonable time limit for management response.
_____ Establish hazard reporting system.
_____ Inform employees of their rights to report workplace hazards.
_____ Evaluate employer accountability system.
_____ Establish procedures of investigating accidents/incidents.
_____ Provide access to safety/health codes.

Inspections:
_____ Establish procedures for inspections.
_____ Conduct quarterly inspection of satellite locations
_____ Train representative in hazard identification
_____ Committee shall review inspection reports
_____ Establish methods for correction and follow-up

Train committee members regarding:
_____ Duties and Responsibilities
_____ Hazard Identification and Correction.
_____ Principles of Accident Investigation.

Glossary

Glossary

B

Braking Distance — The distance the vehicle travels from the time the brakes are applied until the apparatus comes to a complete stop.

C

Candidate Physical Abilities Test (CPAT) — Consists of eight separate events. The test is a sequence of events that requires a candidate for the fire service to progress along a predetermined path from event to event in a continuous manner. In order to pass the test, a candidate must complete all the events in a prescribed manner and in a prescribed time.

Chemical Protective Clothing (CPC) — Clothing designed to shield or isolate individuals from the chemical, physical, and biological hazards that may be encountered during operations involving hazardous materials.

Communication model — The process that all people use when interacting with others on a daily basis.

CPAT — Abbreviation for Condidate Physical Abilities Test.

CPC — Abbreviation for Chemical Protective Clothing.

E

Ergonomics — The applied science of workplace equipment design intended to maximize productivity and minimize potential injuries by reducing operator fatigue and discomfort.

F

Fitness — Defined by the American College of Sports Medicine as appropriate facility in 5 areas: muscular strength, muscular endurance, flexibility or range of motion, cardiorespiratory efficiency, and body composition.

Fitness for Duty Evaluation — By order of the fire chief, an evaluation of a firefighter by a medical authority to determine if the firefighter can physically and mentally perform the essential duties of the position.

Freelancing — Operating independently of the incident commander's (IC's) command and control.

H

Hazard — Condition, substance, or device that can cause injury or loss; the source of a risk.

Health and Safety Officer — Officer designated by the chief of department as defined by NFPA® 1521. Officer in the fire department primarily responsible for the management of the health and safety program.

Health Insurance Portability and Accountability Act (HIPAA) — The *Standards of Individually Identifiable Information* (Privacy Rule) establishing for the first time, a set of national standards for the protection of certain health information. The U.S. Department of Health and Human Services (HHS) issued the Privacy Rule to implement the requirement of the HIPAA act of 1996.

Health Program — Any program designed to assist and encourage employees to adopt and attain a healthier lifestyle.

HIPAA — Abbreviation for Health Insurance Portability and Accountability Act.

I

Intrinsic safety (IS) — A protection technique for safe operation of electronic equipment in explosive atmospheres.

IS — Abbreviation for Intrinsic Safety.

L

Level A Protection — Highest level of skin, respiratory, and eye protection that can be afforded by personal protective equipment. Consists of positive-pressure self-contained breathing apparatus, totally encapsulating chemical-protective suit, inner and outer gloves, and chemical-resistant boots.

Level B Protection — Personal protective equipment that affords the highest level of respiratory protection, but a lesser level of skin protection. Consists of positive-pressure self-contained breathing apparatus, hooded chemical-resistant suit, inner and outer gloves, and chemical-resistant boots.

Level C Protection — Personal protective equipment that affords a lesser level of respiratory and skin protection than Levels A or B. Consists of full-face or half-mask air-purified respirator (APR), hooded chemical-resistant suit, inner and outer gloves, and chemical-resistant boots.

Level D Protection — Personal protective equipment that affords the lowest level of respiratory and skin protection. Consists of coveralls, gloves, and chemical-resistant boots or shoes.

Life Safety Initiatives Program — Recognizing the need to do more to prevent line-of-duty deaths and injuries, the National Fallen Firefighters Foundation has developed 16 Life Safety Initiatives.

Life Safety Rope — Rope that meets the requirements of NFPA® 1983, *Standard on Fire Service Life Safety Rope, Harness, and Hardware*, and is dedicated solely for the purpose of constructing lines to be used for the raising, lowering, or supporting people during rescue fire fighting or other emergency operations, or during training. Also called *lifeline*.

Light Duty — Assignment of a firefighter to a duty position with physical restrictions or limitations.

Line-of-Duty Death — Any injury sustained in the line of duty that proves fatal, any illness that was incurred as a result of actions while on duty that proves fatal, and fatal mishaps involving nonemergency occupational hazards that occur while on duty.

N

Narcotic — Any class of substances or drugs that numbs, paralyses, or causes loss of feeling in the body. Narcotics may be addictive, alter moods or behavior, and induce sleep.

National Wildfire Coordinating Group — Its purpose is to establish an operational group designed to coordinate programs of the participating wildfire management agencies.

Near Miss Reporting — A voluntary, confidential, non-punitive and secure reporting system with the goal of improving firefighter safety.

NWCG — Abbreviation for National Wildfire Coordinating Group.

O

On-the-Job Training (OJT) — System of training firefighters that makes full use of personal contact between firefighters and their immediate supervisor; trains firefighters both physically and psychologically for the position they will perform.

OJT — Abbreviation for On-the Job-Training.

P

Preincident plan — Plan developed from the information gathered during the preincident survey and used during emergency operations and training.

Preincident survey — The act of collecting information on a site prior to the occurrence of an incident.

R

Reaction Distance — The distance the vehicle travels from the time the brakes are applied until the apparatus comes to a complete stop.

Regulations — Rules or directives of administrative agencies that have authorization to issue them.

Risk — Likelihood of suffering harm from a hazard; exposure to a hazard; the potential for failure or loss.

Risk Management Plan — A plan that considers all fire department operations, the duties and responsibilities of members, and policies and procedures. The plan should include goals and objectives to ensure that the risks associated with daily operations of the fire department are identified and effectively managed.

S

Safety and Health Committee — A committee assigned by the fire chief in an advisory capacity to conduct research, develop recommendations, and study and review matters pertaining to occupational safety and health within the fire department.

Situational awareness — The process through which one perceives and comprehends the implications of surrounding conditions and then applies that understanding to help predict how and when the situation may change in order to take appropriate action.

Size-Up — Ongoing mental evaluation process performed by the operational officer in charge of an incident that enables him or her to determine and evaluate all existing influencing factors that are used to develop objectives, strategy, and tactics for fire suppression before committing personnel and equipment to a course of action. Size-up results in a plan of action that may be adjusted as the situation changes. It includes such factors as time, location, nature of occupancy, life hazard, exposures, property involved, nature and extent of fire, weather, and fire fighting facilities.

Standards — Criterion documents that are developed to serve as models or examples of desired performance or behaviors and that contain requirements and specifications outlining minimum levels of performance, protection, or construction. No one is required to meet the requirements set forth in standards unless those standards are legally adopted by the authority having jurisdiction, in which case they become law.

Strategic safety plans — The process that identifies specific goals that support and enhance the safety and risk management activities of the department.

T

Total Stopping Distance — Sum of the driver/operator reaction distance and the vehicle braking distance.

Train-the-Trainer — Training program that relies on training key members of the organization, such as company officers, to train other members, such as company personnel, in a specific topic, procedure, or policy. An effective train-the-trainer program will teach trainers such skills as how to facilitate a meeting, how to deal with students who are having trouble, and how to teach effective note-taking.

U

Utility Rope — Rope to be used in any situation that requires a rope — except life safety applications. Utility ropes can be used for hoisting equipment, securing unstable objects, and cordoning off an area.

V

Visual lead time — The concept that driver/operators should scan far enough ahead of the apparatus, for the speed it is being driven, to assure that evasive action can be taken if it becomes necessary.

W

Wellness — In good health.

Work Capacity Test — A set of 3 physical test developed by the NWCG to determine the fitness and work capacity level of a wildland firefighter and support staff.

Workload management — The Crew Resource Management factor that attempts to reduce stress by reducing a job to its most basic tasks and assigning those tasks to individuals and groups.

Index

Index

Index by Nancy Kopper.